EDWARD VII LAND
SHIRASE COAST
SAUNDERS COAST
FORD RANGES
ROOSEVELT I.
Mt Berlin
3498
Siple I
MARIE BYRD LAND
ROCKEFELLER
EXECUTIVE
Mt Sidley
4181
COMMITTEE
RANGE
BAKUTIS COAST
SIPLE COAST
PLATEAU
CRARY MTS
3677
Mt Frakes
Toney Mtn
3566
Mt Takahe
3398
SEA
GOULD COAST
HOLLICK - KENYON
WALGREEN COAST
WEST
Faure Peak
3941
HORLICK MTS
WISCONSIN RA
OHIO RANGE
PLATEAU
THURSTON ISLAND
HUDSON MTS
EIGHTS COAST
ELLSWORTH
LAND
ANTARCTICA
BELLINGSHAUSEN
SEA
Peter 1 Øy
THIEL MTS
ELLSWORTH MOUNTAINS
Vinson Massif
4897
SENTINEL RANGE
HERITAGE RANGE
ZUMBERGE COAST
BRYAN COAST
PATUXENT RANGE
PENSACOLA MTS
NEPTUNE RANGE
FORRESTAL RANGE
Smyley I
Spaatz I
Latady I
Charcot I
ORVILLE COAST
ENGLISH COAST
ARGENTINA RANGE
RONNE ICE SHELF
FILCHNER ICE SHELF
ALEXANDER ISLAND
BERKNER ISLAND
PALMER LAND
Mt Jackson
3180
LASSITER COAST
FALLIÈRES COAST
ADELAIDE I
SHACKLETON RANGE
BELGRANO II (ARGENTINA)
BLACK COAST
WILKINS COAST
BOWMAN COAST
GRAHAM COAST
GRAHAM LAND
ANTARCTIC PENINSULA
LUITPOLD COAST
COATS LAND
LARSEN ICE SHELF
FOYN COAST
Cape Agassiz
Cape Framnes
OSCAR II COAST
NORDENSKJOLD COAST
HALLEY (UK)
CAIRD COAST
DAVIS COAST
BRUNT ICE SHELF
WEDDELL SEA
NORDENSJOLD'S HUT (1902) Snowhill I
NORDENSJOLD'S HUT (1903) Paulet I
Joinville I
MARTHA KYST
HEIMEFRONTFJELLA
KRONPRINSESSE
GEORG VON NEUMAYER (GERMANY)
AFRICA)
SOUTH ORKNEY ISLA
ORCADAS (ARGENTINA)
SEA

ANTARCTICA

EXPLORING THE EXTREME

400 YEARS OF ADVENTURE

MARILYN J. LANDIS

Library of Congress Cataloging-in-Publication Data
Landis, Marilyn J.
Antarctica : exploring the extreme : 400 years of adventure / Marilyn J. Landis
p. cm.
Includes bibliographical references (p. 375) and index (p. 383).
ISBN 1-55652-428-5
1. Antarctica—Discovery and exploration. I. Title.
G860.L354 2001
919.8'904—dc21
2001042109

COVER IMAGE: Photo by Frank Hurley, dated January 14, 1915, from the Ernest Shackleton lead expedition on the *Endurance*, courtesy Royal Geographical Society, London.

Cover and interior design: Lindgren/Fuller Design

First edition
Published by Chicago Review Press, Incorporated
814 North Franklin Street
Chicago, Illinois 60610
ISBN 1-55652-428-5
Printed in the United States of America
5 4 3 2 1

To all intrepid Antarctic explorers and dreamers, past and future

CONTENTS

ACKNOWLEDGMENTS

This book could not have been written without the help, support, and encouragement of many individuals and organizations. I am indebted most of all to the scholars of Antarctic history, geology, and wildlife listed in the bibliography, for their works allowed me to build upon a solid foundation. The explorers' published diaries, narratives, and logbooks provided great inspiration. These men, drawn to the most desolate place on Earth, had the talent to paint the honorable and the horrible with unforgettable word imagery. The quotations in this book are from these sources.

A good library reference department is an invaluable tool to an author. I was fortunate to have the services of the unfailingly cheerful staff at Ontario, California, who located many obscure, out-of-print books and articles for me. My thanks go to Scott Polar Research Institute's William Mills, librarian and keeper, and Lucy Martin, picture library manager, for their help. I appreciate the efforts of Ewen Smith, director, and Dr. J. W. Faithfull, curator in geology, Hunterian Museum in Glasgow; Simon Bennett and George Gardner, assistant archivists, University of Glasgow; Holly Reed, NWCS-Stills Department of the National Archives and Records; and Amy Kimball, special collections of the Milton S. Eisenhower Library, Johns Hopkins University, were also most helpful.

I am deeply indebted to the professors at the department of geology and Institute of Polar Research at Ohio State University who set my feet on a career path that ultimately led to this book. I thank Gary Hutchinson for hiring and giving me the opportunity to travel the world and expand my skills in technical writing. Friends in California and Washington have listened, commented, and commiserated whenever the need arose, but Dr. Gordon Dower, Barbara Birch, and Donna Larsen were unwavering with their support. I extend my appreciation to Marine Expeditions, Mountain Travel-Sobek, and Heritage Expeditions for their support and encouragement during several trips to Antarctica. New Zealand associates Jeni Bassett and David Harrowfield were also very supportive.

The hardworking and dedicated staff at Chicago Review Press shaped my original manuscript and in the process transformed it into this book. My thanks to all of you: Cynthia Sherry, executive editor, whose enthusiasm and attention to detail assured me that I had found the right publisher; playwright Robert Koon copyedited the manuscript with care and insight; Pamela Juárez proofread the manuscript meticulously and made excellent suggestions; and Lisa Rosenthal, author, playwright, and editor. Her consummate skills, intelligence, and sense of humor helped to refine the final manuscript. She was beside me every day by telephone, fax, and e-mail during the last few hectic months of rewrites, new-writes, and clarifications. I am immensely grateful to her.

My family was my anchor throughout this five-year project. I can only express my love and appreciation to Heather Landis and David Starinieri for their steadfast loyalty and belief in this book; and to Brian Landis, who encouraged me by saying the right things during the rough times with humor and sensitivity.

Every writer should have an intelligent and thoughtful first reader of a manuscript in progress. I am so fortunate to be married to the best—Nick Landis read and reread every section of this book with insight and offered many suggestions for improvement. He inspired me to be a better writer. This book is as much his as it is mine.

ANTARCTIC EXPEDITIONS

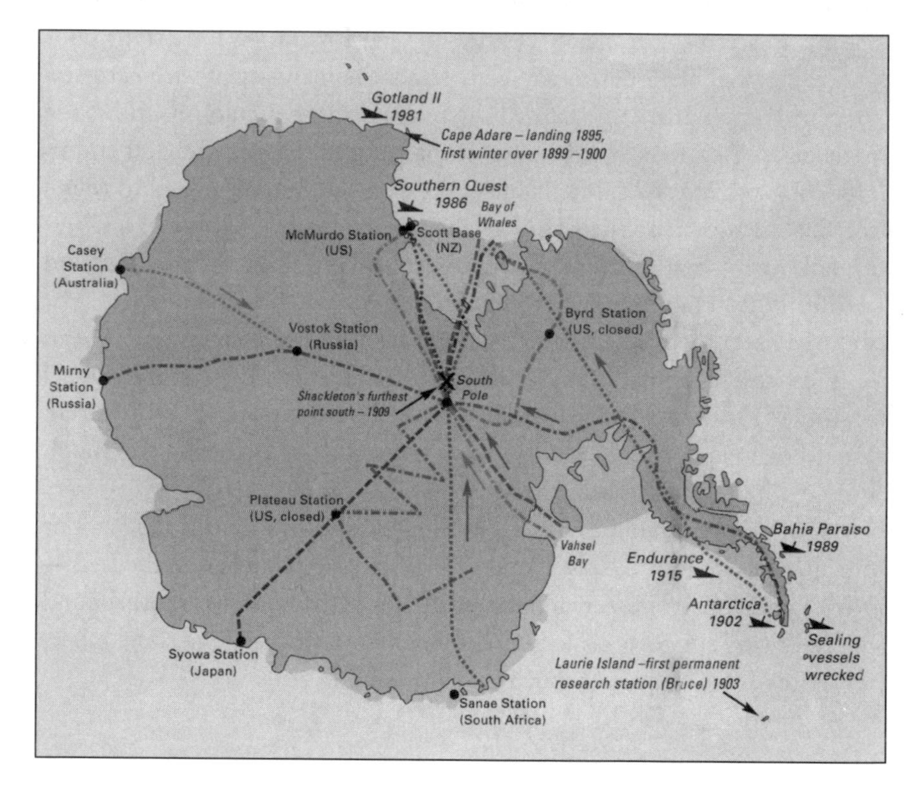

British Antarctic Expedition (Shackleton) 1907–1909 X
British Antarctic Expedition (Scott) 1910–1913..........................
Norwegian Antarctic Expedition (Amundsen) 1911–1912...
First Flight to the South Pole (Byrd) 1929...........................
Trans-Antarctic Flight (Ellsworth/Kenyon) 1935–1936.............
Commonwealth Trans-Antarctic Expedition
(Fuchs/Hillary) 1957–1958...
Tractor train and Byrd Station–South Pole (US) 1960–1961......
Wilkes–Vostok Traverse 1962–1963..
South Pole–Dronning Maud Land Traverse 1967–1968............
Syowa–South Pole 1968...
Trans-Globe Expedition (Fiennes) 1980–1981..............................
International Trans-Antarctic Expedition (Steger) 1989–1990...
Pentland–South Pole Expedition (Fiennes/Stroud) 1992–1993...

TIME LINE

600–330 B.C.: Greek philosophers develop theories that Earth was a sphere with equally distributed land masses and climatic zones, including the North and South polar regions.

A.D. 150: Claudius Ptolemy writes the influential *Guide to Geography* which was translated from Greek to Arabic in the ninth century, and subsequently from Arabic to Latin by Spanish translators sometime before 1300.

1519–1522: Ferdinand Magellan undertakes the first circumnavigation of the world; Magellan sails through the strait that now bears his name in November 1520.

1577–1580: Sir Francis Drake discovers his namesake sea passage in September 1578.

1719–1720: George Shelvocke sails to 61° 30' south on a voyage that inspires Samuel Taylor Coleridge's *Rime of the Ancient Mariner*.

1738–1739: Jean-Baptiste Charles Bouvet de Lozier discovers Cape Circumcision, Bouvetøya Island in January 1739.

1768–1771: Captain James Cook's *Endeavour* expedition explores Tierra del Fuego in January 1769 and searches for *Terra Australis Incognita*.

1771–1772: Yves-Joseph de Kerguélen-Trémarec discovers Îles Kerguélen.

1772–1775: Captain James Cook's *Resolution* and *Adventure* reach 71° 10' south; Cook claims South Georgia and South Sandwich Islands for England.

1777: Captain James Cook publishes his two-volume narrative *Resolution*.

FEBRUARY 19, 1819: William Smith discovers South Shetland Islands.

SEPTEMBER 1819: *San Telmo* wrecks in the Drake Passage with loss of all hands.

OCTOBER 16, 1819: William Smith returns and claims South Shetland Islands.

1819–1821: Thaddeus von Bellingshausen circles Antarctica with the *Vostok* and *Mirnyi*; sights continent on January 27, 1820.

JANUARY 30, 1820: Edward Bransfield and William Smith sight Antarctic Peninsula.

1820: Nathaniel Palmer explores Deception Island with the *Hero* and sights the Antarctic Peninsula on November 17, 1820.

FEBRUARY 7, 1821: John Davis sails the *Cecilia* to Hughes Bay on the Antarctic Peninsula and lands on the continent.

1822–1824: James Weddell travels to Antarctica with the *Jane* and *Beaufoy*; reaches 74° 15' south on February 20, 1823.

1830–1832: John Biscoe expedition with the *Tula* and *Lively* sights Enderby Land and discovers Adelaide Island.

1837–1840: The French Antarctic Expedition, led by Jules-Sebastien-Cesar Dumont d'Urville, sails the *Astrolabe* and the *Zélée* to the Antarctic Peninsula and eastern Antarctica; discovers Adélie Land.

1838–1839: John Balleny discovers Balleny Islands with the *Eliza Scott* and *Sabrina*.

1838–1842: Charles Wilkes heads the United States Exploring Expedition, sailing with the *Vincennes*, *Porpoise*, *Peacock*, *Flying Fish*, *Sea Gull* (Peninsula only), and *Relief* (Tierra del Fuego only) to the Antarctic Peninsula and eastern Antarctica.

1839–1843: James Clark Ross discovers the Ross Sea, Ross Ice Shelf, Victoria Land, and many other Antarctic landmarks with the *Erebus* and *Terror*.

1892–1893: The Dundee whaling expedition, with the *Balaena*, *Active*, *Diana*, and *Polar Star*, sails to the Weddell Sea in search of whales.

1893–1895: Henryk Bull sails the *Antarctic* to the Ross Sea to hunt for whales.

1897–1899: Adrien Victor Joseph de Gerlache heads the *Belgica* expedition, exploring the Antarctic Peninsula and the Bellingshausen Sea.

1898–1900: Carsten Borchgrevink sails with the *Southern Cross* to Cape Adare and spends the winter on the continent.

1901–1903: Erich Dagobert von Drygalski spends the winter in eastern Antarctica on the *Gauss*.

1901–1904: Otto Nordenskjöld heads the Swedish Antarctic Expedition, on the *Antarctic*, to the Weddell Sea and Snow Hill Island.

1901–1904: Robert F. Scott heads the *Discovery* expedition, spends two winters on Ross Island, and treks toward the South Pole.

1902–1904: William S. Bruce heads the Scottish National Antarctic Expedition and sails on the *Scotia* to the Weddell Sea and the South Orkney Islands.

1903–1905: Jean-Baptiste Charcot, heading the French Antarctic Expedition, sails the *Français* to the Antarctic Peninsula and spends the winter.

1907–1909: Ernest Shackleton heads the *Nimrod* expedition, sails to Ross Island, and sledges toward the South Pole.

1908–1910: Jean-Baptiste Charcot returns to the Antarctic Peninsula on the *Pourquoi Pas?* and spends the winter.

1910–1912: Roald Amundsen sails to the Bay of Whales on the *Fram* to trek to the South Pole.

1910–1913: Robert F. Scott heads the *Terra Nova* expedition, sailing to Ross Island for a sledge journey to the South Pole.

1911–1912: Wilhelm Filchner sails the *Deutschland* into the Weddell Sea pack ice.

1911–1914: Douglas Mawson heads the Australian Antarctic Expedition, sailing on the *Aurora*, and spends two winters at Commonwealth Bay.

1914–1916: Members of Ernest Shackleton's expedition sail on the *Endurance* into the Weddell Sea pack ice.

1914–1917: The other half of Ernest Shackleton's expedition, the Ross Sea Party, sails on the *Aurora* to Ross Island.

1921–1922: Ernest Shackleton sails to South Georgia on the *Quest*.

1928–1930: Members of the Richard E. Byrd party spend the winter at Little America on the Bay of Whales, to fly to the South Pole in November 1929.

1933–1935: Richard E. Byrd's second expedition to Little America II; Byrd stays alone at Bolling Advance Station.

1933–1939: Lincoln Ellsworth leads expeditions that explore four Antarctic regions by air.

1939–1941: Richard E. Byrd heads the United States Antarctic Service Expedition to Little America III.

1946–1947: Richard E. Byrd undertakes Operation Highjump.

1957–1958: The International Geophysical Year, an eighteen-month year of Antarctic science, is held.

1956–1958: Vivian Fuchs of the Commonwealth Trans-Antarctic Expedition crosses the continent from the Weddell Sea to the Ross Sea, with help from Edmund Hillary.

DECEMBER 1959: Antarctic Treaty signed.

1964: The Agreed Measures for the Conservation of Antarctic Fauna and Flora is enacted.[1]

DECEMBER 1965: The last whaling station on South Georgia closes.

1967: Researcher from Ohio State University discovers jaw fragments of a four-legged, lizard-like amphibian, proof that Antarctica was part of Gondwana two hundred million years ago.

1978: Convention for the Conservation of Antarctic Seals is enacted.[2]

1979: Fatal Air New Zealand crash on Mount Erebus kills all 257 passengers and crew.

1980: Convention on the Conservation of Antarctic Marine Living Resources is enacted.[3]

1982: Fossil of a rat-size marsupial is found on Seymour Island, near the tip of the Antarctic Peninsula, evidence that marsupials migrated from South America to Australia via Antarctica.

---

[1] The Agreed Measures for the Conservation of Antarctic Fauna and Flora protects Antarctic animals and plants from exploitation and destruction and allows ecologically sensitive areas to be out of bounds to everyone but scientists.

[2] Convention for the Conservation of Antarctic Seals ensures that seals will not be hunted for commercial gain.

[3] The Convention on the Conservation of Antarctic Marine Living Resources protects krill (the foundation of the Antarctic food chain) and other marine resources from excessive harvesting.

1982: Falkland Islands War between Great Britain and Argentina from April 2 to June 14.

1985: British researchers at Halley Station discover a circular layer in the stratosphere depleted of ozone; it is now referred to as the Ozone Hole.

1986: International moratorium on commercial whaling goes into effect.

1987: Massive iceberg B9 calves from the Ross Ice Shelf and carries away Little Americas I–III.

1991: Protocol on Environmental Protection to the Antarctic Treaty (Madrid Protocol) negotiated.[4]

1991: Soviet Union is dissolved; conversion of ice-class research ships into tourist vessels commences.

1996: NASA scientists examine meteorite AH84110, collected in 1984 by Japanese researchers in Antarctica, and detect possible microscopic "worm" burrows.

1998: Researchers find that Greenland and Antarctic ice cores show wildly erratic temperature shifts after the last Ice Age.

1998–2001: Scientists discover Lake Vostok and seventy-eight other warm freshwater lakes beneath East Antarctic Ice Sheet.

MAY 2000: Three massive icebergs calve from Ronne Ice Shelf; largest measures 107 by 21 miles.[5]

---

[4] Protocol on Environmental Protection to the Antarctic Treaty prohibits mining for fifty years and designates Antarctica as a "natural reserve, devoted to peace and science." It also requires environmental impact statements on all activities, including scientific studies and tourist operations.

[5] Instability of the West Antarctica Ice Sheet, a probable product of global warming, is an ongoing concern and focus of many international research projects. The disintegration of the West Antarctic Ice Sheet would increase sea level around the world and devastate coastal areas.

INTRODUCTION

"I think this Southern Land to be a Continent."
—*Seal hunter John Davis,*
February 7, 1821, Hughes Bay, Antarctica

That long-ago February morning was cloudy. Light winds rippled the water in a large bay ringed by snow-covered mountains. Icebergs and floes drifted by in the distance, spawned by glaciers that fanned into the bay. The misty blows of humpbacks may have resounded across the water, breaking the silence with hissing sighs. Whales held little interest for John Davis. He and his crew were searching for fur seals and had sailed from the South Shetland Islands, away from the beaches that swarmed with American and British sealers. He ordered a boat to shore "to look for Seal at 11 A.M. . . . but found no sign of Seal at Noon." Davis did not linger and sailed away to fresh sealing grounds. The sheer magnitude of the then-unexplored land convinced Davis that he and his men had stood at the edge of a vast continent.

The story of Antarctica does not begin with John Davis and his men wading ashore. Antarctica began as a reasoned conjecture twenty-three hundred years before in Athens, Greece. Throughout mankind's intellectual evolution, the idea of an unknown southern land has changed to reflect each era's perspective and knowledge, from Aristotle's lifeless climatic zone to Alexander Dalrymple's tropical paradise populated with friendly antipodes. Antarctica the Imagined has lured the adventurous and ambitious, men who sought this geographic Grail for its promise of wealth and fame.

When Davis and his men stepped ashore for the first time, the dreamscape sharpened into focus. But the complex interplay of ice, water, light, and cold air created a fantasyland of mirages, a Merlin's mix of improbabilities and magic that confounded all those who followed to explore the continent's geography or to exploit its wildlife.

Antarctica's history is a continuum of exploration—from speculative imagination to satellite imagery, from the dim to the defined. Part I of this book reveals the continent through the diaries and narratives of the famous and the obscure, from Captain Cook to Kerguélen-Trémarec, men who searched for the unknown land and respectively found acclaim and ignominy. Those who followed exploited the resource-rich subantarctic islands and ocean. Barefoot sealers witnessed and recorded the brutality of that long-ago age, and in turn exposed the pathos of the human heart. International rivalries spurred three men to seek the South Magnetic Pole in the name of science and proprietorship. Later, Robert Scott and Roald Amundsen trudged toward the South Geographic Pole on separate but parallel journeys to immortality and despair. Their histories and others are rich in detail, told from very different perspectives yet sharing the commonality of hardship and heroics. These are the impressions of the men who hunted the seals and whales, who painted and photographed the landscapes, who studied the wildlife, who trekked across the vast Polar Plateau, and who wrote an innovative treaty based on peace and cooperation.

Part II divides Antarctica into four regions, each with its unique voices and history. The *Belgica* expedition was the first to spend the long Antarctic night trapped in the pack ice along the Antarctic Peninsula. The *Endurance* expedition, Ernest Shackleton's great epic of survival and hope in the Weddell Sea pack ice, illustrates how to face hopelessness while keeping one's humanity. Apsley Cherry-Garrard's terrible journey to collect emperor penguin eggs during the dead of winter in the Ross Sea Region is a tale of scientific curiosity and innocence lost. East Antarctic blizzards set the stage for Douglas Mawson's horrific solo trek back to Commonwealth Bay. The subantarctic islands are intertwined with

Antarctic exploration and exploitation, from the slaughter of the fur seals to the formation of sanctuaries and World Parks.

Part III traces Antarctica's geologic past from Gondwana to the continent with the harshest climate on Earth, inhabited by a diverse wildlife population. More than thirty years ago, the discovery of a fossil imprint of a deciduous plant and fossil jaw fragments of a four-legged amphibian revolutionized earth sciences. Since then, by analyzing ozone hole data, chemical composition of air samples from ice cores, and measurements of unstable ice sheets, scientists have been decoding Antarctic's past to understand its impact on the world's climate.

The edges of this continent are home to an astounding variety of life that has adapted to its hostile climate. From the near-microscopic krill to the 120-ton blue whales that feed on them, Antarctic waters and shores abound in wildlife. Penguins, seals, and birds are Antarctica's living link between what was and what will be.

Three lone huts still stand in the Ross Sea Region. Rough furnishings and memorabilia offer poignant glimpses into the past. Stiff hand-knitted wool socks dangle from a top bunk whose owner will never return; empty stalls retain the scent of ponies, mules, and fodder; and pungent black slabs of seal blubber wait to be cut into squares to burn for warmth. Reindeer sleeping bags, frayed dogskin mittens, and grass-stuffed boots are reminders that to stand here is to step back into an age when cold was an invisible companion who trudged next to the men, killing the unprepared, the careless, or the weary. This book chronicles these and many other human histories—magnificent legacies that endure.

• PART I •

GENERAL EXPLORATION: FROM THE CONCEPT OF ANTARCTICA TO THE ANTARCTICA TREATY

• 1 •

THE SEARCH FOR *TERRA AUSTRALIS INCOGNITA*

IN 600 B.C. ANTARCTICA LAY WRAPPED IN A WHITE FROZEN MANTLE, hidden beyond fierce storms and endless swells of unknown seas, its seasons marked by the advance and retreat of ice. Icebergs, spawned from the flat fringe that ringed the continent, thundered into bays. Strong currents carried the ice northward in concentric circles like ripples on a pond. The colors of cold—shades of white and blue—gave them texture and dimension. Wind and waves whittled the icebergs into gleaming sculptures as they floated into warmer oceans and melted, unseen by human eyes.

Thousands of miles to the north of this realm of ice, Europe entered the Bronze Age. The Babylonians sketched a map of the known world on a clay tablet and charted the constellations. The Persians waged war in distant lands, while the Phoenicians sailed the Mediterranean and beyond the Straits of Gibraltar in search of tin for bronze weapons.

Prosperous Greek colonies dotted the Mediterranean coastline "like frogs around a pond," wrote Plato. Wealth freed the Greek aristocrats from the daily drudgery of work, and they discovered the luxury of time and leisure. A few stared upward into the dark abyss filled with pinholes of light, searching for illumination and answers. Their inquiries started a revolution of ideas that eventually led to the concept of *Terra Australis Incognita*, the Unknown Southern Land.

A Question of Balance

Thales of Miletus (640–550 B.C.), a prosperous ship owner, studied meticulous Babylonian charts of the constellations and concluded that the stars moved with seasonal predictability. But one star in the Little Bear constellation anchored the nighttime sky, never changing its position. The star was Polaris, and its discovery meant that Greek ships now could safely sail beyond the boundaries of the Mediterranean Sea. When Thales predicted the eclipse of 585 B.C., youths flocked to hear his words and later founded schools that further refined his ideas.

Pythagoras (born about 570 B.C.) found harmony in the universe and sought to explain it. He reasoned that Earth and the other planets must be spheres, since the most harmonious geometric form was a circle. The most powerful force in the universe was, he believed, the number ten. Although Pythagoras could only see nine heavenly bodies, his theories of harmony and balance required a tenth. So he imagined *Antichthon*, a planet hidden behind Earth. Pythagoras's followers founded a school based on his principles and influenced one of the greatest thinkers of all time—Plato.

After the death of Socrates, Plato (427–347 B.C.) traveled to southern Italy to study Pythagorean mathematics. When he returned to Athens and established his school, Plato also taught his students that Earth was a sphere. If man could soar high above the clouds, Earth would resemble "a ball made of twelve pieces of leather, variegated, a patchwork of colors." Undiscovered continents to the south perfectly balanced the known lands to the north. If this were not so, reasoned Plato, Earth would wobble and turn over.

Aristotle (384–322 B.C.) was Plato's prize student and "the mind of the school." He observed "there are stars seen in Egypt and . . . Cyprus which are not seen in the northerly regions." Since this could only happen on a curved surface, he too believed Earth was a sphere "of no great size, for otherwise the effect of so slight a change of place would not be so quickly apparent."

The concepts of symmetry, equilibrium, and cyclic repetition permeated Aristotle's work. In *Meteorology* he divided the world into five climatic zones. Two temperate areas were separated by a torrid zone near the equator. Two cold inhospitable regions, "one near our upper, or northern pole, and the other near the . . . southern pole" were impenetrable, girdled with ice. Although no humans could survive in the frigid zones, inhabitants in the southern temperate regions could exist. He called these theoretical people *antipodes* ("feet opposite").

During the next several hundred years, the unknown but theoretical southern continent evolved into a populated landmass, complete with its own folklore. The brilliant astronomer and geographer Claudius Ptolemy was the first to grace this region with a name. He called it *Terra Australis Incognita*.

Claudius Ptolemy lived in Alexandria, the center of scholarship in the second century A.D. A prodigious writer, he published manuscripts on optics, music, mathematics, and astronomy. In about A.D. 150 he produced his eight-volume *Guide to Geography*, which listed the latitude and longitude of eight thousand places between 63° north and 16° south—the limit of the known world. He included in his treatise instructions for drawing *Terra Australis Incognita*, its imaginary dimensions calculated to balance the mass of the known world. From its cocoon of reasoned speculation, the Unknown Southern Land emerged as an accepted reality, complete with its population of antipodes.

As Christianity spread northward, the Greek principles of science devolved into anathema and heresy. For nearly seven hundred years dogma, fear, and superstition ruled intellectual life behind the stone walls of monasteries. Then the Church rallied the masses to wage holy wars against the infidels. The eleventh century Crusades enriched not only the Papacy with expanded territories and the merchants with new trade routes, but intellectual life as well. Greek manuscripts, previously translated into Arabic by Muslim scholars, were now more prized than gold. As these writings were translated into Latin, Europe slowly emerged from darkness and embraced enlightenment.

By the early 1400s cartographers produced maps of the world based on translated instructions in Ptolemy's *Guide to Geography*. The mapmakers also unwittingly perpetuated his errors. Ptolemy had reduced the Greek Eratosthenes's near-accurate estimate of Earth's circumference by 25 percent—a mistake that would later confuse and confound seafaring explorers. Africa and Asia were joined, and the land from Spain through China greatly exaggerated. The dimensions of the seas suffered a reduction, and *Terra Australis Incognita* nearly filled the Southern Hemisphere.

Maps of Ptolemy's world jolted Europe and ushered in the Age of Discovery. Christopher Columbus learned Latin, studied Ptolemy's manuscripts, and saw a world of possibilities for an ambitious seaman. He narrowed his focus to three potential expeditions: (1) to seek new North Atlantic islands; (2) to search for unknown southern lands; and (3) to sail west across the Atlantic to Asia in search of spices. His financial backers, King Ferdinand and Queen Isabella of Spain, ignored the first two possibilities and opted for the third, the route with the most potential economic windfalls. The search for *Terra Australis Incognita* would have to wait.

Magellan's Discovery in the Land of Fire

Ferdinand Magellan was a small, wiry, unassuming man with a severe limp, a permanent reminder of Portugal's war with the Moors. In 1517 he petitioned King Manuel for an increase in his yearly allowance for his wound, but the king refused and pronounced him cured. Although Magellan seethed in private at this unsympathetic turn of events, he courted the King's favor by proposing a new route to the Spice Islands. He would sail west across the Atlantic to search for the rumored waterway through South America. His plan was met with royal indifference. Seventeen years earlier, Vasco da Gama had pioneered an eastward passage around Africa. What purpose could another route serve?

This final rejection spurred Magellan to renounce his Portuguese citizenship. Seeking revenge and retribution, he swore allegiance to Charles I, King of Spain, who agreed to finance Magellan's expedition. For two years vicious court intrigue—Spanish and Portuguese—focused on Magellan. Rumors of treachery, plots of murder, and whispered betrayals permeated palace dining halls and bedchambers. A man of few words, Magellan ignored the ruthless gossip and concentrated on securing provisions for his five ships. But warnings of a planned mutiny persisted.

On September 20, 1519, Magellan sailed from Spain with five ships and 240 men including ten wealthy passengers who paid a small fortune for the privilege of facing hardships and danger. One of these adventurers was Antonio Pigafetta, the son of an Italian nobleman. A keen observer, he soon sensed trouble brewing among the four Spanish captains.

During October the expedition endured severe storms and high seas near the western coast of Africa. Gales shredded the sails and thirty-foot waves broke over the decks, nearly drowning a helmsman. Even the brief calm periods between storms filled the superstitious sailors with dread. Eerie green orbs of light glowed at the tips of masts and outlined the ropes on the rigging. One night the men stared at the light "shining so brightly on the topsail that they were blinded for a quarter of an hour and thought they were dead," wrote Pigafetta.

Magellan skirted the eastern coast of South America from December through February 1520, searching for the elusive waterway across the continent. By March, snow squalls and cold winds battered the ships. Sleet-stiffened ropes cut the sailors' numbed hands and thick plates of ice fell from the sails, shattering like glass near their feet. Below the deck, sea water from numerous small leaks sloshed from side to side in the bilge.

Rather than press farther south, Magellan anchored in San Julian, Argentina for the winter to repair his ships and rest his men. As their supplies dwindled, the crew hunted for food beyond the desolate and uninhabited Patagonian shores, but no one spotted a single animal. Even fish were scarce. When Magellan cut rations, grumbling escalated into a mutinous plot. A trusted member warned Magellan when the attack was imminent. Under the cover of darkness, Magellan surprised the mutineers and bound forty-five men in irons. Punishment for the three Spanish captains responsible for the mutiny was grim. One was drawn and quartered; another was beheaded. The third man was then abandoned alive on shore with nothing but the rotting bodies of the other two.

Magellan wintered near Rio de Santa Cruz and headed south in the late austral spring. On October 21, 1520, the ships approached the massive headlands Magellan later named Cape of Eleven Thousand Virgins, in honor of St. Ursula. He sailed into a large bay that narrowed into a channel flanked by dark, brooding cliffs. Magellan knew he had found the elusive strait, to everyone's joy and relief. "The Captain General was determined to sail as far as 75 degrees south latitude, where there is no night, if he had not found this passage to the sea," wrote Pigafetta.

During their passage through the treacherous strait, they faced strong currents that foamed over hidden rocks and winds that howled down from the stark, snowcapped peaks. Hiding among the stunted birch trees near the shore, curious but shy Patagonians watched the ships maneuver through a maze of channels. In the evenings their campfires twinkled in the darkness. Magellan named the region Tierra del Fuego, Land of Fire.

"Strange black geese" were spotted on two small islands and the ships stopped to investigate. "They are unable to fly," noted Pigafetta, "and they live on fish, and are so fat that it is necessary to peel them." Fresh penguin meat soon filled the ships' larders.

Magellan emerged from the strait that now bears his name on November 28 with three ships intact. The *San Antonio*, carrying much of the expedition's provisions, had deserted. But not even the threat of starvation weakened Magellan's determination to complete his voyage. They sailed for three months across the Pacific, surviving on maggoty biscuit crumbs yellowed with rats' urine. The men gnawed on leather straps and swallowed soup made with boiled sawdust. Rats were stalked and sold to the highest bidder for "half a ducat or a ducat," noted Pigafetta. Scurvy ravaged the initial survivors of the voyage and its perils. "The gums of some men swelled over their upper and lower teeth . . . they could not eat and so died."

Brisk winds and a calm sea favored the desperate men. The three ships reached Guam on March 6, 1521, and the survivors regained their strength. To consolidate their meager resources, Magellan ordered one ship dismantled before sailing on to the Philippines. Shortly after his arrival there, fighting broke out between two tribes on Cebu Island. Magellan intervened and was speared to death on April 27. Only the

Victoria eventually returned to Seville on September 8, 1522, with a crew of just thirty-one men. Antonio Pigafetta was one of the survivors.

Twenty years later, after several wrecks, ships no longer attempted to navigate the treacherous Straits of Magellan. Since Magellan had seen land south of his passage, cartographers modified their maps and joined Magellan's Land of Fire to *Terra Australis Incognita*.

Spanish Gold and English Sea Dogs

By the late 1500s Portuguese ships no longer dominated world exploration nor monopolized the spice trade routes. Fleets of Spanish galleons, loaded with gold and cannons, ruled the seas instead. Although royalty pocketed much of the South American bullion, merchants profited from the sale of spices, silk, and ivory. Even the Spanish Inquisition received a share of the bounty—for torturing foreign sailors and navigators. The seamen had to endure the rack until they relinquished their secret maps or "confessed" their knowledge of the ports, channels, and coastlines of Spain's many enemies.

England was a latecomer to exploring the riches of the New World. However, the British claimed two lucrative niches: the slave market and piracy. Sir John Hawkins was a man of the times. Not only did he establish the notorious slave route triangle from Europe to Africa to the West Indies, he seized every opportunity to loot rich Spanish ships that had the misfortune to sail across his path. These acts of piracy were publicly condemned by Spain—and England—but secretly condoned by Queen Elizabeth I. Her fleet of "gentlemen sea dogs," as she fondly referred to them, basked in royal approval.

Envious and penniless relatives of the prosperous sea dogs competed for favor. Some won jobs as seamen and merchants, following in the illustrious wake of their kin. John Hawkins had many such relatives, including one particular cousin, Francis Drake. Born in poverty, the son of a preacher, Drake's only advantage was this distant family tie. But his knack for navigation caught the eye of Hawkins, who promptly put Drake on his company payroll.

In 1572 Drake plundered Spanish colonies near the Isthmus of Panama, looted Peruvian llama pack trains, and sacked Spanish treasure ships. Fifteen months after his departure, he returned to England with a Spanish bounty on his head and a musket ball in his leg. The bounty only increased England's admiration for his heroism, but the wound permanently lamed Drake.

Wealth opened palace gates for Drake. Queen Elizabeth summoned him for a private audience, and Drake outlined his next grand scheme to her. Since Spanish ships had abandoned the Straits of Magellan, he would navigate the treacherous passage and claim land on the western coast of South America for England. Raiding unsuspecting Spanish colonies along the way would assure profits for investors. Drake would later disguise his true intentions by saying the expedition would search south of Tierra del Fuego for the fabled *Terra Australis Incognita*.

Queen Elizabeth secretly approved his plan. The tenuous peace between England and Spain did not prevent her, or others, from investing heavily in Drake's expedition. As he readied his fleet of five ships, spies for Spain tried to uncover the true purpose of the voyage but were unsuccessful. Even the seamen who signed on thought they were headed to Alexandria to pick up a cargo of currants.

While the crew loaded the ships with supplies, Drake furnished his cabin on the *Pelican* with all the comforts of home—luxurious oriental rugs, brocade cushions, and lavish wall hangings. Meals would be served on silver plates and the finest wines would fill his goblet. Drake hired four musicians to provide dinner music and took his drum to accompany them. But, ever secretive about his intentions, Drake hid a copy of Magellan's *Voyage* in one of his cabin chests.

On November 15, 1577, in the midst of heightened suspicions, Drake and his five ships had just sailed from Plymouth into the Channel when the worst storm in memory struck. Drake was forced to return to port for repairs. The sailors, ever sensitive to ill omens, predicted disaster for the journey, but on December 13 the ships once again sailed. When Drake continued south past Gibraltar, the crew realized that Alexandria was not their destination. Drake ignored their protests and refused to divulge his plans.

The ships skirted the western coast of Africa and then sailed across the Atlantic. They sighted Brazil and continued south into a series of severe storms that nearly sank the ships. Talk among the cold and disgruntled men turned mutinous—too dangerous for Drake to ignore. On June 20, 1578, he anchored at then-uninhabited Port St. Julian, the grisly site of the revolt against Magellan fifty-eight years earlier. Within days, human bones and the wooden gallows were found. The ship's carpenter, untroubled by superstitions and thrifty by nature, sawed the wood into drinking cups for the sailors.

Perhaps the mood of the place, with its history of insurrection, suited Drake, for he chose this site to end the brewing mutiny. On June 30 Drake charged one man, Thomas Doughty, with treason and beheaded him two days later. The grim act quelled the rebellion and Drake regained full control of the expedition.

On August 17, 1578, the fleet left Port St. Julian. Three days later Drake rounded the Cape of Eleven Thousand Virgins, and on August 23 entered the Straits of Magellan. He celebrated the occasion by renaming the *Pelican*; his flagship was now called the *Golden Hind*. The next day the men discovered abundant food on three small islands. "We found great store of strange birds which could not fly at all . . . in body they are less than a goose, and bigger than a mallard . . . in the space of one day we killed no less than 3,000. . . . They are very good and wholesome victuals," wrote Chaplain Francis Fletcher, Drake's nephew. Several Welsh sailors called the birds "pen gwynns" for their white heads.

Drake's voyage through the Straits of Magellan was smooth, but when he emerged on September 6 a ferocious hurricane blew the ships at least one hundred miles south of Tierra del Fuego. Streamers of white spray marked the crests of monstrous waves and glowed with eerie iridescence against the dark sky. "No traveler hath felt, neither hath there been such a tempest so violent . . . since Noah's flood," Drake later wrote. On the night of September 15 the men watched an eclipse of the moon, a sure sign of impending disaster.

Calamity struck on September 30. As another storm intensified, the *Marigold* floundered and sank in dense fog and high swells. Chaplain Fletcher, on watch duty, could do nothing but listen to the muted cries of the twenty-nine drowning men, muffled by the howling winds.

When the storm finally died, Drake steered north, leaving that tempestuous body of water where the Pacific and Atlantic "meet in a most large and free scope." With this discovery, Drake had unintentionally fulfilled his promise to explore the region south of Tierra del Fuego. The stormy waters would later be named in his honor: Drake's Passage.

Steering close to the western coast of South America, Drake looted settlements and Spanish galleons. A final tally will never be known but at least eighty pounds of gold, twenty-six tons of silver, and chests loaded with gems and pieces of eight filled the *Golden Hind*'s hold when the ship lumbered into Plymouth on September 26, 1580.

Queen Elizabeth and Drake's other investors hailed him a hero. King Philip II of Spain demanded full retribution and Drake's head. Queen Elizabeth promptly responded by knighting Drake in 1581. But glory and knighthood didn't change Drake—he continued to plunder Spanish colonies and ships until tensions between England and Spain culminated with the sailing of the Spanish Armada in 1588. Drake and Hawkins helped to defeat the most powerful nation in the world with a resounding victory. Years later, Sir Francis Drake died of dysentery during a raid in the West Indies.

Bouvet's "Unsuitable" Discovery

Interest in *Terra Australis Incognita* waned during the first half of the seventeenth century until a memoir was published in France in 1663. The book, which became a bestseller, detailed Paulmyer de Gonneville's expedition in 1503. During the voyage, a severe tempest near the Cape of Good Hope drove Gonneville far to the southwest—and to a land he described as paradise. He called the region Southern India and stayed for six months, enjoying the tropical climate and the hospitality of the people. His memoirs inspired two later French explorers to search the high southern latitudes for Southern India.

In 1735 Jean-Baptiste Charles Bouvet de Lozier persuaded the East India Company to finance an expedition to rediscover Gonneville's tropical paradise. It took officials three years to reach a decision. Finally, lured by visions of untold profits, the company provided Bouvet with two ships and supplies for eighteen months. He sailed for Southern India on July 19, 1738.

Bouvet reached Brazil in October and then sailed southeast. Temperatures dropped and soon the grim reality of the South Atlantic vanquished dreams of warm sunshine and lush vegetation. Cold, dense fog "wet our clothes like rain," wrote Bouvet on November 26. To help keep track of each other in the thick fog, the men on both ships fired guns at regular intervals during the day and hung lanterns at night.

Icebergs were spotted in December. "We saw some of these islands of ice," wrote Bouvet, "and they presented us with several amusing figures, resembling fortresses, houses, [and] ships." Bouvet optimistically predicted that Southern India must be close since ice formed on "high land." He pressed further south. To keep warm while working the rigging, the men bundled up in blankets and wrapped ragged canvas sacks around their feet. Bouvet ordered two kegs of brandy to be opened and distributed more clothing, but still the men suffered. "The cold was severe.... I saw sailors crying with cold as they hauled in the sounding line," Bouvet wrote in his report.

On January 1, 1739, land was sighted. "It was exceedingly high, covered with snow, and almost hid in fogs," wrote Bouvet, naming his discovery Cape Circumcision. For twelve days he tried to land, but pack ice and strong currents blocked his attempts. Bouvet finally turned his ships north when scurvy struck and food ran short. Although he thought he had discovered the tip of *Terra Australis Incognita*, he was honest in his appraisal, declaring the land to be "completely unsuitable as a staging post" for ships on their way to the Indies.

Kerguélen's Exaggeration

Another French explorer, Yves-Joseph de Kerguélen-Trémarec, was not so honest about his discovery. Also inspired by Gonneville's fabled land, he sailed with two ships, the *Fortune* and the *Gros Ventre*, on May 1, 1771, to discover Southern India. By February 1772, Kerguélen saw penguins diving among the icebergs, while petrels hovered over the ships. Convinced that land must be near, he took his cue from the penguins and steered southeast. On February 13 he caught a tantalizing glimpse of gray cliffs guarded by thick ice and high seas. The *Fortune* and the *Gros Ventre* maneuvered closer, but within five hours the land disappeared behind a curtain of fog. Kerguélen ordered the *Gros Ventre* to steer closer while he remained on board his flagship the *Fortune* at a safe distance. The captain of the *Gros Ventre* reached the shore of "South France" and claimed the land for France. By nightfall the weather deteriorated. When the *Gros Ventre* failed to return, Kerguélen waited less than a day and then abandoned his companion ship.

When Kerguélen reached France, he described his discovery as a land with deep woods, luscious meadows, and a climate favorable for crops—especially grapes. "No doubt wood, minerals, diamonds, rubies, precious stones, and marble will also be found," he added. "If men of a different species are not discovered, at least there will be people living in a state of nature." France rewarded Kerguélen with a second expedition. Meanwhile, the *Gros Ventre* had reached Australia, but the crew's much more realistic version of Kerguélen's discovery did not reach France until after Kerguélen had sailed in March 1773.

Kerguélen headed south with three ships crammed with seven hundred people, including farmers, merchants, and other colonists. By mid-December the dismal archipelago was sighted through a veil of fog, snow squalls, and high seas. For more than a month the ships skirted the islands and now, faced with the true nature of his discovery, Kerguélen renamed the islands Land of Desolation. A court-martial and a jail sentence were Kerguélen's rewards when he returned to France.

Dalrymple's Dream

France wasn't the only country in the 1700s to launch quests to discover *Terra Australis Incognita*. England's argonaut was Alexander Dalrymple, an ambitious astronomer, geographer, and self-appointed expert on the elusive southern continent. He was also a member of the Royal Society.

Earlier in his career, Dalrymple had translated a book of memoirs written by Pedro Fernandez de Quiros. This early-seventeenth-century explorer and mystic wore flowing Franciscan robes and believed he was Christopher Columbus reincarnated. He claimed to have seen the mysterious *Terra Australis Incognita* in 1606, and wrote vivid descriptions of a land teeming with lush vegetation and inhabited by strange but friendly antipodes. The rediscovery of Quiros's nirvana obsessed Dalrymple.

With the fervor of a zealot, Dalrymple soon convinced his Royal Society colleagues that a massive continent must exist between Easter Island and New Zealand, "populated with at least 50 million people." Since an expedition to Tahiti to observe an astronomical event, the transit of Venus, was planned for 1769, why not extend the voyage and claim the Southern Continent for England? The Royal Society agreed and chose him to command the voyage.

But the Royal Navy shattered Dalrymple's ambitions. They wanted a navy man to lead the expedition, and the person they had in mind was a self-taught man who was an expert surveyor and navigator. His name was James Cook.

The Farm Laborer's Son

James Cook was born in 1728 in a small Yorkshire village. Although his father recognized his son's quick mind and intuitive grasp of mathematics, educational opportunities were limited for the son of a farm laborer. So Cook worked as an apprentice in a general store in Strithes, weighing potatoes and measuring yards of cloth. As he listened to Arctic whalers and sailors spin tales of adventure, his own life must have seemed dull and full of drudgery. In the evenings, he hiked the coastal hills near the town and watched the three-masted ships sail by. Then, one day in 1746, he walked away from the store forever and signed on as an apprentice with Whitby, a coal-hauling company.

Cook worked for Whitby for nine years, and then abruptly joined the Royal Navy at a time when few men willingly volunteered. A long sea voyage was a death sentence in the mid-1700s. Scurvy, dysentery, and drowning took their toll on the seamen who had survived floggings, falls from the rigging, and food poisoning. Life expectancy was three to five years for a newly recruited Royal Navy man.

James Cook survived those grim statistics and rose through the ranks, gaining a reputation as a man who exceeded each assignment's objectives. While surveying the coast of Newfoundland in 1766, Cook made detailed measurements of the eclipse of the sun. He submitted his work to the Royal Society, which published his observations. A year later, the Royal Navy recommended Cook to command the expedition to the South Pacific to study the transit of Venus across the face of the sun. The Royal Society quietly agreed and, as a gesture of peace, invited Alexander Dalrymple to participate as an observer. Bitter after losing the role of commander to an underling—and a navy man at that—Dalrymple refused.

CAPTAIN COOK'S SECRET ORDERS

On August 26, 1768, the *Endeavour* sailed from England with ninety-four men, including botanists Joseph Banks and Daniel Charles Solander, two artists, and three astronomers. The gentlemen converted Captain Cook's large chart/library/mess cabin into a makeshift work area littered with plant presses, collection jars, numerous chests, and easels. The room resembled the den of mad alchemists rather than the quarters of a meticulous captain.

On deck, another menagerie vied for space. Dozens of chickens roosted in one of the boatswain's skiffs. Hogs toppled heavily with each pitch and roll of the ship. One feisty goat with good sea legs, gained from a previous circumnavigation, found peace among sacks of coal. On September 1 during a furious gale, the small boat that doubled as the coop was washed overboard, drowning "between 3 and 4 doz'n of our Poultry which was worst of all," wrote Cook. Now there would be no fresh eggs for the officers or the sick.

By Christmas 1768 Cook reached Tierra del Fuego. Snow covered the stark gray peaks. Ominous lenticular clouds, harbingers of violent weather, scudded across the sky. Icy winds numbed the mariners' hands, and sleet soon saturated their government-issue jackets and trousers. The goat was bundled in an extra coat but the hogs and sheep died from constant drenchings of cold spray. Joseph Banks and his entourage explored the botanical offerings of the bleak but beautiful landscape on a fine January morning. Within hours, sudden bitter winds and blinding snow squalls surprised the unprepared men. Huddling for warmth, the group survived—except for Banks's two servants, who had frozen to death during the night.

After rounding Cape Horn, Cook sailed northwest toward Tahiti. By mid-April 1769 the *Endeavour* anchored in Matavai Bay, previously discovered by Captain Willis. The islanders greeted Captain Cook cheerfully from their canoes—until they spotted a certain four-legged critter. The goat that had sailed with Willis had reveled in butting strangers overboard. But once the goat was securely tethered, the Tahitians boarded the ship and traded fresh fruit for hatchets and iron nails.

On June 3, 1769, after the transit of Venus had passed, Cook opened a sealed envelope that contained secret instructions. He was "to proceed southward" to 40° latitude and search for *Terra Australis Incognita*. If he found no land he was to sail westward and "fall in with the eastern side of land discovered by Tasman now called New Zealand." Alexander Dalrymple had suspected that New Zealand was the tip of his massive undiscovered paradise. Cook's mission was to prove him right or wrong.

After weeks of arduous sailing through a good chunk of Dalrymple's theoretical continent, Cook reached 40° south latitude in September. Several weeks later he sighted the mountains of New Zealand. From October 1769 to April 1770 Cook surveyed nearly twenty-five hundred miles of New Zealand's unknown coasts, circumnavigating both islands. At that point Cook could have returned to England, but he decided to sail west over the Tasman Sea to explore New Holland (Australia). The crossing was rough, with high winds and monstrous swells. Sails were "worn to rags" but John Ravenhill, a sailmaker in his seventies, hunched over the endless folds of canvas, stitching and patching the rips. On April 20, 1770, the lookout sighted the southeast corner of Australia. For the next two months, Cook charted the eastern coast of Australia, anchoring often to provide shore excursions for Banks and kangaroo hunts for the crew.

Although the men arrived fit and strong at Batavia, Java, bacteria from open sewers and bites of malaria-infected mosquitoes killed a good portion of the crew during their eleven-week stay. One exception was Ravenhill, the sailmaker, who remained drunk and ship-bound the entire time, stitching canvas shrouds for shipmates. Just before arriving home he died "of old age" in his sleep.

When the *Endeavour* reached England in July 1771, fifty-six of the original ninety-four men clambered down the gangway to greet friends and family. Joseph Banks enjoyed fame to the fullest, lecturing before the Royal Society and dining at the best tables in London. The goat retired to a landlubber's life of green pastures. James Cook presented his report to the Admiralty—and his suggestions for another voyage. Since vast tracts of southern oceans still remained unexplored, he volunteered to search the high southern latitudes for *Terra Australis Incognita.* The Admiralty approved his trip.

THE SNOW AND SLEET CONTINUED

To accomplish his new mission, James Cook scoured English shipyards and settled on two sturdy ships, renamed *Resolution* and *Adventure*. Both vessels met his exact requirements for stability on the high seas; roominess and comfort were not primary considerations.

Cook ordered all supplies, including ice anchors and hatchets, "such as used by Greenland ships." He had studied Bouvet de Lozier's report and knew what to expect: ice. To combat scurvy he ordered twenty-two gallons of carrot marmalade, nine tons of sauerkraut, and three tons of salted cabbage. Hogs, sheep, goats, chickens, and cattle transformed the upper decks into a barnyard. These animals would find new grazing grounds in Tahiti and New Zealand, if they survived the winds, sleet, and swells of the Southern Ocean.

On July 13, 1772, the *Adventure* and the *Resolution* sailed from Plymouth with the usual shouts and curses from the seamen, punctuated with grunts, bleats, clucks, and snorts from the animals. Cook's human entourage included not only the usual officers and seamen but also two Highland pipers and a marine drummer who could fiddle in a pinch. In place of Joseph Banks, the Royal Society substituted a gloomy father-son team of naturalists, the only individuals who applied for the position. Cantankerous and cynical, Reinhold Forster made life miserable for his long-suffering son. George Forster, just seventeen years old, had no choice but to endure his father's daily barrage of complaints about his "damp and moldy" cabin, uncouth seamen, and meals of "salt-horse." Queasiness assaulted the Forsters, but they found "the greatest relief from red Wine mulled with sugar and Pimento."

However, no one found comfort or escape from the violent downpours and high seas as Cook steered for Cape Town, South Africa. "I was suddenly driven out of my bed by a flood of Saltwater rushing down upon my head through the seams.... I was fairly soused," grumbled the elder Forster. The carpenter caulked the deck seams with oakum and tar but still rivulets of seawater continued to douse Forster and his cabin became "a scene of desolation... not one place being dry and secure."

Cook arrived at Cape of Good Hope in October. Rumors of Kerguélen's discovery of "paradise" were rampant, but Cook didn't alter his plans. More sheep and poultry were herded aboard the ships, and soon Cook set his course toward Bouvet's Cape Circumcision, the possible tip of the Southern Continent. To help chart its position should he find it, Cook carried four experimental chronometers, two for each ship. During the voyage, three of the four large clocks stopped working, victims of low temperatures and salty spray. Only the chronometer designed by John Harrison functioned beyond Cook's expectations.

On November 23 the two ships headed south from the Cape of Good Hope. Mountainous waves curled over the bow, knocking down the sheep, drowning the chickens, and soaking the helmsman. "My poor dog wanted to take shelter in my Cabin, against wind, rain, cold, and the Sea," wrote Forster. "He found some relief against the three first inconveniences but he could not escape entirely the last."

The sheep weren't so lucky. Sopping wool coats soon stiffened with ice. "So very cold... that great part of the Sheep brought from the Cape died; served fresh Mutton to ye People," wrote Lieutenant Pickersgill.

The first "ice island" loomed through the fog on December 10, and day by day the number of bergs increased. "Sailing here is render'd very Dangerous, Excessive Cold, thick Snows, islands of Ice... ye people

Numb'd," wrote Pickersgill. Layers of ice coated the sails like plates of transparent armor and thick icicles hung from the rigging and ropes. Cold weather garb was unpacked from tightly sealed chests and distributed to the men. So, too, were the hardy viruses that had survived within the heavy folds of jackets and trousers. Within a week, most of the seamen suffered from colds and influenza. Reinhold Forster wrapped a long woolen scarf around his neck and took to his bed.

Fog, violent squalls, and the ice-littered sea wearied the men. "We cannot see the Length of the Quarter Deck," Lieutenant Charles Clerke jotted in the logbook on December 16. "These 24 hours we've abounded rather too plentifully in Fogs and Ice Islands—either one or the other we can very well cope with, but both together is rather too much."

A few men showed flashes of imagination that transformed icebergs into ruined villages, abandoned cathedrals, or strange temples. "The Ice appeared like Coral Rocks, honey-combed and . . . exhibited such a variety of figures that there is not an animal on Earth that was not in some degree represented by it," wrote the usually laconic Captain Cook.

Soon tangles of seaweed drifted by the ships. Diving petrels were spotted, but frequent sightings of land turned out to be distant icebergs dappled with shadows. Officers tracked shoals of porpoising chinstrap penguins with their telescopes until the black and white flashes were lost among the icebergs. Bouvet's Cape Circumcision continued to elude Cook.

However, penguin watching relieved the tedious hours for the men on duty. "Pass'd an ice island which had 86 Penguins upon it—I counted them," boasted midshipman Mitchell. Pickersgill became a dedicated student of penguin behavior. "We fired two 4 pounders at them but Missed . . . they wheel'd off three deep and March'd down to the water in a Rant; they seemed to perform . . . so well that they only wanted the use of Arms to cut a figure on Wimbledon Common." William Wales, an astronomer, analyzed their call. "Their Note is not much unlike that of a Goose, which some of our people so happily imitated that they could draw them almost close to the Ship's Side."

The men celebrated Christmas with second helpings of pudding and cups of grog. The pipers wheezed soulful tunes on the bagpipes and the fiddler played reels on the violin. "Mirth and good humor reigned throughout the whole Ship," wrote Captain Cook in his journal. "Savage noise and drunkenness," noted the Forsters.

During January 1773 the number and size of the icebergs increased. When the fog lifted, the men stared with astonishment—and terror. Immense flat-topped ice islands surrounded both ships. Waves crashed against the bergs' one-hundred-foot sides and surged beneath Gothic arches into deep chambers that pulsed with indigo blue and reverberated with the roar of the sea. The Forsters thought "the whole scene looked like the wrecks of a shattered world, or as the poets describe some regions of hell . . . sailors' oaths and curses re-echoed about us on all sides." Botanist Andreas Sparrman wished he had heard less swearing from Captain Cook, who "while the danger lasted, stamped about the deck and grew hoarse with shouting." Astronomer Wales just stared in horror as "a very large Island of Ice burst . . . into three large and many small pieces." He well understood that if a ship were too close to an iceberg "paring its skin," no further tales would be told.

At least fresh water was not a problem. When the casks ran low, the men hoisted the boats over the side, rowed to the nearest iceberg, and scooped large chunks out of the sea with baskets. The irony of the situation was not lost on Pickersgill. "I believe this is the first Instance of drawing fresh water out of ye Ocean in hand baskets," he mused. During the next few days they hauled over forty tons of ice on board the ship. "Thus we can say," wrote Forster, "we took up our water, in the midst of the Antarctic Ocean, with bare hands and laid it up in piles on the Quarter deck." Extra rations of brandy helped the men cope with hands numb and white with cold "without flinching."

Freezing temperatures offered the men no respite from the required weekly baths; however, cleanliness alone couldn't halt scurvy. Reinhold Forster peered into one man's mouth and discovered "ulcerated Gums and the Teeth were so loose that they lay quite sideways." Cook promptly ordered "infusion of Malt" and bowls of marmalade of carrots. To Forster's surprise, the man soon chewed tough salt beef with no further problems.

On January 17, 1773, Cook crossed the Antarctic Circle, "a place where no Navigator ever penetrated," wrote Reinhold Forster, ". . . and where few or none will ever penetrate." The next day heavy pack ice blocked any further progress to the south, and Cook turned north and then east after reaching 67° 15' south latitude. Although bells clanged on both ships every few minutes to keep track of each other in dense fog, the *Adventure* disappeared during a blinding snowstorm. Cook searched for three days and then set his course toward Queen Charlotte Sound, New Zealand, their predetermined place to rendezvous in May.

High seas and terrible storms prevented Cook from approaching the eastern side of Tasmania, which he had planned to survey. He then sailed across the Tasman Sea and anchored in Dusky Bay, New Zealand, on March 26. Days were whiled away escorting the Forsters on their "botanizing" expeditions. Seal hunts were frequent, and Cook wrote expansively about the seals whose fur was "comparable to Otter." Their skins of "that most Valued animal" were used for parts of the rigging; the fat yielded oil for their lamps; and the meat tasted much better than their two-year-old salt beef.

On the morning of May 19 the *Resolution* and *Adventure* were reunited as planned in Queen Charlotte Sound, and both ships sailed for Tahiti to winter. Cook then returned to New Zealand the following October but, just as land was in sight, "the Sea rose in proportion with the Wind," wrote Cook, "so that we not only had a furious gale but a mountainous sea... we had the mortification to be driven off from the Land." The *Resolution* labored against monstrous swells, deep hollows, and screaming winds. A sudden violent jerk of the tiller flipped the helmsman over the wheel. "The man resembled much a seal in substance and make," noted Wales, "and his fall on the deck made exactly the same squash that a Bag of Blubber would have done." Helmsmen were a hardy lot and this man continued his turn at the wheel.

Below the deck, the Forsters had problems, too. "A huge mountainous wave struck the ship on the beam, and filled the decks with a deluge of water," wrote George. "It poured through the sky-light over our heads, and extinguished the candle, leaving us for a moment in doubt whether we were not totally overwhelmed and sinking into the abyss."

The *Adventure* disappeared during the tempest and was not seen again during the rest of the expedition. Cook searched for the lost ship, but by November 27 he could wait no longer and sailed south. Ten days later the *Resolution* was exactly 180 degrees opposite London. The Forsters clicked their cups and saluted the "Antipodes at London." Wales, too, toasted his friends "who may now rest perfectly satisfied that they have no Antipodes besides Penguins and Petrels, unless Seals can be admitted as such."

By Christmas 168 ice islands surrounded the *Resolution*, but the men were "cheerful over their grog," joking about how "they would certainly die happy and content on an ice island, with some rescued keg of brandy in their arms," wrote Sparrman, the botanist.

Early on the morning of January 30, 1774, a vivid white band of light stretched across the horizon. As Cook approached the edge of the pack ice, he counted ninety-seven distinct ice hills which "looked like a ridge of Mountains, rising one above another, till they were lost in the clouds... and seemed to increase in height to the south." His position was 71° 10' south latitude and 106° 54' west longitude, opposite what is now Marie Byrd Land in the Amundsen Sea sector. Polar historians still argue whether this "ridge of mountains" was a mirage of the Walker Mountains on Thurston Island, approximately one hundred miles to the southeast, or huge blocks of rifted pack ice.

For a few hours that morning the weather had been clear, but haze and fog obscured the view by noon. Cook was intrigued and several officers were convinced that land was very close. "It was indeed my opinion as well as the opinion of most on board, that this Ice extended quite to the Pole or perhaps joins to some land, to which it had been fixed from the Creation... it is here... where all the Ice we find scattered up and down to the North are first form'd."

Since ice blocked the way south, Cook turned northward with feelings of regret—but also relief. "I, who had Ambition not only to go farther than any one had been before, but as far as it was possible for Man to go, was not sorry at meeting this interruption."

The *Resolution* swept across the Pacific, calling on Easter Island, the Marquesas, Tahiti, and New Caledonia before returning to New Zealand for a brief stay. Then on November 11, 1774, Cook sailed back across the Pacific and reached Tierra del Fuego. "I have now done with the Southern Pacific Ocean and flatter myself that no one will think that I have left it unexplored," he wrote on December 17.

On December 29 Cook steered for the eastern side of Tierra del Fuego, where "the Whales are blowing on every point of the Compass," wrote Clerke. One island was a haven for fur seals. "The whole shore was covered with them," noted Cook, "and by the noise they made one would have thought that the island was stocked with Cows and Calves." Undisturbed for countless generations, the fur seals had no reason to fear Cook or his men as they walked among the vast herds. Cook's unbridled enthusiasm for that "most Valued animal" would soon have disastrous consequences for the fur seals.

Cook then sailed east from Tierra del Fuego. On January 14, 1775, just when the mariners least expected it, they sighted land. Foul weather and high seas prevented Cook from landing until January 17. Early that morning Cook, the Forsters, and Sparrman were rowed to shore. "The head of the Bay, as well as two places on each side, was terminated by a huge Mass of Snow and ice of vast extent," Cook wrote. "The inner part of the Country was not less savage and horrible: the Wild rocks raised their lofty summits till they were lost in the Clouds and the Valleys laid buried in everlasting Snow. Not a tree or shrub was to be seen, no not even big enough to make a tooth-pick."

Cook quickly displayed the flag and took possession of the land he named South Georgia. "A volley of two or three muskets were fired into the air," Forster related, "and the barren Rocks re-echoed to the sound, to the utter amazement of the Seals and Penguins, the inhabitants of these newly discovered dominions."

Most of the men were overjoyed, believing that at last they had discovered *Terra Australis Incognita*. Cook continued southward and surveyed the eastern coast for the next three days. When the land abruptly trended to the northwest, Cook knew the truth. South Georgia was an island, not the elusive continent. He named the southern tip of the island Cape Disappointment.

Cook sailed southeast from South Georgia. By the end of January, domes of ice-covered land appeared on the horizon. Within a few short days, Cook concluded "that what we had seen, which I named Sandwich Land was either a group of Islands or else a point of the Continent, for I firmly believe that there is a tract of land near the Pole, which is the Source of most of the Ice which is spread over this vast Southern Ocean.... I can be bold to say, that no man will ever venture farther than I have done and that the land which may lie to the South will never be explored . . . a Country doomed by Nature never once to feel the warmth of the Sun's rays, but to lie forever buried under everlasting snow and ice."

Cook searched the icy fringes of the southern Atlantic for Bouvet's Cape Circumcision but didn't linger. The long voyage had taken its toll on his men. Meals of moldy biscuits and rancid salt beef almost led to violence. "Confin'd Messiers Maxwell, Loggie and Coglan for going into the Galley with drawn knives and threatening to stab the cook," wrote Clerke, "and Read the Articles of War to the crew." Life was dismal for Reinhold Forster, too. "I do not live, not even vegetate; I wither; I dwindle away."

On March 21, 1775, at Cape Town, South Africa, Cook received a letter from Captain Furneaux of the *Adventure*, dated one year earlier. After the ships were separated near the coast of New Zealand, Furneaux landed just a few days after Cook had sailed south. The Maori ambushed ten sailors and when Furneaux stumbled upon their cannibalized bones, he sailed to England without a backward glance.

The *Resolution* anchored in Plymouth on July 30, 1775, with the loss of only four men—none from scurvy. When Cook presented a paper about his antiscurvy precautions to the Royal Society, he was awarded the Copley Medal, the Society's highest honor. Soon the Admiralty approached him to lead another expedition, and Cook accepted without hesitation. He sailed from England on July 12, 1776, his final voyage. On February 14, 1779, Captain James Cook was stabbed to death on the shore of Kealakekua Bay, Hawaii. The Forsters later published their journals—and never returned to the South Seas.

When James Cook's journals from his second voyage were published in 1777, public and scientific interest in *Terra Australis Incognita* quickly waned. What possible discoveries could justify expensive expeditions across roiling seas "pestered with Ice Islands" to explore desolate lands bombarded by endless blizzards?

So, men of science dismissed Aristotle's "frigid zone" and Ptolemy's Unknown Southern Land for the next fifty years. However, a powerful group of English and American businessmen studied Cook's and Forster's published journals. There, charted and described in meticulous detail, were the locations of a very valuable commodity—the Antarctic fur seal.

• 2 •

IN SEARCH OF SEALS WITH FURTHER DISCOVERIES

During the mid-1700s, waves of sleek, dark bodies glided through stormy Antarctic seas, converging and congregating on the rocky shores of islands with no names. Each October, male fur seals arrived on the windswept beaches and staked out territories above the tide line, but close to the pounding surf. Younger, less experienced males hauled out too, but were chased from the prime parcels and had to settle for turf in the dense tussock grass.

Pregnant females arrived by November and were soon herded into harems, guarded by the strongest males. With oblique glances toward younger males, snouts pointing skyward and guard hairs bristling around their necks, the bulls roared and whimpered as they patrolled the outer boundaries of their territories. Demarcations shifted, increasing or decreasing as females and pups were gained or lost to more dominant males.

Fleets of small ships would soon emerge from the perpetual fog. Weaving their way through fractured channels in the pack ice, the ships would slowly approach those distant shores, the realm of fur seals since time immemorial.

The Rush for Fur

On December 16, 1773, while Captain James Cook battled "ice-islands" and dense fog in the Southern Ocean, American rebels stormed three British ships and tossed seven thousand pounds of tea into Boston Harbor. Samuel Enderby, Inc. of London, the firm that owned the ships and tea, refused to deliver additional cargo to the colonists, opting for less hostile—and more profitable—sealing and whaling ventures. During the American Revolutionary War, British ship owners had difficulty finding enough able-bodied sailors to operate their vessels. Quashing a rebellion was far more appealing to British men than long whaling voyages on the high seas. This manpower problem was solved by successful British blockades along the northeastern seaboard of the colonies. The crews from captured American whaling ships were simply hijacked and forced to serve on British commercial whalers.

After the war, New England whaling industries suffered not only from a shortage of sailors, but also from a decrease in the demand for whale products. Thrifty colonists now lit their homes with tallow instead of expensive whale oil, and bright-burning spermaceti candles glowed only in the parlors of the wealthy. When New England investors heard rumors that sailors from Captain Cook's last voyage had sold tattered and ill-preserved sealskins in China for enormous sums of money, shipping firm owners took note. They studied Cook's maps and detailed narratives. The fur seal islands were remote but accessible, they learned. As an added bonus, elephant seals could also be killed, their blubber then rendered into high-quality oil that burned and lubricated almost as well as the finest whale oil. And hunting seals was far less dangerous to ships—and men—than hunting whales. Hefty profits would soon inflate the bottom line on ledgers.

Americans were quick to capitalize on the trend toward cheap fuel and fine furs. In 1784 the *United States* sailed from Nantucket to the Falkland Islands. The crew slaughtered enough elephant seals for three hundred tons of oil and skinned about seven thousand fur seals. When the ship reached New York two years later, the seal pelts were sold to a New York merchant who, in turn, shipped the merchandise, described as "otter skins," to Canton, China, and sold them for about sixty-five thousand dollars. Whether the thick furs were otter or seal probably didn't matter to the Chinese, since they had already developed a process for removing the stiff guard hairs from seal pelts to expose the soft underlying fur.

Sealing pumped money into the New England economy and revived the shipbuilding industry. More ships soon sailed for New Zealand, the coast of Patagonia, Tierra del Fuego, the Falkland Islands, and South Georgia to search for seals. Within a few years, America had cornered the fur seal market in China and Europe. The monopoly ended in 1796 when an Englishman, Thomas Chapman, perfected a technique similar to the Chinese method for removing the outer guard hairs from sealskins. Ships flying the Union Jack now challenged America's dominance for fur.

In 1800, American Edmund Fanning arrived at South Georgia and found seventeen ships, most of them British, anchored in harbors, their crews knee-deep in pelts and carcasses. Elephant seal blubber sputtered and melted in numerous iron try-pots. As competition for seals accelerated, secretive captains burned logbooks as a safeguard against inquisitive—and talkative—sailors when new beaches were discovered. By 1803 only a few hundred fur seals hauled out on South Georgia's rocky shores and sealers searched for more profitable terrain. During the next ten years, the seal population on South Georgia stabilized and slowly increased. But the sealers returned for a second assault from 1813 to 1821—and the slaughter was relentless not only on South Georgia but on other islands as well.

The work was difficult and dangerous. Sealing crews were stranded on far-flung islands for months—even years—without adequate food, clothing, or shelter. Yet, sealers eagerly wrote their names or made their mark on American and British ships' rosters. Each fur pelt or cask of oil put a promised halfpenny in the sealers' pockets.

HUTS AMONG THE HAVOC

Each year, the sealers arrived at nameless beaches by mid to late October, a time when severe weather often hampered shore operations. Sudden and violent *katabatics*, or gravity winds, were especially dangerous to men and equipment. "The light cedar whale-boat moored at the stern of this ship... turned over and over before striking the water, the same as a feather attached to a thread, and blowing in the wind," wrote Captain Edmund Fanning in 1800 at Royal Bay, South Georgia.

The sealers carried their meager belongings and rowed toward a beach lined with sleek fur seals, or a cove that echoed the protracted roars of bull elephant seals. As soon as try-pots and other tools of the trade were unloaded, the men set about their bloody business. "Seals were perfectly tame," wrote one captain in the early 1800s. "They would come up and play among the men who were skinning their companions." Female elephant seals were whacked on the head and then bled before being flensed. The blubber was carefully cut into thin strips and tossed into the try-pots for rendering. The great bull beachmasters were especially prized. Sealers crept up on them and bludgeoned the males on their sensitive noses. As the seal reared back, roaring with outrage and pain, the sealers struck him in the heart with a lance. If a bull panicked and dashed for the sea, the men blocked his escape and shook an iron bucket filled with pebbles in front of his face. Frightened by the loud rattling noise, the elephant seal would retreat to shore—and death. As the men stripped the carcass, ravenous petrels and skuas circled above, waiting to feed on the remains.

Female fur seals were easiest to kill. A quick blow on the head was all it took—most of the time. "The snuffling of the seal, and the sound of the blood spouting and fizzling into the snow... was hardly nice," wrote W. Burn Murdoch, "but when the red carcass sat up and looked at itself, I looked up to see if God's eye was looking."

When the day's work was done, the men cooked thick slices of elephant seal tongue. Then they crawled underneath their overturned boats, wrapping stiff sealskins around their bodies for warmth. Boots made from king penguin skins with the feathers turned inside protected bare feet from the cold.

For longer stays on the islands, the sealers constructed mud and tussock-grass huts, lining the walls with elephant seal hides. In the middle of the single room, chunks of blubber crackled in an iron pot that rested on a grate of seal bones. Oil from blubber fueled handcrafted lamps, and penguin feathers functioned as wicks. Seal bladders, stretched over holes in the walls for light, soon blackened from the oily smoke.

Hardship, disease, and loneliness broke bodies and minds. Joints and knuckles ached with rheumatism from the cold and damp climate; sores on hands from knife cuts continually festered. Only the ingenious and industrious survived. When the men tired of seal meat, they lit fires at the foot of sheer cliffs to entice returning night petrels to crash into the gray stones. Albatross and penguin eggs were snatched from nests in the tussock grass. The sealers fished with hooks whittled from bones and boiled bitter-tasting plants to disguise the pungent flavor of seal and blubber soups. Blood, they discovered, was a good solvent for dissolving the layers of soot and grease from clothes and bodies.

When no more seals hauled out on the beaches, the men scanned the horizon, waiting for their ships to return for the piles of sealskins and casks of oil. Sometimes the sealers waited for years. But sooner or later, either their ship or another rescued them and their bounty. Only the iron try-pots, dilapidated huts, and wind-scoured seal bones remained on the silent shores.

The South Shetlands and Antarctic Peninsula Revealed

After the War of 1812, the price for sealskins skyrocketed in Europe. Where millions of fur seals had once crowded the shores of the Falklands, South Georgia, Tierra del Fuego, Macquarie, and Campbell Islands, only a fraction returned to their breeding grounds by 1819. The search for new territories escalated.

Early in 1819, a British merchant captain named William Smith agreed to carry a cargo of mining equipment from Montevideo (near Buenos Aires) to Valparaiso, Chile. As he rounded Cape Horn, gales pushed his ship southward. On February 19, 1819, Smith spotted land through a curtain of sleet and snow. He sailed to within ten miles of the unknown coast but did not want to risk his valuable cargo by trying to navigate a maze of floes, icebergs, and strong currents.

After he delivered his cargo, Smith told Captain William Shirreff, a senior officer of the British Royal Navy, about the new islands to the south. Captain Shirreff greeted the news with skepticism and ridicule. Stunned by Shirreff's dismissal of his discovery, William Smith sailed for Montevideo in May—and made a detour southward in June toward the land he had seen. But he could not penetrate the winter pack ice.

By the time Smith returned once again to Montevideo, a group of American sealers had already heard rumors of new lands with abundant fur seals. Several of them tried to bribe Smith with lucrative promises of fabulous wealth, if only he would tell them the coordinates of the land. "Your Memorialist having the Good of his country at heart... resisted all the offers from the said Americans," wrote Smith. In September, Smith again veered to the south from Cape Horn. On October 15, 1819, the weather cleared and he saw a group of islands. The next day, Smith jumped ashore on an island later named King George Island. He hoisted the Union Jack, cheered, and took possession, calling the land New South Shetland.

The British Royal Navy finally acknowledged Smith's discovery and prepared a voyage to the new territory. Captain Shirreff placed Edward Bransfield in command of the brig *Williams*, with William Smith as pilot. They carried a year's worth of supplies, including cattle, as a safeguard against starvation if pack ice should trap the ship. On December 20, 1819, the *Williams* sailed from Valparaiso back to the Shetlands to survey the new islands—and to estimate the number of fur seals that covered the beaches.

Bransfield and Smith reached the Shetlands on January 18, 1820, and took formal possession once more on King George Island four days later. For the next ten days, Bransfield surveyed the islands. Then, on January 30, the fog lifted briefly, unveiling land to the southwest. A curtain of mist descended once more as they sailed toward the land but cleared again just as they rounded three mammoth icebergs. The magnificent mountains of the Antarctic Peninsula were at last revealed in all their glory.

Bransfield and Smith returned to South America with news of their discovery—and a large cargo of pelts gathered from the shores of the South Shetland Islands.

Political Intrigue

United States State Department agents, stationed in several South American ports, quietly noted Smith's discovery. Couriers carried letters to Secretary of State John Quincy Adams, urging him to send a ship to the South Shetland Islands as soon as possible, but Adams delayed informing President James Monroe.

Pressure mounted in the U.S. State Department to protect American interests during the anticipated 1820–21 fur rush. John Quincy Adams wrote a blunt letter to President Monroe, dated August 16, 1820. "The British Government just now have their hands so full of Coronations and Adulteries, Liturgy, Prayers and Italian Sopranos... High Treasons and Petty Treasons, Pains, Penalties and Paupers, that they will seize the first opportunity they can to... make a question of national honor about a foot-hold... upon something between Rock and Ice-Berg, as this discovery must be." Adams' final recommendation was to send a navy frigate "to take possession.... There can be no doubt of the right... for protecting the real object—to catch seals and whales."

In the fall of 1820, bold statements appeared in a number of newspapers and journals claiming that Americans had not only known about the Shetland Islands but had hunted seals there for years. However, no logbooks ever surfaced to justify such claims, and no seal carnage blighted any beaches at the time of Smith's initial discovery. The South Shetland Islands and Antarctic Peninsula were pristine—but not for long.

Claims, Counterclaims, and Controversies

At the same time that Bransfield and Smith were exploring the South Shetland Islands, at least one American vessel explored hidden coves for shelter. The *Hersilia*, captained by James Sheffield, had sailed from Stonington, Connecticut, on July 20, 1819. Nathaniel Brown Palmer, just nineteen years old, served as second mate. When the brig arrived in the Falkland Islands, Palmer heard sealers boast about new islands, flush with fur seals, just south of Tierra del Fuego. The *Hersilia* sailed south on January 11, 1820, to search for the rumored islands.

According to Palmer's logbook, the South Shetland Islands were sighted by the Americans on January 18; January 24–28, sealing gangs gathered pelts on several islands. Palmer made no entries in his logbook from January 29 through February 7—the same days when Edward Bransfield and William Smith charted the tip of the Antarctic Peninsula and landed on Clarence Island. Later historians speculated that Palmer, or any other crew member, may have caught a brief glimpse of the peninsula from their unrecorded position during those ten days, but no definitive records have been found. On February 8, the *Hersilia* sailed north and arrived in Stonington, Connecticut, in May 1820 with over eight thousand fur seal pelts.

Nathaniel Palmer returned to the South Shetlands the following season in charge of the forty-seven-foot *Hero*. While sealing in the Falkland Islands, Palmer learned that many British ships were sailing toward the South Shetlands. Palmer related this disturbing news to Benjamin Pendleton, commander of the Stonington fleet. Pendleton worried that fierce competition for fur seals could trigger a showdown with the British. On November 12, 1820, he sent Palmer on a mission to find a more defensible harbor—and one with more "seals up."

Palmer sailed from Rugged Island on November 14, and by evening he sighted Deception Island. The next day Palmer entered a narrow passage later called Neptune's Bellows and explored a large protected bay. On November 16 he anchored in Pendulum Cove, "an Excellent Harbor secure from all winds." Later that day, Palmer sailed from Deception Island southward in clear weather. Eight hours later Trinity Island, flanked by the precipitous mountains of the Antarctic Peninsula, was in full view. On November 17 Palmer wrote that he "found the sea filled in immense Ice Bergs—at 12 hove to under the Jib Laid off & on until morning at 4 AM made sail in shore and Discovered a strait." The channel Palmer found was later named Orleans Strait. Ice—and the absence of seals—dissuaded him from further exploration, and he headed north to search for a safe harbor for the American fleet. He discovered a perfect one: Yankee Harbor on Greenwich Island.

What had begun as a routine mission to search for a harbor ended with the rediscovery of the Antarctic Peninsula—ten months after Edmund Bransfield first charted it. When Bransfield's official report of his expedition later disappeared, American historians, geographers, and politicians declared Palmer to be the true discoverer of Antarctica. For the next 125 years, controversy raged and slanderous aspersions were hurled on both sides of the Atlantic. Tales of forgery, insidious plots, and character assassinations appeared in journals and newspapers. The achievements of Bransfield and Palmer were ruthlessly maligned—a fate neither man deserved. Eyewitness accounts from Bransfield's expedition were published in 1822 and, more important, his original chart still exists. Nowadays most historians credit Bransfield with the discovery of the Antarctic Peninsula—but not the continent. That honor belongs to Admiral Thaddeus von Bellingshausen.

For the Glory of Russia

On March 25, 1819, in St. Petersburg, Czar Alexander I announced his desire to "help in extending the fields of knowledge" by ordering two expeditions to the ends of the earth, "the higher latitudes of the Arctic and Antarctic Oceans." The purpose of the Arctic voyage was to search for the Northwest Passage near the coast of Alaska. At the time, convoys of wagons and packhorses were forced to haul furs and supplies through thousands of miles of forests and across the barren windswept steppes of Siberia to St. Petersburg, a costly and time-consuming trek. If the expedition discovered the Northwest Passage, then Russian ships could carry the fur pelts from Arctic villages, sail through the Northwest Passage, and then cross the Atlantic Ocean to St. Petersburg.

An expensive alternative was moving cargo from Siberian ports to St. Petersburg by sea, via Cape Horn or the Cape of Good Hope. Unfortunately, the Russians had no established ports along either route and had to rely on foreign vessels to carry their goods. If lands not already claimed by Great Britain, France, or Spain could be found, then the Russians could establish safe ports for their ships. Subantarctic islands and the southern continent itself—if it existed—were strong geographic contenders for potential ports.

The man Czar Alexander selected for this historic voyage was Thaddeus von Bellingshausen, a career naval officer and a veteran of long sea journeys. Born in Estonia in 1779, the year that James Cook died, Bellingshausen entered the naval academy when he was ten. In 1803–1806 he sailed with A. J. von Kruzenstern on Russia's first voyage around the world. Bellingshausen was later assigned to Sebastopol, a Crimean port on the Black Sea. The Czar's summons reached Bellingshausen in late April 1819. By the time he arrived in St. Petersburg after a grueling, one-thousand-mile overland trip, Bellingshausen had just six weeks to organize the Antarctic expedition.

Stacks of official documents had to be read, supplies purchased, and the two ships readied for the arduous sea voyage. Although the Imperial Academy of Science had been contacted to provide Bellingshausen with someone to record scientific observations, no one volunteered for the long voyage. However, science was very important to the Czar, and two German naturalists were included to report on meteorology, geology, astronomy, and oceanography.

On July 26, 1819, two small ships sailed from Kronstadt, a port near St. Petersburg. Bellingshausen commanded the *Vostok*, and Mikhail Lazarev was in charge of the *Mirnyi*. In both ships' holds were odd assortments of gifts—knitting needles, kaleidoscopes, and tambourines—for foreign officials and natives inhabiting any discovered land.

Problems beset Bellingshausen almost immediately. When he arrived in Copenhagen, both German naturalists backed out. "They had refused on the ground that too little time have been given them to complete preparation for the voyage," noted Bellingshausen, "but I . . . cannot help thinking that all that a scientist need bring with him is his scientific knowledge; books were to be found at Copenhagen . . . and all the bookshops in London would have been at their service." Just the sight of the two ships would have been reason enough for the last-minute cancellation—neither vessel inspired confidence. Both were constructed from unseasoned pinewood, and only a thin sheet of copper protected the *Vostok*'s hull from dangerous encounters with ice. The *Mirnyi* didn't even have that small margin of safety—its hull was lined with tarred canvas.

Burdened with two fragile ships and the responsibility for scientific observations to boot, Bellingshausen must have felt overwhelmed. But he rose to the challenge and purchased scientific equipment, books, and charts. He would do the work of the naturalists. As the *Vostok* and *Mirnyi* heaved away from England in early September, one of the most remarkable circumnavigations in Antarctic history was about to begin.

A PUZZLING SIGHT: FIRST VIEW OF THE CONTINENT

As Bellingshausen approached South Georgia in December 1819, dark clouds brooded over the island, whales spouted near ice floes, and penguins raised themselves out of the water "to call to each other as people do in a thicket," noted Bellingshausen. South Georgia, he summarized, "was inhabited only by penguins, sea elephants and seals; there were but few of the latter, since they are killed by the whalers." During the next two days he surveyed the southern coast of South Georgia to complete Captain Cook's charted northern section, and then both ships headed east toward the northern extremity of Cook's Sandwich Land.

Bellingshausen soon realized that Sandwich Land was not an extension of the southern continent but several clusters of islands. He named the northern group of three after the Marquis de Traversay, Russian minister of naval affairs. Although the largest island, Zavodovski, was an active volcano, Bellingshausen allowed a small party to land and collect rocks. "Our travelers," wrote Bellingshausen, "had gone almost half way up the mountain and had found the ground warm. A particularly bad smell from the great quantity of guano from the penguins forced them soon to return." Sulfurous volcanic gas was the real culprit.

Continuing south in heavy mist, the ships passed through bands of bergs and brash ice. One January evening near Candlemas Island a dense curtain of fog descended, hiding nearby icebergs from view. To Bellingshausen, nothing was more terrifying than the sound of thunderous waves slamming against icebergs in dense fog. It was a potential death knell for his wooden-hulled ships. With only their ears to guide them through the danger, the sailors leaned far over the sides of the ship, listening for the telltale sounds of disaster. "The noise of waves breaking on the ice and the screaming of the penguins created a most unpleasant feeling," he wrote succinctly.

A few days later they spotted penguins on a floe and launched a boat. The men managed to stuff thirty into sacks and haul them back to the ships. "I ordered a few to be sent to the mess, a few to be prepared for stuffing, and the remainder were kept on board and fed on fresh pork." The diet didn't agree with the penguins, and soon the men were busy making caps out of the skins and greasing their boots with the fat. Several days later they captured more penguins but had no space to keep them—except in the chicken runs and in a bathtub.

On January 26 Bellingshausen crossed the Antarctic Circle, the only explorer since Captain Cook to do so, yet he made no mention of this achievement in his journal. The next day he was within sight of the Princess Martha Coast at 69° 21' 28" south latitude and 2° 14' 50" west longitude, but snow squalls hid the elusive continent. "We encountered icebergs, which came in sight through the falling snow looking like white clouds," he wrote. "We observed that there was a solid stretch of ice running from east through south to west. Our course was leading us straight into this field, which was covered with ice hillocks." Bellingshausen was less than twenty miles from the ice shelf that extends from Princess Martha Coast.

Meanwhile, about sixteen hundred miles to the west, Edward Bransfield and William Smith were approaching the tip of the Antarctic Peninsula. They sighted it in clear weather three days later on January 30.

The barometer continued to fall, and for the next five days Bellingshausen sailed eastward in abominable weather, skirting the coast of eastern Antarctica without realizing it. On February 2 Bellingshausen took advantage of the first good weather since leaving the South Sandwich Islands. Clothing and bedding were hauled to the deck to be aired and dried. The men carried red-hot cannon balls in iron kettles through the lower quarters and the officers' cabins to dry the damp moldy wood. This method of airing the ship could only be done during very calm weather. If the pots were dropped during a sudden lurch or deep roll, fire would sink the ship.

On the morning of February 17, Bellingshausen noticed "a vivid brightness" southward on the horizon. By midafternoon, he "observed a great many large high flat-topped icebergs, surrounded by small broken ice.... The ice toward the south-south-west adjoined the high icebergs.... Its edge was perpendicular and formed into little coves whilst the surface sloped upwards toward the south to a distance so far that its end was out of sight even from the mast-head." Although he did not realize it, Bellingshausen was gazing at the Antarctic continent. This sector, later named the Princess Astrid Coast, has steep vertical ice cliffs that gradually slope higher to the Polar Plateau.

Perhaps Bellingshausen expected the coast of Antarctica to look like South Georgia or the South Sandwich Islands, with dark-colored cliffs and glacier-clad mountains, for he hesitated to call the vast icy terrain "land." Yet, this spectacle haunted him. If it wasn't land, then what was it? A year later he formed his hypothesis. "Huge icebergs, which as the pole is approached rise in size into veritable mountains, I call mother-icebergs," he wrote. Bellingshausen believed these massive bergs extended to the South Pole and were attached to land. They also spawned the colossal tabular icebergs that drifted northward. Almost a hundred years would pass before concepts such as *ice sheet*, *ice shelf*, or *Polar Plateau* would become accepted terminology. However, Bellingshausen's mother-iceberg theory encompassed all these terms.

The next day Bellingshausen could push his ships no farther into the dense brash ice. "In the further distance we saw ice-covered mountains similar to those mentioned above and probably forming a continuation of them." Bellingshausen suspected land was near. Flocks of Arctic terns hovered near the ice. "We had observed similar birds near South Georgia Island," he wrote. "The reappearance of these birds... gave ground for the suggestion that there might be land fairly near, as we had never seen these birds out in the open sea." But Bellingshausen had to turn to the north, since both ships were low on wood and he didn't want to "resort to breaking up our water or wine barrels for firewood."

During his clockwise trek along the eastern side of the continent toward Australia, Bellingshausen encountered "appallingly high breakers...the sides of waves...were almost perpendicular. The vessel coming up on such a wave, encounters a great mass of water on one side, whilst from the other side, the same wave forms a deep hollow, into which the vessel tends to heel over.... It seems shameful to feel fear, but even the hardiest man prayed."

Although danger was ever present, there were moments of extraordinary beauty. One night in March in the southern sky "there appeared two columns of a whitish-blue color like phosphorescent fire, flashing from clouds along the horizon with the rapidity of lightening. Then this amazing spectacle spread along the horizon...and the whole sky was covered with similar pillars.... The light was so strong and penetrating that opaque objects threw shadows exactly as in daylight...and it was possible to read even the smallest print without difficulty." Bellingshausen watched the aurora australis all night. The light illuminated a wide belt of icebergs, and the ships glided through the maze with ease.

On March 9 the most violent storm of the expedition struck with a fury that shredded their sails—and the replacements. "We remained now under bare poles at the mercy of the furious gale. I ordered some of the sailors' hammocks to be hoisted on the mizzen shrouds in order to keep the vessel to the wind," wrote Bellingshausen. Then he heard the cry of "Ice Ahead!" Horror spread across the men's faces as they headed on a collision course with the bergs. Just as they braced themselves for a stove side or damaged bow, "a huge wave, rising under the vessel, carried the ice some yards away and rushed it past the quarter galley to leeward."

All morning the storm battered the helpless ships. A lone albatross rode out the gale "in the hollow of the waves and maintained itself there with outstretched wings whilst it paddled with its feet." By midnight the storm intensified to hurricane force. Waves like steep dark hills lifted the ships and then plunged them down into deep hollows. "I do not ever remember having seen seas so massive or of such excessive height," wrote Captain Lazarev of the *Mirnyi*. "It seemed...as if we were hemmed in by hills of water." The ships listed to port and then to starboard. Worse than the perpetual motion was "the groaning of all parts of the ship." The men furiously stitched and patched the damaged canvas while the ships pitched and rolled. By the end of the next day, the men hoisted the sails and at last gained control of the ship. The nine-day gale finally ended on March 17. Less than a month later, the storm-battered *Vostok* and *Mirnyi* dropped anchors in Sydney Harbor.

"WE SAW A BLACK PATCH THROUGH THE HAZE"

On November 11, 1820, the *Vostok* and *Mirnyi* sailed from Sydney to battle gales, roiling seas, and swarms of icebergs—the staunch guardians of the Antarctic coast. Bellingshausen carried with him eighty-four birds from Australia for entertainment, including cockatoos, parrots, and one parakeet. "They all made a great deal of noise; some of the parrots knew a few English words," noted Bellingshausen. Another unusual passenger—a kangaroo—received nothing but praise from Bellingshausen. "It was very tame and clean, and often played with the sailors, required very little care and ate everything that was given to it."

Although Bellingshausen had intended to sail to the Auckland Islands, he decided instead to stop at Macquarie Island. "To our great surprise, we found that Macquarie Island was beautifully green.... Through our glasses we could see that the shore...was covered with huge sea animals, called 'sea elephants,' and penguins; sea birds flew about the shore in great numbers," noted Bellingshausen on November 29, 1820. No fur seals were found. "The abundance of fur seals had caused many vessels to come direct from New South Wales to trade in their skins, for which there was such a demand in Britain that the skin of a seal had risen to one guinea; but the unbounded greed of the sealers had soon exterminated the animals."

The next day he went ashore and visited with sealers "who were engaged in melting down the blubber of sea elephants. One party of them had been on the island nine months and another six." Although the sealers lived in mud hovels, he believed conditions here were much better than on South Georgia. "They live on the same sort of sea birds, the flippers of young sea elephants and eggs of penguins and other birds, but the traders here enjoy, besides a better climate, the advantage of natural remedies to scurvy. The so-called 'wild cabbage'...grows abundantly over the whole island." Before leaving the island, Bellingshausen gathered cabbage and purchased a parrot from one of the sealers for three bottles of rum.

They sailed south on December 1 from Macquarie Island, which soon faded into the early morning mist. From December 10 to 14, both ships had to pick their way through a wide belt of icebergs at 64° south latitude and 173° west longitude. The band was "not less than 380 miles [east to west] in extent and . . . consisted of ice blocks blown together . . . into various positions and shapes. Some had the appearance of Gothic pointed roofs of large buildings and others resembled the ruins of ancient towns."

On December 24, 1820, Bellingshausen crossed the Antarctic Circle at 164° west longitude and passed through several wide streams of broken ice, numerous large bergs, and flat floes. Four days later, the fog momentarily lifted and Bellingshausen counted nineteen massive tabular icebergs within a distance of one and a half miles. The fog thickened and cut visibility to a few yards. All around them they heard the tremendous roar of waves breaking against the icebergs. "I posted the men all around the ship, on the gangway, ladders, on the lowest steps near the very water's edge," wrote Bellingshausen, "for the lower the eye or ear is placed, the more easily is the whiteness of the bergs picked up . . . and the more quickly can the noise or sound be detected at that level because the fog is thinner." All night they listened to the roar of the sea thundering into caves on the sides of the bergs. Sometimes the small boats were launched to tow the ships slowly past the icebergs, some of which were ten to fifteen miles long and over one hundred feet high. For two days no one slept. On the third day the fog lifted, and the boats were hoisted on board.

Bellingshausen pressed further south, but the number of icebergs increased and no one rested for long. "We were sitting at dinner when suddenly the *Vostok* began to roll and the sails flapped; we all ran out on deck and beheld a majestic and terrible spectacle. Ahead of us lay a single narrow track between crowding icebergs, obliging us to close the bergs to windward in order to clear the other that lay to leeward. The tops of these icebergs were so high that it took the wind even out of the topsails." For several hours their position was perilous as the circle of icebergs tightened. Whenever the wind shifted, the fog lifted and Bellingshausen saw a narrow path around the bergs for just minutes at a time. The ships slipped between iceberg after iceberg, slowly advancing and then retreating in another direction. Bellingshausen knew the danger had passed when the swell of the sea increased, no longer dampened by the bulk of bergs. The ships continued to sail eastward; the western coast of Antarctica lay behind the girdle of ice.

On January 21, 1821, Bellingshausen reached the southernmost point of his voyage: 69° 53' south latitude and 92° 19' west longitude. In the midafternoon "we saw a black patch through the haze.... Words cannot describe the delight which appeared on all our faces at the cry of 'Land! Land!'," wrote Bellingshausen. Early the next morning, the sun dispersed the mist. "We saw a high island covered with snow.... I then signaled from the *Vostok* [to the *Mirnyi*] ordering a glass of punch to be served to each of the crew, who all drank to the health of the Emperor." Bellingshausen named the small island Peter I. A hundred years would pass before the first man would step onto its volcanic shores.

Signs of land again appeared on January 28. The sea grew darker, snow petrels and skuas wheeled overhead, and penguins porpoised southward through the swells. The next morning the men sighted high mountainous land. Ice floes prevented Bellingshausen from getting any closer than forty miles but "it was the most beautiful day that could have been desired in high southern latitudes." He named this discovery Alexander I Land. Although Bellingshausen assumed he had discovered part of the continent, it was not until 1936 that Alexander I Land was shown to be an island, separated from the continent by a narrow, two hundred-mile-long channel.

Bellingshausen now headed northeast toward the South Shetland Islands, having heard about the discovery of the islands from the captain of an East Indian ship in Australia. On February 5, 1821, "there was a hail from the forecastle: 'Land in sight above the clouds!'" Bellingshausen had reached the South Shetlands. The next day, near Deception Island, one of the most controversial meetings in Antarctic history occurred between Bellingshausen and an American sealer, Nathaniel Palmer.

SOUTH SHETLAND ISLANDS INVASION

The 1820–21 Antarctic summer season was a grim one for fur seals hauling out on the shores of the South Shetland Islands. At least fifty

British and American ships vied for prime seal territories on a first-come, first-served basis. Even William Smith, the discoverer of the South Shetland Islands, was horrified at the congestion. "To your Memorialist's surprise, there arrived from 15 to 20 British Ships with about 30 sail of Americans," he lamented. The situation was volatile, and gangs of sealers strictly enforced their own version of property rights.

While sailing near Greenwich Island after his voyage to the Antarctic Peninsula in November 1820, Nathaniel Palmer discovered Yankee Harbor, a protected cove with beaches swarming with fur seals. Captain Pendleton moved his fleet to this less competitive—and safer—location, but by mid to late January, all seals in the vicinity had been butchered. Either Captain Pendleton, Nathaniel Palmer, or both may have searched for seals along the Antarctic Peninsula as far south as Marguerite Bay, but logbooks and firsthand accounts are sketchy and inconclusive.

A few weeks after this voyage, Palmer continued shuttling supplies to crews scattered throughout the islands. On February 6, Palmer was surprised to see two Russian vessels. He pulled his sloop, the *Hero*, alongside the *Vostok* and hailed to come aboard. According to Edmund Fanning, Palmer's unofficial biographer, the twenty-one-year-old sealer told Bellingshausen that he had seen land much further to the southwest. Bellingshausen "arose much agitated" and requested to see Palmer's logbook.

After reviewing Palmer's entries, the Russian commander stated, according to Fanning, " 'What do I see and what do I hear from a boy... that he has pushed his way toward the Pole through storm and ice and sought the point I... have for three long, weary, anxious years, searched day and night for.' " Then Bellingshausen supposedly placed his hand on Palmer's head and said " 'I name the land you discovered in honor of yourself, noble boy, Palmer's Land.' "

Although Bellingshausen wrote briefly in his journal about this meeting with Palmer, he made no mention of scrutinizing the sealer's logbook—or bestowing any honors on Palmer. Yet, by focusing on the sequence of events—and ignoring Fanning's exaggerated rendition of Bellingshausen's out-of-character speech—there may be an element of truth in Palmer's version.

Just days before his encounter with Palmer, Bellingshausen had discovered Alexander I Land at 68° south latitude. It would be reasonable for Bellingshausen to ask to see Palmer's logbook—if only to reassure himself that the young American sealer had not discovered Alexander I Land earlier. After reviewing Palmer's charts, Bellingshausen may have felt enormous relief—Alexander I Land was a full two degrees further south than Palmer's limit—yet Bellingshausen did not disclose his own discoveries to Palmer. Although Bellingshausen was known as a man of few words and cautious with praise, he could now afford to be generous toward the bright-eyed and enthusiastic American who had so willingly shared information with him. He may well have suggested that the land Palmer had seen be named for the young Stonington sealer.

Palmer and Bellingshausen also discussed sealing—and the numbers of skins taken (eighty thousand). Bellingshausen later confided in his journal that "there could be no doubt that round the South Shetland Islands, just as at South Georgia and Macquarie Islands, the number of these sea animals will rapidly decrease. Sea elephants, of which there also had been many, had already moved from these shores further out to sea."

The First Small Step

About a week before Palmer and Bellingshausen met on the *Vostok*, American sealer Captain John Davis approached Cape Shirreff on Livingston Island in his small sloop, the *Cecilia*. Sixty gun-toting British sealers quickly discouraged his attempt to land. He concluded that it would be best "to go on a cruise to find new Lands as the Seal is done here." On February 1 Davis tried to land on Smith Island, but a hostile party of Australian sealers convinced him to keep searching. In the distance he sighted an uncharted small island and sailed toward it.

The next day Davis named his discovery Low Island and remained there until seven hundred seals had been slaughtered. On February 6 Davis headed for Hoseason Island and then toward the Antarctic Peninsula. "Commences with open cloudy Weather and light winds," wrote

Davis in his logbook on February 7. "Standing for a Large Body of Land . . . high and covered entirely with snow." He sent a boat to shore "to look for Seal at 11 A.M. the Boat returned but found no signs of Seal. . . . I think this Southern Land to be a Continent."

Davis did not record whose boot first touched the rocks on that deserted shore in Antarctica's Hughes Bay. It was the first time a human had stepped upon the Antarctic Peninsula. Within hours, he turned north toward Deception Island to continue his search for seals.

The End of the 1820–1821 Season

Captain Bellingshausen finished charting the South Shetland Islands by mid-February 1821 and returned to Kronstadt the following August. The Russian Admiralty was not impressed with Bellingshausen's discoveries and showed no interest in the high southern latitudes. Ten years elapsed before his journal was published, only to lie buried in dusty archives for the next one hundred years.

Edward Bransfield retired from the Royal Navy after his return to England in 1820 and faded into obscurity. William Smith filled his ship with fur seal pelts during the 1820–21 season. On the north corner of Livingston Island, he found planks of wood strewn on the shore and anchor cases from the galleon *San Telmo*, wrecked during violent storms in early September 1819. He salvaged enough of the timber for a coffin for himself. Returning to England with a full cargo of sealskins in 1821, he learned that his partners were bankrupt. Smith requested compensation for his discovery from the Royal Admiralty, but his petition was denied and for the rest of his life he lived in poverty. Whether he was buried in his custom-made coffin is not known.

Nathaniel Palmer was luckier in business than Smith. He continued sealing in the Shetlands and discovered the South Orkney Islands in December 1821 with George Powell. A few years later, Palmer founded a successful shipbuilding business and enjoyed wealth and fame to the end of his life.

By the end of 1822, beaches that had once reverberated with the expressive whimpers and bellows of fur seals were silent. A few stragglers hauled out in secret coves but soon they, too, returned no more. One year later a Scottish captain revisited the South Shetlands and was angered by the wanton destruction. "When these Shetland seals were first visited, they had no apprehension of danger, from meeting men; in fact, they would lie still while their neighbors were killed and skinned.... The quantity of seals taken off these islands... during the years 1821 and 1822, may be computed at 320,000.... The system of extermination was practiced on the Shetlands." The writer was James Weddell, the first of a small group of sealers who would find fame but not fortune in the realm of ice.

• 3 •

THE END OF AN ERA

By the end of March 1822, no ships disturbed the solitude of the South Shetland Islands. Snow buried dilapidated huts, piles of seal bones, and a few lone unmarked graves. The surrounding sea cooled and congealed into a viscous skin of ice that crept over sunken shipwrecks near deserted harbors and bays. Like a white carpet embedded with random icebergs, the pack ice spread across known and unknown seas, linking discovered and undiscovered regions into one impenetrable mass.

Gradually, as winter shifted into spring, the ice that covered an unknown sea deteriorated faster than normal. The pack ice, riddled with deep fractures, split into a network of floes and bergs that moved northward, their paths dictated by the whims of current, storm, and swell. Perhaps for the first time in centuries, a wide expanse of sea lay exposed to sunlight, its icy cover dispersed for a month or two. This unique event opened a path for a record-setting event when Scottish sealer James Weddell followed it into the heart of an undiscovered sea in 1823. Seven years later, John Biscoe circumnavigated Antarctica on a voyage that degenerated into a nightmare, and in 1838, John Balleny found land but no fur seals during his disastrous journey.

The Year the Ice Disappeared: James Weddell's Incredible Voyage

James Weddell, the orphaned son of a Scottish upholsterer, enrolled in the Royal Navy in 1804 when he was just eight years old. Later, he joined the merchant marines and hauled cargo to and from the West Indies. In 1819, lured by reports of unlimited wealth in the far south, he led his first sealing expedition to South Georgia, the Falklands, and the South Shetland Islands. Weddell was not a very successful sealer, for he returned to England in 1821 with a half-empty hold. Still, he persuaded shipbuilder James Strachan and several other investors to finance another sealing expedition.

On September 17, 1822, Weddell sailed from England on the brig *Jane* with twenty-two men, accompanied by Matthew Brisbane and thirteen men on the *Beaufoy*, a small cutter. The ships resupplied at Madeira, crossed the equator on November 7, and continued southward. After a two-week stay in the Falklands, the two ships reached the eastern edge of the South Orkney Islands on January 12, 1823. With his expensive array of navigational equipment, some of which he purchased with his own money, Weddell carefully charted the coasts of the islands. He also filled his journal with descriptions of wildlife, harbors, ocean currents, air and sea temperatures, and ice conditions. Although he and his men scoured the beaches, no fur seals were seen. During one of those forays ashore, Weddell killed seven seals of an unknown species, one of which was later delivered to the Edinburgh Museum. This seal species was later named *Leptonychotes weddelli*—the Weddell seal.

Since the South Orkney Islands offered no economic promise for the sealers, Weddell headed northeast, hoping to find land south of Captain Cook's previously discovered Sandwich Land. Although Bellingshausen had determined in 1820 that Sandwich Land was a group of islands rather than an extension of the southern continent, Weddell had no knowledge of his voyage. Bellingshausen's charts and journals weren't published until 1831.

Heavy fog soon swirled around both ships. Weddell offered a reward to the first man who spotted land, but "this proved the cause of many a sore disappointment," he wrote, "for many of the seamen, of lively and sanguine imaginations, were never at a loss for an island." On January 27 the man on watch reported passing a dark brown object that looked like a large rock. "The lead was immediately thrown out, but finding no bottom, we continued lying to." The chief mate concluded that the "rock" was a very swollen dead whale. "Such objects, seen imperfectly in the night, are often alarming," noted Weddell.

His generous ration of three glasses of rum a day for each man may have contributed to unusual sightings and enthusiastic shouts of "Land ahead!" Many times Weddell changed directions but "after the usual practice of pursuing all such appearances, we discovered it to be one of our delusive attendants, the fog banks."

However, on February 10, the chief mate, a reliable and sober man, reported that land was ahead, in the shape of a "sugar loaf." The sight even convinced a skeptical Weddell, but four hours later he admitted the unusual iceberg with its layers of black soil had fooled him—and encouraged him to continue southward. Land, he reasoned, must be near.

By February 14 the ships were surrounded by large flattopped icebergs at 68° 28' south latitude. "Sixty-six were counted around us," Weddell recorded, "and for about 50 miles to the south we had seldom fewer in sight." Two days later, however, almost all the icebergs had disappeared and the weather improved dramatically. Although Weddell had moved the ship's stove below deck for the crew's comfort, since they suffered from "colds, agues, and rheumatisms," the men now had no need to huddle around the fire. Weddell described the temperature as "certainly not colder than we had found it in December in latitude of 61 degrees."

On February 18 Weddell reached 72° south latitude. Many whales spouted about the ships, and the sea "was literally covered with birds of the blue petrel kind," he noted. "NOT A PARTICLE OF ICE OF ANY DESCRIPTION WAS TO BE SEEN." Weddell's capitalized statement emphasized his amazement at this absence of ice on the sea that now bears his name. Warmer than normal air and sea temperatures during the previous winter may have inhibited the growth of the pack ice. This climatic anomaly, coupled with an unusually mild spring, may have produced the exceptional ice-free conditions in 1823.

The fine weather continued and the men remained on deck, stitching canvas sails, airing bedding, and washing clothes. Two days later both ships reached 74° 15' south latitude at 34° 16' west longitude, 214 miles further south than James Cook had achieved. Nearly one hundred years would pass before any explorer would match Weddell's southward penetration in this sector.

Three icebergs—one crowded with penguins—loomed on the horizon in the afternoon. When the wind freshened from the south, Weddell decided to turn northward. One thousand miles of "sea strewed with ice islands, with long nights and probably attended with fogs" still faced Weddell and his men.

When he announced his decision, the crews' dreams of new lands flush with fur seals evaporated. Since their income depended on a small share of a full cargo, the men had reason to feel disgruntled, but Weddell turned disappointment into a celebration. "Our colors were hoisted," he wrote, "and a gun was fired, and both crews gave three cheers. These indulgences, with an allowance of grog, dispelled their gloom, and infused a hope that fortune might yet be favorable."

Weddell's luck held until March 6, the day he lost sight of the *Beaufoy* during a heavy fog and "a most distressing sea." He could only hope that Captain Brisbane would guide the *Beaufoy* toward South Georgia, their place of rendezvous. The next day a severe gale struck just as Weddell was approaching a thick cluster of icebergs. Although he tried to reduce the ship's speed, Weddell was forced to run the iceberg gauntlet at a hair-raising ten miles per hour. "The chief mate was stationed in the foretop to look out for low ice in the hollow of the sea," noted Weddell. The mate could do little but hang on as gargantuan waves "cleared the decks of almost everything movable." The next day the sky cleared, and Weddell sailed for South Georgia.

On March 10 both ships were reunited at Adventure Bay. "Our sailors had suffered much from cold fogs and wet," wrote Weddell. "Our vessels, too, were so much weather-beaten, that they greatly needed refitting." Although he searched hidden coves and bays along the coast for seals, both crews were dismayed at the meager pickings. "These animals are now almost extinct," Weddell wrote. He estimated that since the 1780s the total number of fur seals killed at South Georgia was 1.2 million. After satisfying himself that no seals remained on South Georgia, Weddell and Brisbane sailed the *Jane* and the *Beaufoy* for the Falkland Islands to winter.

Forty-two years earlier, the beaches of the Falkland Islands had been "lined with sea elephants," Weddell wrote. Now not one hauled out. Although he estimated that he had killed over two thousand elephant seals on his previous voyages, the sight of barren beaches saddened him. The crew managed to keep busy despite the absence of seals. They cleaned and caulked both ships, often working in their bare feet, naked to the waist. Weddell remarked in his journal that he believed the climate had significantly warmed since the 1780s, and noted the unusual absence of any icebergs in the vicinity of the Falklands.

Refreshed and rested, both crews sailed from the Falkland Islands in October 1823 for the South Shetland Islands, but heavy pack ice blocked their paths about ninety-five miles from their destination. Weddell continued to skirt the pack for several weeks, searching for an opening. At midnight on October 28, Weddell recorded that they faced hazardous conditions: dense fog, heavy ice, and a high swell. The west wind strengthened to gale force. "We fell in with many ice islands, some of which, by the heaviness of the sea around us, were rolling with the noise of an earthquake," wrote Weddell. Worse was to come.

By 8 A.M. Weddell faced a full-fledged hurricane. Everything movable was swept off the deck and replaced with thick layers of ice that also coated the hull. "The vessel rose to the sea but sluggishly, and the fear of falling in with ice, kept us constantly on the alarm," wrote Weddell. Then the rudder froze, rendering the ship helpless. The crew could do nothing but wait for daylight.

The next morning the winds and sea moderated, but many men had suffered injuries after being hurled about on the lurching decks. Hands, faces, and feet were frostbitten. Since most of the men owned no extra stockings, Weddell ordered the crew to cut up blankets for leggings. Miserable as they were, "no dastardly request to reach a better climate was ever hinted at," wrote Weddell. Both ships continued to search for a lead in the impenetrable ice.

As the ships skirted the edge of the pack ice, collisions with drifting icebergs were an ever-present threat. On the morning of November 5, an iceberg struck the *Jane* with such violence that "the people were thrown out of their cabins. I was on deck in an instant," noted Weddell. He then checked the bilge for leaks but found none—a situation that surprised not only Weddell but also the carpenter. Both men knew that just two-and-a-half inches of wood planking covered the bottom of the *Jane*.

Weddell and Brisbane sealed near Cape Horn and Tierra del Fuego for several months and then returned to England in May 1824. Weddell's search for seals had led him deep into the heart of an unknown sea—and further south than any man. When the *Jane* was dry-docked for repairs, Weddell had every reason to feel lucky. During the ship's violent encounter with the iceberg, part of the bottom planking had been shoved inward and lodged between two thick timbers, plugging a hole that would have sunk the ship.

John Biscoe's Circumnavigation: "Only three of the crew can stand"

When James Weddell's account of his extraordinary voyage was published in 1825, the book may have piqued Samuel Enderby's interest. Owner of the shipping firm that was the focus of the Boston Tea Party in 1773, he was one of those rare individuals who actually read his captains' logbooks and journals, especially of those men who shared his love of geography and adventure. Enderby encouraged his captains to explore seldom-visited islands and lands in the high Arctic and southern oceans while they pursued whales and seals.

When Samuel Enderby died in 1829, his sons inherited controlling interest in the firm. Charles Enderby, who shared his late father's love of geography and exploration, helped to establish the Royal Geographical Society in 1830. One of the earliest papers presented to the Society was a report by Captain Horsburgh about the massive icebergs in the southern Atlantic during the 1828 austral summer, which had disrupted trading routes. Such a vast quantity of ice, he reasoned, must have originated from undiscovered land between the Greenwich meridian and about 20° east longitude, an Antarctic sector now known as Princess Astrid Coast.

Captain Horsburgh's presentation convinced Charles Enderby to search for new Antarctic lands by outfitting the brig *Tula* and the cutter *Lively*. He then placed John Biscoe in command with orders to seek new lands in the high southern latitudes—and to procure enough sealskins to offset the expedition's costs.

On July 14, 1830 both vessels left England to begin the third circumnavigation of Antarctica. In late November, Biscoe reached the Falkland Islands and continued eastward toward the South Sandwich Islands. A series of gales and thick fog hindered the ships, but on December 21, 1830, Biscoe sighted the islands. The next day he sent several men to explore the beaches but they soon returned, "not having observed the least trace of seal or elephant, or even finding a landing place," wrote Biscoe. He searched for more productive sites, but on December 26 icebergs surrounded both ships. "We were utterly prevented from steering on any one course for more than a few minutes at a time," noted Biscoe, "and never at any time had we less than fifty to a hundred ice-islands round us." Three days later Biscoe escaped the deadly ring of bergs, but after another futile seal hunt he left the islands and sailed southeast toward Captain Horsburgh's theoretical Antarctic landfalls.

On January 8, 1831, he wrote, "I found myself at the head of a bay of firm ice, with a view of it from the mast-head of at least 20 miles to the southward... any person might have walked upon it without difficulty." Although he suspected that land might be near, the absence of wildlife astonished him, as did the temperature on January 16. "Thermometer in the air 45 degrees F... but in the sun it immediately rose to 77 degrees F; in fact it appeared more like a summer in England than what I had expected in these latitudes," he recorded.

Biscoe crossed the Antarctic Circle on January 22, 1831. One week later he reached 69° south latitude at 10° 43' east longitude, north of Princess Astrid Coast, described eleven years earlier by Bellingshausen. For the next four weeks, Biscoe pushed through drift ice well within the Antarctic Circle as he continued eastward. On February 24 he had his first glimpse of "an appearance of land." On that day, the weather briefly cleared, and Biscoe saw a steep icy cliff that "bore the marks of icebergs having been broken from off it... it ran away to the southward with a gradual ascent, with a perfectly smooth surface, and I could trace it in extent from 30 to 40 miles from the foretop with a good telescope." He also observed "two or three lumps which had the appearance of land from the irregularity of their surface." Biscoe ordered a boat to be lowered but after a half-hour of arduous rowing, he made little headway through the icy rubble. He convinced himself that he had seen nothing more than a solid field of ice.

On February 28 Biscoe saw "several hummocks to the southward, which much resembled tops of mountains, and at 6 P.M. clearly distinguished it to be land, and to considerable extent." His position was just north of White Island and Casey Bay, a stretch of coast later named Enderby Land. For two days Biscoe tried to maneuver the ship around bergs and floes toward the coast. The men endured tedious hours of endless thumps, jolts, and bone-jarring blows, but the ice refused to yield a passage.

On March 3 an unearthly spectacle presented a new threat: distraction. "The Aurora Australis showed the most brilliant appearance," noted Biscoe, "at times rolling itself over our heads in beautiful columns, then as suddenly forming itself as the unrolled fringe of a curtain, and again suddenly shooting to the form of a serpent." Icebergs still surrounded the ships, but Biscoe "could hardly restrain the people from looking at the Aurora Australis instead of the vessel's course."

As the splendid display faded with dawn, a cold breeze from the south stiffened into a gale. The wind grew fiercer, and by the next day it "blew a perfect hurricane." Biscoe lost sight of the *Lively* among the heaving swells and the *Tula* became encased in ice from the freezing spray. "It was impossible for the people to hold anything in their hands for more than a minute or two at a time," wrote Biscoe. By the time the wind moderated on March 8, "it left us almost a wreck." The force of the storm had driven Biscoe 120 miles to the north. "I am under much apprehension for the cutter, as I think this is the hardest blow I have ever known with the exception of the hurricane of 1814."

Biscoe sailed south to the area where he had last seen the *Lively*. On March 16 he saw "land bearing south by compass, which was a very high mountain... but am sorry to say could see nothing of the cutter." The ship's disappearance depressed Biscoe; he wrote in his journal that he was ready "to give up all further pursuit" for new but inaccessible lands. The hardships his men had endured also weighed heavily on his mind. "Heavy gales frequent every day, some of the people getting sick, the carpenter for some time past having lost the use of his legs... and two or three more under medicine for the same complaint." The symptoms of scurvy were unmistakable.

Gales, snow squalls, and high seas battered the *Tula* while scurvy stalked the lower decks. "Only three of the crew can stand," Biscoe wrote on April 6. Although he placed the entire crew on "fresh provisions," scurvy affected nearly every man. On April 23, the carpenter died after "a long illness, which commenced with pains and swellings in his legs," noted Biscoe. "Afterwards his mouth became much affected, the gums swelling over his teeth." Three days later "only one of the crew able to stand, together with one boy, the officers (two mates) and myself, I find it absolutely necessary to make the best of my way for Van Diemen's Land [Tasmania] to save the vessel and the lives of those who remain."

Eleven days later Biscoe reached Tasmania, but a heavy swell prevented him from landing. "I endeavored all in my power," wrote Biscoe, "to keep up the spirits of those on board, and often had a smile on my face with a very different feeling within." He continued to battle the swells and finally arrived at Hobart on May 10. The dying men were carried to a makeshift hospital for treatment.

The crew of the *Lively* had suffered even worse catastrophes after the hurricane separated the two ships. Starvation and scurvy reduced the crew to Captain Avery, one seaman, and a boy with a crushed hand. Somehow these three starved and exhausted men managed to pull into Port Phillip (near Melbourne, Australia) in April. For the next few months, the three survivors rested and repaired the cutter with help from the townspeople. Still weak and sick, they sailed to Hobart, arriving on September 3, 1831—just as Biscoe was leaving for his second expedition south. Biscoe returned to Hobart to resupply the cutter and wait for the three men to recover. A month passed before both ships sailed once more for Antarctica.

THE NIGHTMARE CONTINUES

In early October 1831 the *Tula* and *Lively* left Hobart to hunt seals for three months. Although the men searched New Zealand's shores and several subantarctic islands, few seals were found. Biscoe then headed southeast, following Captain Cook's trek toward Cape Horn, hoping to find land southwest of the South Shetland Islands. On February 4, 1832, Biscoe's position was 65° south latitude, opposite what is now named Marie Byrd Land. Although there were few icebergs in sight, he was reluctant to push further south. Any land would be inaccessible and, more importantly, the effort would "disable the crew for their necessary duty, should we find any land in the neighborhood of Palmer's Land or South Shetland." The "necessary duty" was, of course, sealing. Biscoe concluded that to dawdle longer than necessary in the high southern latitudes "would be throwing the remainder of the season away to little purpose."

Yet, Biscoe crossed the Antarctic Circle and sighted land on February 15, 1832. "It has a most imposing and beautiful appearance, having one very high peak running up into the clouds," he noted. Biscoe named the island Adelaide, after the British queen. Four days later, he discovered a cluster of small islands that now bears his name. He named one island Pitt when he noticed how much a nearby iceberg resembled the British statesman. Although Biscoe searched the island, no seals were found.

On February 26, the South Shetland Islands appeared on the horizon. Biscoe steered "in that direction in hopes of better success," but the South Shetland Islands offered no financial windfalls for the crew. "I was in hopes . . . I might still be in time to load the vessels with elephant and oil, as I now expected the March bulls would be coming up," noted Biscoe. Only thirty elephant seals were slaughtered on March 7.

Biscoe continued his search until April 10, and then decided to hunt sperm whales. But before he raised his anchor in New Plymouth Harbor on Livingston Island, the wind shifted, bringing a monstrous swell on its back. The *Tula* struck ground in the deep troughs between the waves. So forceful were the blows to the ship, the rudder broke on impact and the vessel floundered helplessly in the heavy swell. Biscoe ordered all hands to abandon the ship but "the dip of the swell was not less than 17 feet . . . and that added to the breakers made it extremely difficult to get at the boats." After a hair-raising trek to the shore with provisions, Biscoe could do nothing but watch the breakers curl over the *Tula*'s decks.

When the swells had subsided, Biscoe rigged the rudder to the ship with ropes, and on April 15 sailed from the South Shetland Islands. "Had this misfortune not happened," Biscoe noted, "I had still some hopes of making a tolerable voyage. But the rudder being in this state renders it absolutely necessary to make for some port." On April 29 Biscoe arrived in the Falkland Islands "after a very rough passage." Many men deserted. Biscoe finally arrived in England on January 30, 1833, with a near-empty hold and a skeleton crew.

Charles Enderby presented John Biscoe's log to the Royal Geographical Society, which awarded Biscoe their highest medal of honor. Although Biscoe commanded several more expeditions for Enderby Brothers, all were unsuccessful. In 1837 he moved to Sydney and then to Hobart, Tasmania, where he lived in abject poverty. A public appeal raised enough money for his return to England, but he died in 1843 while en route.

New Islands and a Puzzling Iceberg

John Balleny joined Enderby Brothers in 1838 and was given command of the *Eliza Scott*, a sealing schooner. That July, Balleny's flagship and the cutter *Sabrina* with Captain Freeman in command, sailed from England. Their orders were to explore the high southern latitudes and return with enough sealskins and oil to offset the expedition's costs. By early December both ships had anchored in Chalky Bay, South Island, New Zealand. Gales, heavy rains, and thunderstorms with horrific lightning displays lasted the entire month. But worse than the weather were the stinging hordes of black flies. "The bite of a mosquito is not to be compared to it for severity and effect," noted Balleny.

The ships left New Zealand on January 7, 1839, much to the relief of the men. Three weeks later, Balleny crossed the Antarctic Circle at 172° east longitude and reached 69° south on February 1. Balleny now sailed eastward through "dirty-green looking water," littered with copious penguin feathers. On February 9 he spotted the mountainous islands that were later named after him.

The next day Balleny maneuvered the *Eliza Scott* closer to the central island but heavy ice blocked his path. Dense fog soon hid the islands, but on the eleventh the weather cleared. He estimated the height of one mountain at twelve thousand feet. On February 12 John Balleny and Captain Freeman rowed toward a narrow scrap of rocky beach visible only during the ebb of the waves. "Captain Freeman jumped out and got a few stones," wrote Balleny, "but was up to the middle in water." As they rowed back to the ships, the sopping wet Freeman and Balleny paused to watch smoke rise from several ice-covered volcanoes.

During the next two weeks, Balleny sailed eastward with good winds. On March 2, squalls of snow and sleet assaulted the ships throughout the day, but by 8 P.M. the weather had settled. "Saw land to the southward," wrote Balleny confidently. The next day immense icebergs surrounded both ships, but Balleny and first mate McNab recorded two more sightings of land. Neither man noted any details or descriptions. Balleny named the section of coast he presumably saw Sabrina Land.

On March 13, the ships passed within a quarter mile of a three-hundred-foot iceberg with "a block of rock attached to it," wrote Balleny. McNab grabbed his pen and paper, quickly sketching the berg and boulder before it disappeared into a snow squall. His drawing and recollections would later pique the interest of Charles Darwin—and would lead to a theory of the origin of large, out-of-place rocks called "erratics."

Balleny now turned the *Eliza Scott* northward. On March 24, the west wind increased to hurricane force. At midnight, his companion ship, the *Sabrina*, burned a blue light, a signal of distress. Balleny tried to reach the cutter but high, chaotic waves thwarted his attempts. The flickering blue light soon disappeared. "No signs of the poor cutter in sight," wrote Balleny at daybreak. "I trust she may be safe." The *Sabrina* was never seen again.

On September 17, 1839, Balleny arrived in England with a paltry cargo of just 178 sealskins—a severe financial loss for Enderby Brothers. Charles Enderby presented Balleny's discoveries to the Royal Geographical Society later that year, but his firm stopped funding expeditions to Antarctica.

Although Weddell, Biscoe, and Balleny made important geographical discoveries, their expeditions were financial failures because of the lack of seals. The few fur seals that had managed to escape cudgels and lances now tottered on the brink of extinction. Commercial interest in Antarctica—the prime motive for past exploration—disappeared for the next fifty years.

Scientific interest would briefly fill the gap. In 1831, while John Biscoe and his men were recovering from scurvy in Hobart, James Clark Ross led a well-financed expedition to northern Canada to claim the North Magnetic Pole for Great Britain. In 1835, German mathematical physicist Johann Karl Friedrich Gauss created a formula for determining magnetic deviations anywhere on earth. The South Magnetic Pole, he predicted, would be found at 66° south latitude and 146° east longitude. Now, attention shifted southward as three nations—France, the United States, and Great Britain—vied to be first to claim the South Magnetic Pole and either confirm or refute Gauss' calculations, so critical for accurate navigation.

• 4 •

THE SEARCH FOR THE SOUTH MAGNETIC POLE

FROM 1837 TO 1839 FRANCE, THE UNITED STATES, AND GREAT Britain launched government-sponsored expeditions to the high southern latitudes. The men who led these voyages shared one goal—to reach the South Magnetic Pole. Distinct ambitions, molded by politics and history, shaped each man's focus. American Charles Wilkes wanted to rekindle national pride in his countrymen and gain respect for the fledgling nation from the world community. James Clark Ross had claimed the North Magnetic Pole for Great Britain and now hoped for victory at the opposite end of the earth. Jules-Sebastien-Cesar Dumont d'Urville sailed for the glory of France—and the opportunity to complete his ethnological research.

Each expedition had its own unique origin. A strange theory and a well-publicized plea before Congress spurred the Americans; an anonymous pamphlet, brimming with nationalistic pride, motivated the British; and Dumont d'Urville's proposal prompted his king to respond with a proposition of his own. The three men who searched for the South Magnetic Pole each made discoveries of such magnitude that the veil of mystery was lifted forever from the face of Antarctica.

Weird Science and an Impassioned Speech

In April 1818, American John Cleves Symmes declared, in writing, that "the earth is hollow and habitable within; containing a number of concentrick [*sic*] spheres; one within the other, and that it is open at the poles twelve or sixteen degrees." He sent this letter—along with a certificate of his sanity—to over five hundred universities and governmental agencies in the United States and Europe. The fact that no one responded to his letter or his plea for scientific expeditions to the polar regions didn't discourage Symmes. Having spent a lifetime and a small fortune refining his theories, he set out on a lecture tour across the American Midwest to enlighten the public. He discussed how the earth's crust was much thicker at the equator than at the poles because of the planet's rapid rotation. The thin polar crusts had cracked into a web of massive fissures, he explained, and vast crevasses had gradually merged into gaping holes—four thousand miles wide at the North Pole and six thousand at the South Pole. The auroras that hovered over the polar regions were reflected sunlight, emanating from deep inside these hollows. The magnetic needle's mysterious dip, or deviation from true north, was easily explained: it pointed to the holes' entrances.

Symmes made little headway in gaining support for his theory until he met Jeremiah N. Reynolds, a newspaperman from Ohio, in 1825. A man with good stage presence and a talent for public relations, Reynolds teamed up with the taciturn Symmes and soon dominated the presentations. Less than a year later, the two separated. Reynolds argued passionately for exploration in the southern latitudes, while Symmes campaigned for an expedition to the far more accessible northern regions. The Russians offered Symmes their full support for a Siberian trek to find the northern entrance to the center of the earth.

Reynolds dropped the hollow-earth theory as a reason for a government-funded Antarctic expedition. Instead, he focused on the need to protect American fishing interests in the Southern Hemisphere. Hadn't the British already outmaneuvered the Americans, he asked politicians, in the race for fur seals, subantarctic whaling ports, and new territory? It was time, he declared, for America to "open channels for commercial enterprise" in the high southern latitudes. Backed by the powerful New

England whaling and sealing industries, Reynolds sailed with a fleet to the South Shetland Islands in 1829. There he saw the effects of the mass destruction of the fur seals, for not a single animal was spotted. What angered him was not the relentless slaughter, but the fact that American sealers had lost a valuable commodity to the British. When he returned, he dedicated himself to the task of whipping up public and congressional support for a naval fleet to protect any natural resources near Palmer's Land.

News of the British conquest of the North Magnetic Pole in 1831 had little impact on Reynolds's Antarctic campaign but destroyed Symmes' hollow-earth theory. The Russians withdrew their offer to fund a northern expedition, and Symmes retired quietly to his farm in Kentucky. However, reports of John Biscoe's Antarctic circumnavigation intensified Reynolds's crusade for an American expedition. In 1836 he pleaded with Congress to expand America's fisheries by exploring the "extensive group of islands lying north of the coasts of Palmer's Land." He then vented his anger toward the English, who had renamed several islands. "A British vessel," he sputtered, "touched at a single spot in 1832, taking from it the American and giving it a British name." With this statement Reynolds had tapped the core of a fledgling country's pride. American discoveries—and commercial interests—must be protected.

Congress responded swiftly. Funds were appropriated in May 1836 for the first official United States Antarctic Expedition commanded by Thomas Ap Catesby Jones. Riding on the wave of enthusiastic public support, the Navy Department tapped the Philosophical Society of Philadelphia for scientific input, formed committees to supervise equipment purchases, and outlined the expedition's goals. Lieutenant Charles Wilkes, a brash and temperamental New Yorker, traveled to London to purchase scientific equipment and charts. There he consulted with James Clark Ross, the man who had planted the British flag at the North Magnetic Pole in 1831.

When Wilkes returned to the United States, the first inkling of trouble surfaced and escalated into a storm of verbal brawls, backroom intrigues, and public humiliations that quickly soured public opinion. Newspaper editorials declared the expedition a failure before it even sailed.

A Public Appeal to the Royal Geographical Society

When news of America's planned Antarctic expedition reached Great Britain in 1836, an anonymously published pamphlet urged the Royal Geographical Society to pressure the government into funding an official expedition. How could England, the author wondered, just stand by and watch a "foreign and in some points a rival nation . . . step in and bear away the palm of glory" so arduously won by Cook, Bransfield, Weddell, and Biscoe? How much better it would be to build on their magnificent achievements than to allow "a nation . . . in her infancy to snatch laurels that have been planted and watered by the toils of our seamen."

After this nationalistic appeal, the author argued that magnetic research alone presented enough justification for an Antarctic expedition. "The safety of our ships, the value of our commerce, and the lives of our fellow creatures are all risked by the unknown agency of this mysterious power which seems to baffle investigation," the author stated. The discovery of new sealing grounds, just "a single spot for British enterprise and capital to exert itself" would help recoup the expedition's expenses—and revive a doomed industry.

The pamphlet had little influence on the cash-strapped Royal Geographical Society; members pointedly ignored it. In France, however, the pamphlet had a profound effect on Jules-Sebastien-Cesar Dumont d'Urville.

D'Urville's Dilemma

Dumont d'Urville was born in 1790 in the Calvados district of Normandy. When he was twelve years old, he demonstrated his formidable linguistic abilities by translating the Latin works of Cicero and Virgil. Languages became one of his lifelong passions, and he immersed himself in studies of Hebrew, Greek, German, English, Spanish, and Italian. At seventeen, he joined the French Navy, where his insatiable curiosity led him to the fields of astronomy, entomology, ethnology, geology, and his most consuming interest, botany.

In 1819 d'Urville sailed on the *Chevrette* to survey eastern Mediterranean regions. While the ship was anchored off Melos Island, d'Urville headed for the hills to collect plants but heard rumors that a six-foot marble statue had just been unearthed. Ever curious, he investigated the site. The beautifully sculpted female face and body fascinated d'Urville. "It represented a naked woman, whose raised left arm was holding an apple and the right arm was skillfully holding a draped sash that fell casually from her hips to her feet," he wrote. His sketches stunned the French ambassador in Constantinople, who sent his secretary to Melos to acquire it. French sailors carelessly dragged the statue over the rocky terrain to the ship. As they hauled it aboard, they noticed several body parts were missing, but did not search for them. The Venus de Milo—minus both arms—now resides at the Louvre Museum. For his role in the acquisition of this treasure, d'Urville was honored with a promotion to the rank of lieutenant by King Charles X. In 1821 he joined with other prominent intellectuals to form the Paris Geographical Society.

From 1822 through 1829, d'Urville sailed on two voyages around the world with extended stops at South Pacific islands, New Zealand, and Australia. During this time he developed his prodigious intellectual gifts by pursuing many scientific interests with aplomb and dedication. Robust and blessed with indefatigable stamina, he tramped over dangerous terrain for days at a time, indifferent to hardships.

He was equally indifferent to his shipboard appearance and refused to don his navy uniform even for foreign dignitaries. The British in particular were offended by d'Urville's choice of uniform: torn trousers, an unbuttoned twill coat, and an old straw hat full of holes. Even worse, wrote Rene-Primavere Lesson, who sailed with him in 1822–1825, "M. d'Urville spoke English with a whistling sound through his teeth... but with the tone of one well aware of his superiority and his position." When the guests mistook d'Urville for an uppity sailor, Lesson invariably whispered the word "lieutenant" to the visitors. The shocked looks and raised eyebrows never failed to delight d'Urville.

In 1830 King Charles X fled France during the July Revolution, and the Duke of Orleans, Louis Philippe, assumed power. Although d'Urville's naval career languished for the next seven years, he submitted legions of scientific articles for publication, completed the narrative of his 1826–29 voyage, and delved into the ethnology and linguistics of the South Sea islands. By 1836 he had found gaps in his research that only a third expedition to the South Seas could resolve. And persistent dreams about Cook's three voyages haunted him during the nights. "What was strange in these dreams," wrote d'Urville, "was that they were always taking me nearer the Pole."

In January 1837 he wrote a proposal to the minister of the navy, outlining a voyage to the South Seas. When his plan was delivered to King Louis Philippe for final approval, the king countered with a proposal of his own. Would d'Urville be willing to push beyond Weddell's tracks toward the geographical South Pole? "I was dumbfounded and irresolute," wrote d'Urville. He debated whether to accept or reject the king's proposal.

D'Urville resolved this dilemma with characteristic self-confidence. "After examining all sides of the business, I finally recognized that an attempt to get to the South Pole would have the character of novelty, of greatness and even of wonder in the eyes of the public," he noted. By this time, d'Urville had received a copy of the anonymous pamphlet and knew that at least the United States and Britain were planning expeditions to Antarctica—an even better reason to accept the King's challenge.

D'Urville presented a translated copy of the pamphlet to the Paris Royal Geographical Society in 1837. At that same meeting, he announced his own plans for a three-year expedition to the South Pacific islands and Antarctica. Although his instructions did not authorize him to pursue magnetic studies, d'Urville reasoned that if his Weddell Sea attempt on the geographical South Pole was unsuccessful, he could at least try to reach the Magnetic Pole ahead of the Americans and the British. Several members privately doubted his suitability for such a long and arduous journey, for d'Urville was forty-seven years old and plagued with frequent attacks of gout. However, the Society respected his prolific contributions to many scientific disciplines and offered d'Urville its full support.

D'Urville immediately set to work. In London, he purchased charts and scientific equipment and developed a passion for phrenology, the "science" that linked the bumps, knobs, and shape of the human skull to intelligence and personality traits. D'Urville presented his head to one of Great Britain's most popular phrenologists. The accuracy of the very flattering report convinced d'Urville to recruit French practitioner M. Dumoutier for the expedition. Officials approved d'Urville's request, and Dumoutier was put on the payroll as Phrenologist and Natural History Assistant.

The two ships commissioned for the expedition were the ninety-four-foot *Astrolabe* and the smaller *Zélée*. One day, as his crew was overhauling the ships, several of the men paused. "They saw me walking slowly and heavily owing to a recent attack of gout," wrote d'Urville, "and they appeared surprised to learn that I was their commander. Several men mumbled, 'Oh, that old gaffer won't take us far!'" D'Urville would soon prove just how wrong they were.

D'URVILLE'S FIRST VENTURE INTO THE ICE

The last several months had been difficult for d'Urville. Although the *Astrolabe* and the *Zélée* had been completely overhauled during the summer, the ships were unsuitable for ice navigation. Neither hull was reinforced, and the large square gun holes that pierced the sides of the corvettes would soon subject men and supplies to frequent drenchings from the roiling swells of the Southern Ocean. Finding enough men to work the ships was also a problem for d'Urville. At first he had to settle for the dregs of sailor society, but the tenacious explorer convinced King Louis Philippe to offer two incentives. First, a reward of one hundred gold francs was to be given to each man if the ships surpassed Weddell's record 74° 15' south latitude. And second, every member of the crew was promoted in grade or class at the start of the expedition. D'Urville's staffing problems were quickly solved.

On September 7, 1837, Dumont d'Urville hobbled up the gangway to board the *Astrolabe*. Suffering from another attack of gout, he held no illusions about his chances of success. Twice before he had participated in long voyages. "Then I was young, strong and full of hope, optimism... But in 1837 I was old, subject to attacks of a cruel malady, completely disenchanted... So I left behind all that was dearest to me." He said his farewells to family and gave the order to sail from Toulon. Captain Jacquinot followed in the *Zélée*.

The ships sailed southward to Tenerife, one of the Canary Islands, to take on fresh supplies and water. Officials, however, quarantined the men for four days because of rumors of plague in d'Urville's hometown. Once the men were allowed to leave the ship, they made up for lost time ashore. "They recklessly imbibed the strongly fortified wine of the region," wrote a disgusted d'Urville, "taking such copious libations that they lost the feeble drop of sense that nature had endowed them with."

At least the wine was digestible—which was more than could be said about the food on the ships. As an experiment, the navy had stocked large canisters of preserved meat. "When the tin was opened a foul smell came from it," wrote d'Urville. After a month the crew's opinion was unanimous: salt beef was the meal of choice. No more tinned meat was served, but not necessarily because of the men's likes or dislikes. "It is lucky we are not carrying much," noted Officer Duroch in his journal. "Three of the drums had exploded."

D'Urville sailed across the Atlantic to Tierra del Fuego. On December 13 he entered the Strait of Magellan and headed for Port Famine. Expecting a desolate region devoid of anything edible, d'Urville was pleasantly surprised. The men feasted on wild geese, fish, and fresh celery. There was also plenty of wood and water—and even a small post box for incoming and outgoing letters. The men celebrated New Year's Day with a picnic on the sand; nine days later, they headed south toward Weddell's 1823 route.

On January 13, 1838, the men found themselves in a twisted maze of ice floes and bergs. "The crew, unaccustomed to this sight, appeared quite moved," wrote d'Urville. "This was but the advance guard of the fearsome enemy with which they would soon join battle." By January 20 d'Urville had intersected Weddell's path and confidently predicted that he would reach 65° south latitude in just a few days. The next morning a vast seamless barrier of ice barred their way. D'Urville gazed at the evocative panorama. "It is a new world unfolding before our eyes, but a world that is inert, mournful. Profound silence reigns over these ice plains and life is only represented by a few petrels, gliding soundlessly, or by the whales whose loud and ominous spouting occasionally breaks this sad monotony."

For the next three days they skirted the pack ice, always on the lookout for an opening. The tireless M. Dumoulin endured stiff hands, numbed with cold, as he charted the pack ice and took magnetic readings on the floes. On January 24 d'Urville admitted defeat. Where were those wide avenues of water that had carried Weddell so far south fifteen years earlier, d'Urville wondered. "Either Weddell struck an exceptionally favorable season, or he played on the credulity of his readers..."

D'Urville sailed northward toward the South Orkneys, where he hoped to land, but heavy swells and high winds held the ships at bay. Horizontal sheets of sleet pounded the ships for two days. When the weather improved, he tempted fate by pushing toward the south once again. "I learn that their lordships, the sailors of the *Astrolabe*... have suddenly become enthusiastic about the Pole and their only fear is that I may give up too soon. They need not worry. When I give up, none of them, I believe, will have any desire to push on any further!" He was as good as his word.

From the crow's nest came the dreaded yell of "Ice field ahead!" Fearlessly, d'Urville smashed a path through the fringes of the field into a two-mile open area within the pack ice. Sailors and officers from the *Zélée* visited with the *Astrolabe*'s crew and "emptied a bottle of punch amid mutual congratulations that we had dared what no other navigator had so far done... we had sailed boldly into the Antarctic ice," d'Urville wrote. He happily fell asleep. "Ominous noises" awakened him several hours later. "Going up on deck I saw that on all sides we were hemmed in by serried ranks of huge blocks of ice filing past." Both vessels were trapped. He decided not to wake the men, hoping that the situation would soon improve.

Heavy fog made matters worse the next morning. D'Urville retreated blindly toward the channel that had led him through the pack ice, but it had closed during the night. The *Astrolabe* rammed through the ice for several hundred feet, but then rattled to a bone-jarring halt. The men scrambled down onto the ice with picks, axes, and mattocks in their hands. All morning they hacked at the thick ice that barricaded their path to open water. Hands, faces, and feet turned white with frostbite. The men tied ropes around the larger bergs and then boarded the ship to haul on the lines, heaving the vessel forward inches at a time. For the next five days the crews battled ice, snow blindness, and frostbite while d'Urville directed operations and prayed for a strong wind to fill their sails. "I am not afraid of death," he wrote, "but I am tormented by the tragic prospect of seeing my two crews forced to seek precarious shelter on the icebergs . . . awaiting, in despair and horror, for the end." Three miles of ice separated life from death.

Then on February 9, the wind suddenly freshened. Just as the ship began to crunch through the few remaining floes toward the open sea, d'Urville shouted at the men to jump aboard. Every man leaped for the ropes and clambered up the sides to the deck to safety—except Aude, the caulker. Delayed by widening gaps between the ice floes, he now dashed toward the ship, leaping across the narrower cracks and running long zig-zagged detours around the wider channels. "Despite anything I tried, the corvette was moving smartly forward. . . . I was afraid that I might have to leave this unfortunate man on the ice," wrote d'Urville. But a lingering death among the floes was not what Aude had in mind. With extraordinary strength and stamina, he raced after the *Astrolabe*. "At last, to my joy and relief, he reached the ship and was dragged aboard more dead than alive." Aude survived his ordeal, but a year would pass before he fully recovered.

D'Urville now headed again toward the South Orkneys. On February 15 they landed on Weddell Island and filled their stomachs with boiled, minced, and stewed penguin meat. Hunger must have silenced all criticism, for the men vowed that the dark, oily meat was better than chicken. During long days ashore they collected rocks, moss, lichen, and more penguins for the galley. The ships left the South Orkneys on February 22 and steered toward the South Shetlands, skirting Clarence and Elephant Islands and continuing toward the Antarctic Peninsula. Although he rediscovered land first seen by Palmer and Biscoe, he still charted islands, channels, harbors, and coastline until March 4. On that day Lieutenant Dubouzet on the *Zélée* wrote, "The whole night we had freezing, penetrating rain that exhausted our sailors. Several of them were starting to get sore mouths, pain and fever, the early symptoms of scurvy." D'Urville immediately turned north toward Tierra del Fuego. "I myself openly confess that this first attempt was a complete failure," he wrote two days later.

Contrary winds buffeted the ships. Progress was slow and the men grew sicker. By March 27 the *Astrolabe* had twenty-one cases of scurvy, and the *Zélée* was little more than a floating infirmary. When the ships arrived in Talcahuano, Chile, nearly a dozen men deserted, and the most seriously ill were left to either die or recover. In May 1838, both ships arrived at Valparaiso, where d'Urville learned that rumors were circulating in the town stating that the expedition was a failure. Furious at this attack, he confronted the critics—officers from other French ships in port and the French Consul—with a smile and an elaborate lunch. Their opinions slowly shifted in his favor when d'Urville discussed his charts and observations.

He next headed toward the South Seas to survey potential whaling ports and to complete his ethnological research. Within twenty months he would try again to penetrate the Antarctic ice to search for the Magnetic Pole—at the same time as American Charles Wilkes.

Delays and Disputes in America . . . Solidarity in England

In 1837, preparations for the United States Antarctic Expedition bordered on chaos. Secretary of the Navy Dickerson had altered the expedition's objectives by allotting more time and resources to the South Pacific regions. An infuriated Jeremiah Reynolds, the man who had lobbied for the expedition, unleashed a torrent of bitter letters and pamphlets to the public. The Navy responded by dropping him from any further involvement. The scientists, who had put their careers on hold in order to sail with the expedition, seethed with anger and disgust at each delay. Rumors of fraud and high-level corruption surfaced daily and made good stories for newspapers at first; however, even the most jaded reporters soon tired of the steady stream of scuttlebutt. "Success alone can efface the recollection of the bickerings and the heartburnings, and delays and blunders, which have marked the progress of this expedition from its inception to the present time," noted one exasperated editorial in October 1837. Less than four months later, the scientists resigned en masse.

Secretary Dickerson, under siege for indecisiveness, listened to the officers vent their anger and then accepted their resignations. One of the first to depart was Captain Thomas Ap Catesby Jones, commander of the expedition, followed by many sympathetic officers. Finding replacements was a monumental chore for Dickerson. His list of qualified men dwindled daily, as officer after officer listed excuses for refusing assignment with the expedition. But on March 28, 1838, he found a man willing to accept command: Lieutenant Charles Wilkes. His appointment caused another storm of protest from higher-ranking officers, but Wilkes persevered.

As Wilkes grappled with bureaucrats during the summer of 1838, members of the British Association for the Advancement of Science held a meeting to draw attention to the problems of terrestrial magnetism. The participants composed a series of formal resolutions that, if adopted by the government, would advance the science of magnetism. One resolution in particular stressed the importance of observations in the high southern latitudes and strongly recommended "the appointment of a naval expedition expressly directed to that object." Parliament granted funds for the expedition and officially named James Clark Ross to command the operation in April 1839.

No one was surprised at Ross's appointment, for he was by far Britain's most experienced polar officer. Although the anonymous pamphlet two years earlier had caused widespread interest and enthusiasm for an Antarctic expedition, it was the British Association that provided the sound scientific justification for such a voyage that led directly to funding the project. In September 1839, just ten months after Parliament appropriated funds, James Ross led his ships southward.

A GLOOMY DEPARTURE FOR THE AMERICANS

While Dumont d'Urville was recuperating in Valparaiso, Chile, Lieutenant Charles Wilkes inspected his squadron of six ships. The sight didn't impress him, but there was no time to overhaul the vessels. The *Vincennes*, *Peacock*, and *Porpoise* were warships with large uncovered gun ports lining their sides. The *Sea Gull* and *Flying Fish* were old New York pilot boats that had outlived their usefulness. The *Relief* was a small cargo ship. None of the hulls were reinforced with copper or an extra layer of wood planking. Every one of the vessels leaked. Not even the barest human necessities—a dry place to sleep, warm clothing, adequate food—were supplied. Unscrupulous navy contractors and lax bureaucracy had seen to that.

Wilkes was aware of the problems but had little choice. Another postponement was out of the question—the expedition had already been delayed a year. If he resigned at this critical point, he reasoned, the whole project would be scrapped and America would face humiliation in the eyes of the world. With single-minded determination, focus, and perseverance, he readied his ships and men for the long voyage.

On August 18, 1838, Wilkes and his fleet of six dilapidated ships were about to begin the most ill-equipped and controversial expedition in the history of Antarctic exploration. Aboard his flagship, the *Vincennes*, Wilkes watched the shore disappear with a sense of foreboding. "It required all the hope I could muster to outweigh the intense feeling of responsibility that hung over me. I may compare it to that of one doomed to destruction." Later that day, Wilkes summoned the men for prayers and a sermon. The men congregated around Chaplain Elliot and listened attentively. Meanwhile, Captain William H. Hudson, who doubled as the preacher on the *Peacock*, felt gratified that "scarce an eyeball rolled during the service—a stillness reigned throughout."

Sailing southward, the fleet experienced few storms but lost days waiting for the sluggish *Relief* to catch up. Captain Hudson also had his hands full—the *Peacock* leaked "buckets full." Wilkes was forced to stay at Rio de Janeiro until January 6, 1839, while the *Peacock*'s crew tarred and patched the worst leaks, caulked the rotting deck and sides, and dried out the lower quarters. The ships then steered south along the South American coast and arrived in Orange Harbor, Tierra del Fuego, on February 17. Morale problems soon beset Wilkes, and he acted decisively. Eleven disgruntled officers were quickly relieved of their duties and dispatched with the *Relief* for survey work in the Strait of Magellan before returning home.

The Antarctic summer season was now dangerously advanced, forcing Wilkes to modify his plans for his first Antarctic voyage. He divided the six ships into three groups. The *Vincennes* and *Relief* would stay behind. He would take the *Porpoise*, accompanied by Lieutenant Johnson in the *Sea Gull*, and penetrate the Weddell Sea on the eastern edge of the Antarctic Peninsula. Hudson on the *Peacock* and Walker on the *Flying Fish* were to try to surpass Captain Cook's record 71° 10' south latitude in an area now known as the Amundsen Sea. On February 25 the four ships left Orange Harbor.

WILKES'S SHORT SOJOURN

As the *Porpoise* and the *Sea Gull* headed toward Drake's Passage, the weather was clear and mild. "The whole scenery around us was viewed to great advantage," Wilkes wrote, "but a dense bank of cumuli in the southwest foretold that we were not long to enjoy the moderate weather." High seas and gale-force winds thrashed the ships that afternoon. Wilkes viewed this as an opportunity, however, and measured the height of the waves (32 feet) and the speed of the running sea (about 26 MPH). Albatross, he noted, effortlessly kept pace with the ships during the gale, hovering above the masts as though tethered with invisible lines.

On March 1 the first icebergs emerged from a curtain of snow near the South Shetland Islands. The next day, Wilkes watched smoke billow from a snow-covered dome on Bridgeman Island. Determined to make a landing, Wilkes ordered the boats lowered, but within minutes a dense fog enveloped the ships. On March 3 the mist dispersed as the ships neared the northern tip of the Antarctic Peninsula. "Penguins, uttering their discordant screams, seemed astonished at encountering so unusual an object as a vessel in these frozen seas," Wilkes wrote. The soft evening light washing over countless icebergs mesmerized him. "I have rarely seen a finer sight," he wrote, "some were of pure white, others showing all the shades of the opal, others emerald green."

The next morning, ominous signs heralded an approaching storm: the barometer fell, the size of the swells increased, and all wildlife disappeared. The icebergs, so magnificent the past evening, now vanished behind a shroud of fog. The men could do nothing but wait—and listen to the waves smash against the massive bergs. The savage storm struck before noon. As the men wrestled with ice-encrusted ropes and sails, cold wind and sleet cut through their shoddy, government-issue clothing. Their suffering convinced Wilkes to cease any further explorations. When the gale ended on March 6, the ships headed back to the South Shetlands. The *Sea Gull* anchored for a week at Deception Island, and Wilkes returned to Tierra del Fuego. His first venture into the Antarctic had lasted just nine days.

"LIKE THE TICKING OF A DEATH-WATCH"

The captains of the *Peacock* and the tiny *Flying Fish* faced far worse perils. The second day out from Orange Harbor, they lost contact with each other during the gale. Captain Hudson fired his guns and burned a blue light continuously on the *Peacock*'s poop deck. After fourteen fruitless hours, Hudson had no choice but to continue his voyage, hoping to intercept Walker in the *Flying Fish* at the first of four rendezvous points. Abominable weather prevented him from reaching it and he now feared that the *Flying Fish* had been wrecked.

Hudson had his hands full just keeping the *Peacock* afloat, for the hastily patched holes and caulked seams were no match for the high, roiling swells. Water poured down the main hatch and sloshed across the lower deck with each roll of the ship. Titian Ramsay Peale, the only scientist to sail on this first venture south, lost all his notes and drawings in the deluge. Yet, even in the most rambunctious seas, Hudson still conducted Sunday services while the men clung to the bulwarks or stretched out prone on the pitching deck.

The *Peacock*, coated with thick layers of ice, struggled across the Antarctic Circle in mid-March. The water that had flooded the lower deck now froze, adding more miseries for the crew. After several days of dense fog, the mist slowly lifted on March 22, revealing glimpses of massive two-hundred-foot icebergs in all directions. Hudson retreated to the northwest. Three days later, the lookout spotted the *Flying Fish*. All hands on the *Peacock* gave three cheers as the seventy-foot vessel hauled up beside the ship. Lieutenant Walker then boarded the *Peacock* and told Hudson astounding news: against all odds, the fragile *Flying*

Fish had sailed to within one degree of Captain Cook's record. Walker explained that after the storm had separated the two ships, he sailed to each of the four rendezvous points without encountering the *Peacock*. Gale-force winds split the sails and snapped the yards while mountainous waves exploded over the deck, splintering the planking in places. The helmsman's ribs were fractured, but he carried on his duties while girdled in a harness of tarred canvas. In spite of the horrific conditions, Walker guided his crippled craft southward. On March 19 the vessel wound through fields of floes and icebergs, "whose pale masses just came in sight through the dim haze, like tombs in some vast cemetery; and, as the hoar-frost covered the men with its sheet, they looked like specters fit for such a haunt," wrote James C. Palmer, the ship's surgeon.

Three days later, on March 22, Walker reached 70° south latitude at 101° 16' west longitude, just north of Thurston Island in the Amundsen Sea, and recorded "an appearance of land" in his journal. Since pack ice completely blocked any passage further south, he turned northward. Later that day thick fog swirled around the ship. "Believing that we were getting into clear sea, I stepped below to stick my toes in the stove," he wrote. While Walker was warming his numb feet, the fog suddenly dissipated and the lookout shouted a warning. Ice floes surrounded the *Flying Fish*. Walker was back on deck within a minute. As he searched for a passage through the ice, he noticed "sutures" of weak ice binding the firmer floes together. If he could fracture the ice along one of those sutures, he reasoned, he might get through.

But the wind deserted them, and a blanket of cold air hovered above the water. "The waves began to be stilled... and a low crepitation, like the clicking of a death-watch, announced that the sea was freezing," wrote James Palmer. The men huddled on deck, praying for a breeze, "whose breath was now their life... and many an eye wandered over the helpless vessel, to estimate how long she might last for fuel." A few hours later the wind returned and Walker set his plan in motion. "I got about six knots on her... and let fly the head sheets," he proudly wrote. The weak ice joining the floes cracked and Walker pushed through to open water—and survival. On March 25, he saw the *Peacock*'s sails in the distance, and the two ships returned to Tierra del Fuego.

At different times during April 1839, Wilkes's six ships headed toward Valparaiso, Chile, but only five ships arrived at the port. The *Sea Gull* disappeared during a nine-day gale and was never seen again.

THE ARCTIC EXPLORER DEPARTS FOR THE ANTARCTIC

While Dumont d'Urville and Charles Wilkes explored islands in the South Pacific, James Clark Ross was finalizing preparations for Great Britain's expedition. Born in London in 1800, Ross entered the navy when he was eleven years old, under the guardianship of his uncle, the famous Arctic explorer Sir John Ross. From 1818 to 1827 he participated in five Arctic expeditions. Recognized for his intelligence and leadership capabilities, Ross was promoted to Commander and elected a Fellow of the Royal Geographical Society. In 1829 he sailed with his uncle to hunt for the Northwest Passage. Although that goal eluded them, James Ross located the North Magnetic Pole on the Boothia Peninsula in northern Canada on May 31, 1831.

His experiences with scurvy in the Arctic convinced Ross to pay particular attention to provisions. Holds were filled with twenty tons of tinned meat, vegetables, soups, and gravies; fresh carrots, potatoes, cranberries, onions, and pickles added another eight tons to the total amount of food.

His two ships, *Erebus* and *Terror*, were the strongest and best-built vessels for ice navigation. Benefiting from a long history of Arctic whaling, the ships' decks were constructed from two thick layers of planking with waterproof canvas sandwiched between. The outer hulls were lined with another skin of wood and then double-sheathed with copper. Inside, the bows and sterns were reinforced with massive timbers; tight-fitting hatches protected the crew's quarters from deluges.

Since the expedition was under the control of the Royal Navy, civilians—including scientists—were banned from participation. Ross, however, had the foresight to ignore protocol. He filled the surgical posts with geologist and ornithologist Robert McCormick and naturalist John Robertson. Botanists Joseph Hooker and David Lyall filled in as assistant surgeons.

Ross received his final instructions in September 1839. His first duty was to conduct a series of magnetic readings en route to Hobart, Tasmania, where he was to winter. Then he was to proceed "to the southward in order to determine the position of the magnetic pole, and even to attain to it if possible." On September 17, sealer John Balleny arrived with news of his Antarctic discoveries: the Balleny Islands and Sabrina Land. Ross carried with him extracts of Balleny's journal and copies of his charts, which later proved to be most important. At the end of September the *Erebus* and the *Terror* sailed down the Channel toward the Atlantic. As the Cornwall coast sank below the horizon on October 5, the *Erebus*'s crew cheered with "joy and light-heartedness," noted Ross. Four and one-half years would pass before they would again see the coast of England.

Dumont d'Urville's Decision

As James Ross sailed across the equator, Dumont d'Urville anchored in Hobart, Tasmania, on December 12, 1839, after twenty months in the South Pacific. Scurvy, combined with rampant dysentery, afflicted most of his men. His first priority was to remove the sick and dying from the ships to a makeshift hospital in town. Next, both ships needed extensive work. Able-bodied sailors had to be found to paint, caulk, clean, and remove "the pestilential smell that the sick had left behind."

He appointed Lieutenant Dubouzet, barely recovered from dysentery, to handle recruitment responsibilities. "There were, at that time, a lot of deserters from French whaling ships... but as soon as they learned of our arrival of our corvettes they cleared out so as not to be caught by us," Dubouzet noted with exasperation. He managed to attract a few deserters by promising to forget the past. Another time-honored tactic—shanghaiing sailors—didn't appeal to him. "It would have been necessary to do the rounds of the taverns and get them to sign on while drinking with them, like they do in England, but I did not have the strength for that."

While Dubouzet scoured the town for sailors, d'Urville scrounged for supplies. Critical shortages had occurred because the number of whaling ships calling at Hobart had increased. When d'Urville wasn't bartering for food, he visited the wharf to renew acquaintances—and to obtain any news about the Americans' plans for returning to Antarctica. D'Urville learned from several whaling captains that Wilkes's officers had been instructed "to remain absolutely silent" about their upcoming expedition. Perhaps his competitive spirit—or the secretiveness of Wilkes's crew—convinced d'Urville to try once more to reach the south polar regions, even though the decision meant disregarding his written orders. He was determined, "come what may, to make it as profitable as possible for the physical sciences. There was one important discovery yet to be made; the position of the Magnetic Pole."

A TRIUMPHANT LANDING: WINE, PROCLAMATIONS, AND PENGUINS

By the end of December Dubouzet had found enough men to work the ships, and d'Urville had purchased "enough sheep and pigs for fresh meat for the crews for twenty days, antiscorbutic lime juice, highly regarded by the English, and potatoes." On January 2, 1840, the *Astrolabe* and *Zélée* headed out to sea.

Flocks of albatross followed the ships from Hobart, but disappeared when d'Urville crossed 50° south latitude eight days later. On January 18 d'Urville crossed 64° south latitude. "We found ourselves suddenly surrounded by five enormous masses of ice, flat-topped like tables," d'Urville wrote. He suspected that land must be near because the icebergs showed no signs of decomposition, as though they had been "detached quite recently from a not very distant frozen shore."

On January 19 the wind died as the ships hovered near the Antarctic Circle, surrounded by icebergs. "The sun was shining brilliantly, and its rays, reflecting on the crystal walls surrounding us, produced a ravishing and magical effect," noted d'Urville. Petrels circled overhead, while penguins porpoised through the calm sea and whales blew in the distance. At 3 P.M. the lookout shouted to Dumoulin that land was ahead. Dumoulin quickly clambered up the rigging and reported to d'Urville "the appearance of land, much more clear cut and distinct."

Both men, however, remained skeptical. All doubt vanished at 10:50 P.M. The sun, low on the horizon, showed the land's "high contours with utmost clarity. Everyone had rushed on deck to enjoy the superb spectacle before our eyes... the sun disappearing behind the land, and yet leaving behind a long trail of light."

Although dead-calm conditions the next day thwarted d'Urville's efforts to sail closer to the coast, the men took full advantage of the lax time. As the ships slowly drifted toward the Antarctic Circle, "Father Antarctica" paid a visit to the *Astrolabe* with songs, skits, and a costumed parade. The men modified that favorite rite of passage—obligatory drenchings at the equator—to fit Antarctic conditions. Cups of wine were substituted for buckets of water—and drunk rather than tossed.

A breeze once again filled their sails on January 21 and d'Urville steered toward the land through a labyrinth of massive icebergs. "Their sheer walls were much higher than our masts; they loomed over our ships, the size of which seemed dramatically reduced in comparison with the enormous masses. The spectacle... was both magnificent and terrifying. One could imagine oneself in the narrow streets of a city of giants." The shouts of the officers echoed and re-echoed from the steep walls of ice. Strong eddies swirled in front of the entrances to vast icy caverns, where waves rushed in with thunderous roars.

Once through the maze of bergs, d'Urville gazed at the snow-covered land, gently sloping upward to about thirty-three hundred feet and stretching from the northwest to the southeast as far as he could see. "We noticed that the snow covering the ground had a furrowed and ruptured surface. We could distinguish actual waves, like those made by the wind in the desert sands... especially in the least sheltered parts." This ice formation, common to windy regions, would later be named *sastrugi*. D'Urville searched until early evening for a place to land, but the shore was a vertical cliff of ice.

Rocky islets were spotted about seven miles in the distance and two boats were immediately launched. After two hours of arduous rowing, the men clambered ashore, hurled aside the astonished penguins, and unfurled the tricolor flag. "Following the ancient custom, faithfully kept up by the English," wrote Dubouzet, "we took possession of it in the name of France.... The ceremony ended, as is mandatory, with a libation. To the glory of France . . . we emptied a bottle of the most generous of her wines. . . . Never was a Bordeaux wine called on to play a nobler role; never was a bottle emptied more appropriately." If France's new territory appeared "worthless and faintly ridiculous," at least the short, squawking inhabitants "will never start a war against our country," mused Dubouzet.

The men quickly searched for shells, plants, lichen, and inhabitants other than penguins. A single dried-out strand of seaweed was their only prize. "We had to fall back on the mineral realm," Dubouzet wrote. Picks and hammers rang out in competition as the men attacked the nearest granite boulders. Dubouzet examined a small stone—an exact replica of the pebbles he had found in a penguin's gizzard the day before, "which could have given an accurate idea of the geological formation of these lands, had we not been able to get ashore." Thirty minutes later, their boats were loaded with enough samples to supply many museums. Before leaving their islet the men carried off "living trophies of our discovery"—a few penguins. Then they saluted the islet and cheered. "The echoes of these silent wastes, disturbed for the first time by human voices, repeated our shouts and then returned to their habitual brooding and awesome solitude." D'Urville called the islet Geology Point in recognition of the men and their souvenirs. He then named the mainland—and the new penguin species—in honor of his wife, Adélie.

A PUZZLING ENCOUNTER

D'Urville explored the Adélie coast in fine weather for the next three days, the ships weaving their way through iceberg obstacle courses, but the good weather did not last. On January 24 a blizzard with gale-force winds struck. The *Astrolabe*'s mainsail was shredded as the vessel foundered in waves that broke over the deck from all directions. "It was impossible to hold on to the rigging, which was covered with sharp icicles," wrote d'Urville, " it was only with considerable difficulty that our sailors could stay on the deck that was being swept constantly by the waves. . . . dreaded every moment hearing the terrible cry of 'pack ice to leeward!' " The men battled seas and winds, steering blindly in the snow and losing all sense of direction. By the time the storm had ended, both ships had sustained more damage in eighteen hours than in the previous twelve months. The ferocious storms and sustained sixty-knot winds near the coast of Adélie Land were later shown to be the rule rather than the exception. Another explorer, who survived in the region for two years, aptly titled his 1915 book *The Home of the Blizzard.*

For another three days d'Urville charted the pristine coast of Adélie Land, the solitude broken only by the cries of penguins, the squawks of petrels, and the occasional thunder of a calving iceberg. Iceberg flotillas fascinated the imaginative Ensign Duroch. "We are in fact, sailing amidst gigantic ruins... here temples, palaces with shattered colonnades and magnificent arcades; further on, the minaret of a mosque, the pointed steeples of a Roman basilica; over there a vast citadel with many battlements, its lacerated sides looking as if they have been struck by lightning; over these majestic ruins there reigns a deathly stillness, an eternal silence."

On January 29 fog settled around both ships. When an iceberg appeared in the distance, the crew unfurled more sail in order to bypass the berg at a safe distance. Suddenly the gray bow of a warship cut through the mist. "The ship was moving fast," wrote d'Urville, "We were able to recognize its flag.... It was an American brig." The men hoisted their colors and d'Urville issued orders to maneuver the *Astrolabe* into position to receive the American ship, the *Porpoise*, alongside. Then, without the customary signals and greetings, the *Porpoise* abruptly turned and disappeared into the fog. D'Urville believed the Americans' rude actions demonstrated their tight-lipped "secrecy about their operations." Lieutenant Ringgold, captain of the *Porpoise*, would later interpret d'Urville's maneuvers very differently, however.

The character of the coast changed the next day. Instead of a jumble of huge tabular icebergs blocking the shore, a 120-foot high solid barrier of ice now jutted into the sea. D'Urville was convinced that "this formidable band of ice was a sort of envelope or crust of ice over a solid base of either earth or rocks, or even of scattered shoals round a big landmass." He named the coast Clarie Land, after Captain Jacquinot's wife. Despite having taken few magnetic readings, d'Urville ended his second Antarctic foray on February 1, 1840, and announced his discoveries in Hobart seventeen days later.

Inquiring Minds

Charles Wilkes and his battered hulks arrived in Sydney in November 1839, after seven months in the South Pacific. To his dismay, the Australians had heard rumors about Ross's elaborately outfitted vessels and were now waiting anxiously to question Wilkes about his ships and equipment. "They inquired whether we had compartments in our ships to prevent us sinking? How we intended to keep ourselves warm? What kind of anti-scorbutic we were to use? And where were our great ice-saws? To all of these questions I was obliged to answer, to their great apparent surprise, that we had none... most of our visitors considered us doomed to be frozen to death."

Wilkes did the best he could to improve conditions on his ships. Rotting bulwarks and gun ports were covered with tarred canvas and sheet lead, weights and pulleys were rigged on the doors so that they swung shut, and charcoal stoves were slung in metal cradles to warm the men and dry their clothes. Although the *Peacock* was in a deplorable state, Wilkes could not afford to wait two months for a complete overhaul. He and Hudson decided to take the ship in spite of its "rotted upperworks."

On December 26, 1839, the Australians cheered the Americans as sails were hoisted and the ships lumbered away from the harbor. Wilkes commanded the *Vincennes*; Hudson, the *Peacock*; Ringgold, the *Porpoise*; and Pinkney, the tiny *Flying Fish*. Wilkes's plan was to head toward 160° east longitude, near the Balleny Islands, and then continue westward to longitude 105° east, reaching as high a latitude as possible. During the voyage, the men were to record magnetic, astronomical, and meteorological observations. Wilkes also stressed to his captains that they should do everything in their power to sail together as a squadron and not to separate.

THE FIRST VIEW OF LAND

On January 1, 1840, the *Flying Fish* was the first to lose contact with the fleet when high winds whipped away part of the ship's rigging. Delayed for repairs and slowed by the roiling swells, the handicapped pilot boat had little chance to catch up with the others. Still, the determined fifteen-man crew doggedly trailed after their companions toward the Balleny Islands. Two days later the *Peacock*, a floating sieve of a ship, separated from the *Vincennes* and *Porpoise* during dense fog and drifted beyond the noise of Wilkes's guns, horns, bells, and drums. Captain Hudson, however, took advantage of the delay to construct housing over the hatches and caulk the deck seams.

Meanwhile, the *Vincennes* with Wilkes and the *Porpoise* with Ringgold sailed toward a flotilla of icebergs on January 10. By the next evening, pack ice had halted any further progress to the south, to the men's disappointment. "We hove to until full daylight," wrote Wilkes. "The night was beautiful, and everything seemed sunk in sleep, except the sound of the distant and low rustling of the ice..." The next morning both ships, enveloped in fog, lost sight of one another but continued westward, close to the edge of the pack ice. On January 15 Captain Hudson in the *Peacock* reached the pack ice and joined the *Porpoise* as the ships continued to skirt the ice barrier.

The next day, January 16, men on each of the three ships sighted land. Although notations were recorded in private journals and in one logbook, Wilkes, Hudson, and Ringgold reserved judgment and tempered the crews' enthusiasm. Fog soon obscured the view, and Wilkes headed southward where he thought the sea lay open beyond the ice floes. With a favorable wind the ship moved rapidly, tossing and rolling with the high swells. Then, in an instant, "all was perfectly still and quiet; the transition was so sudden that many were awakened by it from sound sleep," noted Wilkes. This cessation of motion meant only one thing: a solid field of ice. The men hurried up the hatches and peered into the dense fog, ready to scramble into the rigging to work the sails should the order be given." The feeling is awful," wrote Wilkes, "to enter within the icy barrier blindfolded by an impenetrable fog." Suddenly a dozen voices yelled, "Barrier ahead!" Wilkes shouted orders and the men leaped to activity as the rasping sounds of the ice grew more distinct in all directions. The ship turned away from the ice barrier with not a moment to spare; after two tension-filled hours, the *Vincennes* reached open water.

The three ships continued their westward explorations independently, since too much time had been lost trying to keep together. By now Wilkes, Ringgold, and Hudson had lost all hope of meeting the *Flying Fish*. The sailors were also pessimistic, and speculated that perhaps they would spot pieces of the vessel's wreckage on their homeward trek. Only then would they know for certain their shipmates' fate and mourn their loss.

Gloom was quickly replaced with joy on January 19. "Land was now certainly visible," noted Wilkes. This discovery energized the men and "gave an exciting interest to the cruise." Complaints of fatigue evaporated and now the men were ready to tackle any danger to reach the distant coast. The snow-covered, sloping land with its immense dimensions convinced Wilkes it was not an island but a vast continent. Later that night the sun "illuminated the icebergs and distant continent with its deep golden rays."

Meanwhile, the men on the *Peacock* saw something "in the distance peering over the compact ice . . . very much resembling high craggy land covered with snow," noted Lieutenant George Emmons. " 'Discovery Stock' ran high—Spy Glasses were in great requisition." The land rose to about three thousand feet and looked like a snow-covered amphitheater. Wilkes later named the point Cape Hudson.

As Wilkes continued westward, large tabular icebergs dominated the seascape. The sides of the bergs were chiseled into "lofty arches of many-colored tints, leading into deep caverns. . . . The flight of birds passing in and out of these caverns recalled ruined abbeys, castles, and caves, crowned with pinnacles and turrets, resembled some Gothic keep. . . . If an immense city of ruined alabaster palaces can be imagined . . . with long lanes winding irregularly through them, some faint idea may be formed of the grandeur and beauty of the spectacle."

The crew on the *Peacock* enjoyed another kind of beauty—slate-colored mud and a chunk of rock. The irrepressible Hudson had worked the ship deep into a bay. A boat was launched with Midshipman Eld at the helm. The men clambered on top of a small iceberg to take magnetic readings and measure the depth of the water. When the first task was completed, they tossed a weighted line with a copper cylinder attached into the water. It struck the bottom at 320 fathoms (1,920 feet). When Eld hauled up the line, the cylinder was filled with mud and a piece of granite—proof that what everyone had seen was land, covered with a thick skin of ice and snow. A curious emperor penguin paused to watch these proceedings and was promptly captured by Eld. Once back on the *Peacock*, the crew cheered the men and Hudson ordered everyone to "splice the main brace" (have a double round of rum). The penguin was sent to the galley, where the cook discovered thirty-two pebbles in its crop. Recognizing a financial windfall when he saw one, the cook proceeded to sell the stones to the highest bidders—until Hudson halted the transactions.

Mirth turned into panic the next day as ice in the bay crept toward the *Peacock*, blocking their escape route. As the ship retreated from the encroaching ice, the Peacock's stern struck pack ice. The force of the crash threw the men, who were eating breakfast, to the deck. A hasty survey of the damage showed that the rudder was twisted but repairable, but then came a second blow that smashed the bracings supporting the rudder. The men hauled the rudder on deck and the carpenters worked frantically to repair it while the ship drifted toward the immense wall of a tabular berg. An ice anchor was attached to the nearest floe but another jolt ripped the rope out of the men's hands. The crew tried again and this time the anchors held—until the wind freshened. The anchors broke loose and now the men, powerless to guide the ship, faced imminent destruction against the wall of ice. The surge of a swell turned the ship just before impact. The *Peacock*'s stern slammed into the ice with such force that the rebound gave the men time to work the sails. A sudden breeze carried the ship beyond the iceberg.

Free of one danger, the men now faced another: the weather. The breeze that had saved them now freshened into a gale, and mighty swells pushed the floes and smaller bergs toward the land—and the *Peacock*. With each surge something was carried away—chains, eyebolts, bobstays, and lashings—and "every timber seemed to groan." The unflappable Captain Hudson stuck to shipboard routine and ordered the men to eat their dinner, perhaps to dispel the melancholy mood. As the men listened to the ice grinding and thumping against all sides of the ship, they debated whether it would be better to freeze to death, one by one, or to perish as a group on a dissolving iceberg.

Neither alternative was acceptable to Captain Hudson. He set the men to work. Throughout the long night they repaired stove timbers, launched the gunboats for another attempt to anchor the ship to a floe, and constructed new parts for the demolished rudder. The swell grew more forceful by midnight. "The ice and water were foaming like a cauldron," wrote Hudson. By midmorning Hudson knew the only way to escape was to force the *Peacock* through a narrow channel between the floes and bergs, "or grind and thump the ship to pieces in the attempt." The wind favored the men and the sails were unfurled. With the rudder once again in place but "hanging by the eyelids," the mauled *Peacock* broke through the ice into open sea. Hudson now wisely decided not to push his luck. The *Peacock* headed toward Sydney and arrived safe, but not sound, on February 22, 1840.

RINGGOLD'S COMPLAINT

Meanwhile, Wilkes on the *Vincennes* continued to explore the Adélie Coast, "tracing out the position of the icy barrier and following along the newly discovered continent." On January 28 he sighted land behind the same rocky islets where d'Urville's men had landed one week earlier. As Wilkes continued westward, broad clusters of tabular icebergs loomed ahead. Suddenly, the barometer started to drop precipitously and by evening a blizzard buffeted the ship. "At midnight I had all hands called," wrote Wilkes. "The gale at this moment was awful.... At a little after 1 A.M., it was terrific." He ordered the men to furl the sails. One sailor climbed aloft, and just as he was crawling on the yardarm, a sudden gust of wind blew the sail over the yard, trapping the man. "He appeared stiff," noted Wilkes, "and was almost frozen to death." The sailor was hauled down and placed in front of the charcoal stove to thaw. Miraculously, he survived his ordeal.

The circle of massive tabular icebergs tightened around the *Vincennes* and damped "the roar of the wild storm that was raging behind, before, and above us." Then Wilkes noticed a clear channel between two large bergs and "a glimmer of hope arose." He steered the ship toward the opening. Their passage between the two walls of ice was short but intense. Wilkes listened with satisfaction as the "whistling of the gale grew louder and louder before us, as we emerged from the passage."

By noon the next day the weather had cleared and Wilkes threaded his way back through the iceberg gauntlet he had passed through the night before. How they had avoided being "smashed into atoms" astounded the men. As they again approached d'Urville's Point Geology, Wilkes saw the dark rocky islets for the first time. "Now that all were convinced of its existence, I gave the land the name of the Antarctic Continent," he wrote.

The men's triumphant mood vanished as the barometer once again "declined rapidly." The blizzard raged around the ship, and the driven snow was "armed with sharp icicles or needles." When the storm ended, he received a signed letter from the two surgeons requesting that Wilkes end the voyage since many men were exhausted and sick. Wilkes's response was blunt: it was his duty to proceed "and not give up the cruise until the ship should be totally disabled."

On the same day that Wilkes battled the second blizzard, heavy fog enveloped the *Porpoise*. Lieutenant Ringgold had a fleeting glimpse of two sets of sail in the distance. Assuming that the ships were the *Vincennes* and the *Peacock*, he approached with full sail, quickly closing the distance between the ships and himself. By the time he was almost on top of one ship, he recognized the pennants, indicating the presence of French Commodore Dumont d'Urville. Ever sensitive to the nuances of navy protocol, Ringgold maneuvered to approach the stern of the *Astrolabe*, with its superior ranking commander. When d'Urville turned broadside to the *Porpoise*, Ringgold believed the French commander was rebuffing him by turning away, so Ringgold hauled down his colors and promptly hove to the south.

When Wilkes heard about the incident, he reacted furiously. "By refusing to allow any communication... [d'Urville] not only committed a wanton violation of all proper feeling, but a breach of the courtesy due from one nation to another." D'Urville later responded—and Wilkes replied. Before long, politicians, ambassadors, and government officials from both countries penned faultfinding missives to each other. A pronounced cooling in diplomatic circles between the Americans and the French lasted for several years.

During February, Ringgold and Wilkes separately explored the continent, each sailing westward from Adélie Land and skirting the coast as close as the icebergs permitted. Glimpses of the continent were now recorded routinely—at least during periods of clear weather. But hardships had taken their toll on the *Vincennes*. By February 3, the sick list had soared to thirty names. The symptoms of scurvy—mouth ulcers and skin boils—ravaged the men. Charcoal fires, which burned continuously on the gun deck, released poisonous fumes that added to the crew's misery. Wilkes collapsed on February 4 from breathing the noxious air and was unconscious for thirty minutes. Still, he turned a deaf ear on his officers' requests to end the voyage.

THE FATE OF THE *FLYING FISH*

Whereas Wilkes ignored the complaints of his men and continued his exploration, Lieutenant Pinkney of the *Flying Fish* listened to the pleas of his crew. He turned northward on February 5, having survived almost insurmountable perils.

The *Flying Fish* had reached the ice barrier west of the Balleny Islands on January 21, ten days after Wilkes. Four of the ten sailors were very sick, probably the result of poorly prepared meals. The cook, a surly man full of a self-importance that belied his minuscule culinary skills, executed his duties with minimum effort and maximum indolence. Rancid, undercooked salt beef, weak grog, and moldy biscuits were the usual fare.

At least the weather was good for the next few days. Then came a series of fierce gales and blizzards. By February 1, few men were physically able to work the sails or take a turn at the helm. The sick and exhausted men were moved into the officers' ward, which flooded less than their own quarters. Two nights later, another storm unleashed its fury on the ship. "It was a truly horrible night," wrote Lieutenant Sinclair, especially for the men on watch duty. Lieutenant Pinkney lashed himself to the foremast to keep from being flung overboard as the ship plunged, rolled, and pitched among the turbulent swells. Sinclair followed Pinkney's example but, as the storm further intensified, he was swept off his feet whenever the bow plunged underwater. One of the three sailors still capable of working stood at the helm shoeless, his feet too swollen with cold to be squeezed into leather boots. The only man who probably enjoyed the storm was the cook. Water continually gushed down the stovepipe and doused his charcoal fire, thus relieving him of his duties.

Water also poured in through the seams and drenched every article of clothing. The men reached the limit of their endurance on February 5 and petitioned Captain Pinkney—who was also sick—to end this nightmare voyage. "We, the undersigned... wish to let you know that we are in a most deplorable condition... we have no place to lie down in; we have not had a dry stitch of clothes for seven days... we hope you will... relieve us from what must terminate in our death." Pinkney, in true military fashion, requested a written response from his officers to verify the conditions on board. Their lengthy reply concurred with every grievance, and Pinkney "deemed it his duty to steer for the north." The *Flying Fish* arrived at the Bay of Islands, New Zealand, on March 9.

JOURNEY'S END AND A LETTER TO ROSS

The *Vincennes* and *Porpoise*, widely separated, continued to explore the Antarctic coast to the west of Adélie Land. On February 12, Ringgold noticed a dark brown piece of ice "resembling a large piece of dead coral" that floated by the *Porpoise*. He launched a boat to bring it alongside the ship. The chunk was composed of alternating layers of red clay, brown sand, gravel, and ice. Later that afternoon, as he steered toward what is now called Vincennes Bay, many of the icebergs "assumed a dark color from the clay and sand blown upon them," he wrote. A large black area on another iceberg intrigued him, and he again dispatched a boat to bring back samples. "It proved to be a large mass of black, red, and mixed-colored earth, resting upon a base of snow and ice." As Ringgold gazed toward the undulating, snow-covered edge of the continent, he was "at a loss to account for these frequent signs of land." More than one hundred years would pass before the Bunger Hills, ice free in summer and the origin of the sediments, would be discovered.

Ringgold continued his trek until he reached approximately 100 degrees east longitude. He then turned the *Porpoise* eastward to retrace his route and headed north to the Auckland Islands and New Zealand.

Wilkes, meanwhile, also collected many rock specimens from an iceberg on February 14. "It was amusing to see the eagerness and desire of all hands to possess themselves of a piece of the Antarctic Continent," he noted. In the middle of the berg he found a pond filled with the most delicious water he had ever tasted, and shuttled five hundred gallons back to the *Vincennes*. Their work done, the men raced to the top of steep hummocks and then slid down, using bunched-up jackets, planks of wood, or shovels for makeshift sleds.

On February 17, a massive north–south barrier of ice ended Wilkes's westward trek at 97° east longitude. He fittingly named the barrier Termination Land, now called the Shackleton Ice Shelf. That night the men watched a magnificent display of the aurora. Pencil-thin rays of light flashed across the sky like the opening and closing of a brilliantly colored fan. As Wilkes noted, "the best position to view it was by lying flat upon the deck, and looking up."

Wilkes backtracked for the next four days. On February 21 he mustered the crew. When he announced that the *Vincennes* would now head north, he remarked, "I have seldom seen so many happy faces, or such rejoicings." He and his exhausted crew arrived in Sydney on March 11, 1840. Later the *Vincennes* and the *Peacock* sailed for the Bay of Islands, New Zealand, to join the *Porpoise* and the *Flying Fish*. While there, Wilkes learned that James Clark Ross would soon arrive in Hobart, Tasmania, to prepare for his expedition to the same region that the Americans had just explored. Wilkes wrote Ross a friendly letter that summarized his data on currents, meteorology, and ice conditions. He also included his calculated position for the Magnetic Pole: about 230 miles south from the coast of Adélie Land. He enclosed a copy of his chart and a newspaper article that announced his discovery of the "great Antarctic Continent."

Although Wilkes's orders forbade him to disclose any details of the expedition, he ignored regulations and sent the information to Ross as a gesture of gratitude. When Wilkes was in London in 1837, Ross had shown an interest in America's expedition and had been generous with suggestions and advice. Now Wilkes returned the favor. His letter and chart triggered a strong reaction from Ross, but not the friendly one he had anticipated.

DISTURBING NEWS FOR ROSS

Captain James Clark Ross arrived in Hobart, Tasmania in August 1840. He had spent the previous ten months constructing temporary and permanent magnetic observation stations on several islands, including Kerguélen. The entire town welcomed him with cheers, music, food—and news about d'Urville's and Wilkes's discoveries in the very region that he had intended to explore. Their audacity at preempting his own plans piqued Ross to such a degree that he never acknowledged Wilkes's friendly letter. "I should have expected their national pride would have caused them rather to have chosen any other path in the wide field before them, than one thus pointed out," he later wrote, choosing to ignore the fact that Wilkes had received his orders at least one year earlier than Ross. Since the south Magnetic Pole's position was believed to be at 66° south latitude and 146° east longitude (near the coast of Adélie Land), the route which d'Urville and Wilkes had followed was the shortest and most sensible one.

However, the good news for Ross was that neither Wilkes nor d'Urville had located the Magnetic Pole. Measurements that both men had recorded indicated that the pole's position was much farther south—conclusions that agreed with Ross's own series of magnetic readings. Proceeding to Adélie Land would serve no useful purpose, he reasoned. The impenetrable landmass blocked further progress to the south, and any geographical discoveries would benefit the Americans and the French—not the English—because of proprietorship. However, sealer John Balleny's journal and charts, which Ross had carried with him, offered a glimmer of hope. In 1839 Balleny had seen a clear sea beyond the pack ice near the islands he had discovered. Ross believed that if he could penetrate the pack ice to this unknown sea and then sail southwest, he could reach the Magnetic Pole—if no new lands barred his path.

He now scrapped his original plan and plotted a new route. "I considered it would have been inconsistent," he later wrote, "if we were to follow in the footsteps of any other nation. I therefore resolved at once to avoid all interference with their discoveries and selected a much more easterly meridian [170 degrees E]." Because of this drastic change of plans, James Ross would soon sail toward John Balleny's ice-free sea and begin a voyage of discovery unmatched by any previous explorer.

BREAKING THROUGH THE PACK

On November 12, 1840, Captain Ross on the *Erebus* headed toward the sea from Hobart. Behind the flagship was the *Terror* with Captain Francis Crozier, Ross's second-in-command. Eight days later, the men anchored at Rendezvous Harbor near the eastern tip of Auckland Island. Two stout planks, planted in the peaty soil, were spotted. Their painted messages announced the arrivals of the French and Americans the previous March. Although d'Urville mentioned the discovery of Adélie Land in bold letters, the Americans had simply stated that they had sailed "along the Antarctic Circle." Ross expressed ironic surprise that there was no mention of the "great Antarctic Continent." In a *Sydney Herald* article, Wilkes had declared that the lands he had discovered were part of a massive continent—a contention that had irked Ross. Based on his extensive Arctic experience, Ross believed that Antarctica was a series of ice-covered archipelagos strung around a polar sea. Wilkes's grand assertion, Ross reasoned, was presumptuous at best.

While the crew erected a magnetic observation station, botanist Joseph Hooker tramped through gnarled rata forests, waded across peat bogs, examined fields of giant herbs, and climbed grassy slopes to collect more than two hundred plant species. Naturalist Robert McCormick also trundled the island's width and breadth with gun and haversack in hand, adding to the expedition's wildlife collection with unbridled zeal. Anything that lumbered, waddled, or flitted across his line of fire was dispatched, skinned, and tagged. Wandering albatross fascinated him, and he enjoyed sitting in the midst of a colony, watching their graceful courting displays and clumsy take-offs at the top of a steep slope. The bird, he noted, "waddles off in the most grotesque manner, floundering with outspread wings among the long grass, often rolling over and over if any ruggedness of the ground obstructs her progress."

The expedition arrived at Campbell Island on December 13 and left two days later for the journey into the pack ice. Although no one underestimated the dangers, the men were cheerful, even during periods of boisterous winds and monstrous swells, for they were warm, dry, and well fed. The first iceberg was spotted on December 27; three days later, the ships crossed Bellingshausen's 1821 path at 173° east longitude. New Year's Day was celebrated by sailing across the Antarctic Circle and with extra plates of food and grog. Even the *Erebus*'s goat enjoyed a generous libation.

Several snow petrels, harbingers of pack ice and bergs, flew above the masthead, much to McCormick's delight. After four miscalculations, he finally shot one that dropped to the deck instead of spiraling down into the sea. The next day a brilliant line of light radiated on the horizon: the edge of the pack ice. On January 5, 1841, a strong breeze filled the ships' sails, and Ross guided his heavily reinforced ships toward the frozen scourge that had defeated all previous explorers.

Chunks of loose ice bobbed in the swells, and wedge-shaped floes drifted past the ships. With each mile more floes and bergs consolidated into a white mass, until only a network of ribbonlike channels offered passage further south. Ross skirted the edge of the pack ice and then signaled to Captain Crozier to follow him into the ice. The *Erebus* struck the pack at full sail and the ice cracked and heaved, rasping against both sides of the ship. "After about an hour's hard thumping," the *Erebus* and *Terror* broke through the outer ring of heavy pack ice into more open water. Scattered floes carried many members of the indigenous penguin population, who monitored the ships' progress. The men cupped their hands around their mouths and called to the startled penguins with gooselike noises. Their honks must have struck the right chord with the penguins. With flippers whirling and feet pushing against the ice, the birds tobogganed across the floes, but the ships outpaced them. The penguins then dove into the sea and porpoised beside both vessels, much to the men's delight.

For the next four days, Ross zigzagged among the floes and icebergs. On January 9 the weather deteriorated and soon the winds had risen to gale force, with thick squalls of snow reducing visibility to zero. The next morning the winds died and then the mist dissolved. "We had a most cheering and extensive view," wrote Ross, "and not a particle of ice could be seen in any direction from the masthead." It was a triumphant moment for Ross, for he had accomplished what no other explorer had dared: he was the first to force his way through the pack ice and survive. He now emerged into the sea that bears his name.

"BEHOLDING WITH SILENT SURPRISE..."

The course was set southward toward the Magnetic Pole, which their dip compass indicated was very near. Images of sailing to victory on the open sea flashed through the men's minds, and spirits were high. Then at 2 A.M. on January 11, the lookout reported land straight ahead. Ross's hopes were shattered as he gazed at the distant mountain peaks. "A severe disappointment," he wrote. Others, however, were enthralled. "The Morning was beautiful and clear," wrote Cornelius Sullivan, the blacksmith on the *Erebus*, "...Not a cloud to be seen in the firmament but what lingered on the mountains.... The snow topped mountains Majestically Rising above the Clouds. The Pinguins [*sic*] Gamboling in the water, the reflection of the Sun and the Brilliancy of the firmament Made the Rare Sight an interesting view."

By evening Ross's disappointment dissolved as the beauty of the Admiralty Range worked its magic. Soft golden light washed over the eight-thousand-foot peaks and spilled down the glaciers, illuminating every crevasse and hummock. The utter stillness was broken only by the "breathing of whales," the whisper of brash ice against the wooden planks of their ships, or the sharp report of a calving iceberg. A promontory with dark-colored vertical volcanic cliffs jutted toward the ships. Ross named the feature Cape Adare, after one of the expedition's most enthusiastic supporters. Girdled with a thick fringe of ice, Cape Adare offered no possibility of a landing.

Their prospects improved the next day. Ross, Crozier, and other officers launched small boats and rowed toward the largest of a group of islands just south of Cape Adare. They landed on a scrap of pebbled beach that was strewn with glistening chunks of ice. There the flag was planted, and Ross proclaimed British sovereignty over the Possession Islands and Victoria Land, with only the clamorous penguins serving as the audience. Meanwhile, the irrepressible McCormick tramped across a nearby guano field. "It had attained such a depth as to give an elastic sensation under the feet, resembling a dried-up peat bog," he noted. Thousands of nesting Adélie penguins flanked the field, and "it was like a thistle-bed to pass though their ranks." The penguins squawked "such harsh notes of defiance, in which the whole colony united in concert, that we could scarcely hear each other speak." He hastily crammed a few rocks and one penguin into his haversack, and joined the other expedition members in a toast to Queen Victoria.

During the following week Ross hugged the coast of Victoria Land as he charted mountains that rose to fourteen thousand feet and glacial valleys that swept into ice-choked inlets. Frequent seafloor dredgings revealed a wealth of marine life, including a living specimen of coral. On January 19 Coulman Island was discovered—and another range of mountains. A strong breeze filled their sails, and the men crowded on deck to watch the spectacular scenery change with each mile they sailed further south. Few men closed their eyes that night, afraid to miss the next wonder. On January 22 Ross passed James Weddell's furthest southern record, and "an extra allowance of grog was issued to our very deserving crews," he noted. Five days later Ross discovered an island, which he named after arctic explorer and governor of Tasmania, Sir John Franklin. A very wet landing on ice-covered rocks almost cost Joseph Hooker his life when he slipped from the boulders into the thundering surf. He was hauled into the boat, "benumbed with the cold."

From Franklin Island, Ross continued southward toward a distant speck of land he called High Island. As twilight brightened into dawn on January 28, a thirteen-hundred-foot peak loomed on the horizon. What appeared to be layers of drifting snow ringing the crater-shaped summit were tiers of smoke. Red flames from a ragged fissure near the peak flashed against a background of snow and ice. "The flame and smoke appeared to issue from a monstrous iceberg," wrote Ross. The men were stunned as they gazed at the erupting volcano. Sullivan, the blacksmith, captured its essence: "This splendid Burning Mountain was truly an imposing sight.... Vast Clouds of Smoke when Scattered about with the wind form a Cloudy Surface of Smoke along the Surface of the mountain...the Sun Shone so brilliant on the Ice and Snow it completely Dazzled our Eyes."

Ross named the volcano Mount Erebus and the smaller extinct cone to the east Mount Terror. By midafternoon Ross saw "a low white line" that extended eastward as far as the eye could see. As the ships sailed toward the mysterious line, the men made their second astonishing discovery of the day: a wall of ice rising to two hundred feet, "perfectly flat and level at the top, and without any fissure or promontories on its seaward face. What was beyond it we could not imagine," wrote Ross. Disappointed that the icy barrier blocked his way south, Ross wrote, "we might with equal chance of success try to sail through the Cliffs of Dover, as penetrate such a mass." He named the wall of ice the Victoria Barrier; later, it would be called the Ross Ice Shelf.

When Ross was about four miles from the edge of the barrier he turned eastward, skirting the barrier's perpendicular face for 250 miles. On February 9, a large bay within the barrier was discovered. The *Erebus* entered the indentation through a quarter-mile-wide gap. The ice lining the inlet was just fifty feet high; from the top of the masthead, the barrier's upper surface "appeared to be quite smooth, and conveyed to the mind the idea of an immense plain of frosted silver," wrote McCormick. Gigantic icicles fringed the top edge of the barrier, sparkling in the sun with all the colors of a prism. All hands came on deck and "stood Motionless for Several Seconds before anyone Could Speak to the man next to him," Sullivan noted. "Beholding with Silent Surprise the great and wonderful Works of nature...I wished I was an artist instead of a blacksmith."

By mid-February a thin veneer of slushy ice, a signal that the brief Antarctic summer would soon end, coated the sea. Ross turned back toward Mount Erebus and entered a deep inlet that he named McMurdo Bay. Although he still hoped to claim the Magnetic Pole, located 160 miles inland from the mountainous coast of Victoria Land, he found no suitable landing site. With winter quickly advancing, Ross surrendered his dream and turned northward. He wrote that "few can understand the deep feelings of regret...to abandon the perhaps too ambitious hope I had so long cherished...of being permitted to plant the flag of my country on both the magnetic poles of our globe."

During the first week of March, Ross searched for the land Wilkes had charted to the west of the Balleny Islands. Although the men crowded the mastheads during clear weather, hoping for a glimpse of the "Antarctic Continent" that Wilkes's chart had indicated, no land was seen. Ross suspected that the American had simply mistaken cumulus cloudbanks looming on the horizon for land. He later declared in his published narrative that he had "sailed over" most of Wilkes's eastern discoveries. His accusations cast a long shadow over Wilkes's achievements for the next ninety years.

A UNIQUE NEW YEAR'S CELEBRATION

On April 6, after a voyage of nearly five months, the men arrived in Hobart, Tasmania, "unattended by casualty, calamity, or sickness of any kind," wrote Ross. The people of Hobart welcomed them with endless rounds of parties and good cheer, culminating with a play based on the expedition's discoveries. The following July the ships sailed to Port Jackson, Australia, where the men were treated to another rush of celebrations during their monthlong stay. In August Ross arrived at the Bay of Isles, on the north island of New Zealand. For the next three months the men refitted the ships, loaded supplies, and installed a magnetic observatory. On November 23, 1841, the *Erebus* and *Terror* sailed south once more. Within three weeks the first icebergs were spotted; on December 17 Ross entered the pack ice at about 150° west longitude. Whales swarmed around the ships, curious but unafraid. Ross noted that they barely moved aside to allow the *Erebus* to pass and were so tame that any number could have been killed with ease.

Christmas Day was "anything but a pleasant one," noted Sergeant Cunningham. While the officers dined on fresh goose, the crew ate their meals on watch. Thick fog and even thicker pack ice surrounded both ships. For the next few days the ships advanced and retreated to the whims of ice and wind. Even though the ice extended much further north than during the previous season, Ross was confident that he could plow through the pack ice to the clear sea that lay further south. On December 30 he crossed the Antarctic Circle, about fourteen hundred miles east of his position the year before.

On December 31, the *Erebus* and *Terror* were anchored to opposite sides of the same thick slab of ice. The day "was spent by our people in the enjoyment of various amusing games on the ice," noted Ross. Joseph Hooker helped to carve an eight-foot-high figure of a seated woman and named her "Venus de Medici"; others excavated a large ballroom and sculpted two thrones for Ross and Crozier; a carved refreshment table dominated one end of the room. A few minutes before midnight, the sounds of horns, gongs, bells, and guns welcomed the New Year. The men mustered on the ice and marched across the ballroom in their new red shirts, thick jackets, and trousers. A few carried pigs under their arms, bagpipe fashion, and rhythmically squeezed their "instruments" for the proper squealy tones. On New Year's Day Ross and Crozier were escorted to their thrones to oversee the ball. The men stomped in time to country reels and waltzed across the slippery "floor" in their heavy boots. The formalities quickly degenerated into a free-for-all snowball fight, fueled with generous libations from the refreshment table.

On January 4, 1842, Venus and the ballroom drifted northward as the ships crunched southward through narrow channels between the floes. One tedious day followed another as the vessels zigzagged through the pack ice, only to lose the few miles gained during sudden gales. On January 19 a violent northerly gale descended upon the ships. Monstrous swells broke over the icebergs and the ships pitched in a sea of "rolling fragments of ice, hard as floating rocks of granite, which were dashed against them with so much violence that their masts quivered," wrote Ross. Destruction seemed inevitable. Timbers creaked and groaned; bone jarring blows from the ice "filled the stoutest heart with dismay." The mountainous waves heaped ice floes on top of one another, "dashing and grinding them together with fearful violence."

The men could do little but watch and wait for the shuddering masts to topple. Twenty-eight hours later the storm abated. One of the officers wrote that "the usual smile had gone from Captain Ross's countenance and he looked anxious and careworn." Both the ships' rudders had been heavily damaged. After three anxious days, the repairs were completed. The rudders were rehung and the ships steered south.

On February 2 Ross emerged from the pack ice into open sea. His plan was now to sail toward the eastern sector of the great barrier he had discovered the preceding year. Three weeks later they spotted the high walls of ice, but the season was late. The ships' bows and decks were soon heavy with layers of frozen spray. One wave slammed a fish against the *Terror*'s bow where it was entombed in a block of ice, much to the surgeon's pleasure. He hacked out his prized specimen and carefully placed it aside for later observations. The ship's cat, ever on the prowl for a tasty morsel, swiftly sacrificed science for a fresh meal when the surgeon turned his back.

Plummeting temperatures glazed the sea with ice as Ross steered to within two miles of the barrier, reaching 78° south latitude—a record that would stand for almost sixty years. The ships traced the wall of ice to the east where it rose only one hundred feet above sea level. The view from the masthead was magnificent, "presenting the appearance of mountains of great height, perfectly covered with snow, but with a varied and undulating outline, which the barrier itself could not have assumed." Although most of the officers were convinced that the irregularities represented land beneath a cloak of snow, Ross marked his chart with an "appearance of land" near the region now called Edward VII Land.

COLLIDING SHIPS AND NARROW ESCAPES

By the end of February Ross turned his ships north toward Cape Horn to winter in the Falkland Islands. They recrossed the Antarctic Circle on March 6 and continued their northbound trek with few problems—until March 12. Late that night, thick sheets of falling snow hid a cluster of tabular icebergs, one hundred to two hundred feet high. Suddenly, a massive icy wall loomed in front of the *Erebus*. The ship turned, right into the path of the *Terror*. The force of the collision threw nearly everyone from bunks and hammocks. The men gave no thought to the bitter cold as they raced to their stations, barefoot, their nightshirts flapping about their knees. The ships hung together with their riggings entangled, grinding and crashing against each other with each rise and fall of the swell. The two locked ships headed toward "the weather face of the lofty berg under our lee, against which the waves were breaking and foaming to near the summit of its perpendicular cliffs," wrote Ross. After several more brutal crashes, the ships broke free.

"'Thank God, she [the *Erebus*] is clear' cried I, as she passed under our stern, snapping our spanker boom in two as if it had been a straw," wrote John Davis, second master on the *Terror*, in a letter to his sister. "But my joy was of short continuance... we had this immense berg under our lee, and so close that we already appeared to be in the foam... there was no alternative but to run for the dark place we had seen before, which might be an opening, or be smashed on the face of the cliff." Captain Crozier yelled a series of orders and the men flew to the ropes in the damaged rigging. Somehow the *Terror* was turned, and the ship "passed through an opening between two bergs... the foam and spray dashing over us on each side." Once through the passage, Crozier burned a blue light on the deck. "I looked round me," wrote Davis, "to see the ghastly appearance of everyone's face, in which horror and despair were pictured, the half-naked forms of the men thrown out by the strong light, oh! it was horrible, truly horrible."

Meanwhile, the disabled *Erebus* drifted so close to the berg "that the waves, when they struck against it, threw back their sprays into the ship." Ross shouted orders to turn the ship so that its stern faced the iceberg. Minutes later, just as the stern scraped against the cliff of ice, the strong undertow sucked the ship away from the berg, saving the *Erebus* from catastrophe. Ross then safely guided the ship through the same narrow passage between the two bergs that the *Terror* had just negotiated.

The damaged ships continued toward Cape Horn and on April 6 arrived in the Falklands, where they stayed for five months, repairing and refitting the ships. In November 1842, Ross received permission from the Admiralty to sail on a third and final voyage to Antarctica.

"NIGHTS OF GROG AND HOT COFFEE"

Ross planned to chart the eastern coast of the Antarctic Peninsula, then penetrate the pack ice to surpass Weddell's record south. The men were filled with high spirits as they sailed from the Falklands on December 17, 1842. Convinced that they would succeed where Wilkes and d'Urville had failed, they looked forward to spotting the edge of the pack ice, their gateway to victory. Or so they hoped.

Nine days later the ships skirted the pack ice at 52° west longitude and followed its edge westward toward the eastern side of the Antarctic Peninsula. On December 28 the rugged mountains of Joinville Island were spotted and surveyed. A small cluster of rocky outcrops, difficult to see among the many grounded icebergs, was named the Danger Islets. Whales spouted near the ships and Ross, once again, was struck by their tameness. "Thus within ten days after leaving the Falkland Islands," Ross wrote, "we have discovered not only new land [Danger Islets], but a valuable whale-fishery well worthy of the attention of our enterprising merchants."

On January 6, 1843, Ross, Crozier, and other officers landed on Cockburn Island, taking possession of it and the surrounding land. Nesting snow petrels were found in the nooks and crannies of the precipitous cliffs. Joseph Hooker, once again in his element, carried nineteen species of lichens, mosses, and algae back to the ship in triumph.

For the next six days the pack ice defeated Ross's attempt to penetrate further west or south. He then followed its edge to the east, hoping to find a clear passage to the south, just as Weddell had done in 1823. "It was the worst season of the three, one of constant gales, fogs, and snow storms," recalled Joseph Hooker in 1893. "Officers and men slept with their ears open, listening for the look-out man's cry of 'Berg ahead!' followed by 'All hands on deck!' The officers of *Terror* told me that their commander [Crozier] never slept in his cot throughout the season in the ice, and that he passed it either on deck or in a chair in his cabin. They were nights of grog and hot coffee." But the pack ice didn't relent for Ross. This expedition would not be successful. On March 5, he signaled the *Terror* to head for Cape of Good Hope, homeward bound at last. After an absence of four years and five months, the men set foot on English soil on September 4, 1843.

Homecomings

Public interest had waned by the time Ross returned to England. The spectacular discoveries from his first trip had been reported and praised, but his next two voyages generated little enthusiasm with the public. Nevertheless, Ross was knighted and honored with medals from the London and Paris geographical societies. Although he was asked to lead an expedition to find the Northwest Passage, Ross declined; not even an offered baronetcy induced him to tackle another trek to polar regions. Command of that expedition was given to Sir John Franklin, with Francis Crozier as his second. In May 1845 Franklin and Crozier sailed on the *Erebus* and *Terror* to search for the Northwest Passage—and were never seen again. The disappearance of these men and ships finally convinced Ross to venture one last time into the Arctic in 1848 to search for the lost Franklin expedition. He was not successful.

While Ross and his crews wintered in the Falkland Islands in 1842, Dumont d'Urville worked on his narrative of the 1837–40 voyage. Ill health had plagued him since his return to France, but with persistence and tenacity he completed three volumes. Just when the fourth one was almost ready for the publishers, d'Urville treated his family to a day in Versailles on May 8, 1842. The family boarded the returning train late that afternoon—and never arrived home. Dumont d'Urville, his wife, and his son were incinerated with fifty-six others when the speeding train jumped the rails and burst into flames. Two days after the tragic train wreck, Dumoutier, the phrenologist who had measured his friend's skull many times, performed the sad duty of identifying d'Urville's charred head. The fourth volume was published posthumously.

One month after the death of Dumont d'Urville, Charles Wilkes returned to the United States. After his Antarctic voyage, he traversed the Pacific Ocean, surveyed the western coast of North America, and explored the Philippines before heading home via the Cape of Good Hope. Instead of the warm reception he had anticipated from his navy cohorts and the public, Wilkes was subjected to ridicule, loathing, and outright hostility. Rumors that James Ross had sailed across regions that Wilkes had charted as "mountainous land" had preceded his arrival. Worse still were the charges that Wilkes had secretly plotted for command of the expedition. Several of his officers lost no time filing charges of misconduct against him. Within days, newspaper articles gleefully reported the possibility of a court-martial for Wilkes.

In an effort to vindicate himself, Wilkes made an official call on Secretary of the Navy Upshur to report the completion of his orders. "His reception of me was very cold," wrote Wilkes. "He never offered to shake hands with me nor requested me to take a seat. I felt indignant at such treatment and took a seat unasked." Wilkes then launched into a heated defense of himself and the expedition. "He [Upshur] had not a word to say and showed his agitation by putting on and taking off repeatedly his spectacles, particularly when I told him I was not to be crushed by his machinations." Tact was not one of Charles Wilkes's strengths. The Secretary ordered the court-martial to proceed.

The trial dragged on until September 1842 and Wilkes was found guilty of just one charge: the excessive punishment of a seaman who had generously helped himself to bottles of liquor during the expedition. Wilkes had ordered the man punished with more than the customary twelve lashes. For this he was publicly reprimanded by the Secretary of the Navy. Bitter but still loyal to the navy, Wilkes remained in service. By 1845 he had completed his five-volume narrative of his expeditions.

Even early twentieth century explorers dismissed Wilkes's sightings of land and his fifteen-hundred-mile survey from the Adélie Coast to the Shackleton Ice Shelf. However, aerial photography performed during the late 1950s finally vindicated his work. The results that he achieved with his bedraggled fleet are now recognized not only for accuracy but also as a testament to human endurance.

. . .

Although no country launched a major Antarctic expedition during the next fifty years, American sealers once again assaulted the South Shetland Islands in the early 1870s. The few scattered colonies of fur seals that had survived the 1819–1822 slaughter had propagated to the point of profitability. Other subantarctic haunts were also revisited. A few years later, the islands were once again silent and barren of fur seals. Meanwhile, whalers turned away from the overhunted Arctic waters and steered toward the last holdout for the great leviathans—the Antarctic seas.

• 5 •

ANTARCTIC WHALING: DIMINISHING RETURNS

CAPTAIN JAMES COOK'S UNBRIDLED ENTHUSIASM FOR FUR SEALS had launched a fleet of ships from Great Britain and the United States toward the abundant seal colonies of New Zealand, South Georgia Island, and Tierra del Fuego. Whale sightings, too, had been just as thoroughly documented. In December 1774, near Tierra del Fuego, second lieutenant Charles Clerke wrote, "there are a greater abundance of Whales and Seals rolling about . . . than in any part of the World." Samuel Enderby and Sons, the London whaling firm, took note. In 1788 Enderby commissioned the *Amelia* to sail to Tierra del Fuego to search for sperm whales. The whaler returned in 1790 with enthusiastic reports and a full cargo of sperm oil. Six American ships set out for Cape Horn the following year. By 1800 American, British, and French whaling ships had invaded the sperm whales' breeding grounds along South America's western coast.

As the rush for sperm whales accelerated in the Southern Hemisphere, British seamen found slim pickings in the Arctic, the domain of the Dutch until the 1720s. Bowhead whales, prized for their long plates of baleen, retreated deeper within the pack ice but were still pursued by the tenacious British whalers. Profits plunged as the number of ice-wrecked ships soared, but the dwindling number of baleen whales—and ships—caused whalebone prices to skyrocket. In 1798, a small Dutch fleet attempted to hunt bowheads in the Arctic and was destroyed by British ships. No country now challenged British dominance in the Arctic, but by then the peak whaling period in the Atlantic far north had passed.

At the other end of the earth, the seas surrounding Tasmania, New Zealand, and Australia teemed with thousands of southern right whales. This species was slow-swimming, floated when killed, and carried a fortune in baleen in its mouth for buggy whips, chimney sweep brooms, ladies' corsets, and men's collar stays. Although obliterated in the north, right whales in the Southern Hemisphere offered enterprising firms economic bonanzas. Captains of convict ships returning from Tasmania to Botany Bay on Australia's eastern coast during the first decades of the 1800s were the first to take advantage of this whale windfall. The ships were outfitted with whaling gear, so once the human cargo was unloaded the sailors went a-whaling, filling the holds with rendered blubber and baleen.

The discovery of the South Shetland Islands in 1819 infused the cash-strapped American and British whaling industries with short-term profits from sealing until the fur seals were almost exterminated. For the Americans, who had dominated the fur seal market, it had provided the funds to rebuild whaling fleets destroyed during the War of 1812.

America's whaling industry soon entered its golden era. When Charles Wilkes sailed toward New Zealand at the end of his epic Antarctic voyage in 1840, he noted that "there were at least 100 whale-ships cruising in the neighboring seas." Since his primary objective was to survey "the great Southern Ocean . . . as well as determine the existence of all doubtful islands and shoals" for American whaling interests, his 241 charts and maps fulfilled the expedition's goal. The accuracy of his work enabled whalers to tackle new territories and to find safe ports for repairing and provisioning ships during long voyages. Within a few years at least five hundred American ships hunted sperm and southern right whales each season.

The untapped wealth in the seas surrounding Antarctica impressed Sir James Ross. On January 14, 1841, he wrote that "a great number of whales were observed . . . wherever you turned your eyes, their blasts were to be seen. . . . A fresh source of national and individual wealth is thus opened to commercial enterprise, and if pursued with boldness and perseverance, it cannot fail to be abundantly productive."

The First Subantarctic Whaling Station: The Enderby Settlement

In London, Charles Enderby heeded James Ross's admonishment for "boldness and perseverance" to establish whaling in Antarctic waters. Ross met with Enderby to encourage him to finance a settlement on Auckland Island. Ross's glowing report—and the decline of bowhead whaling in the Arctic—convinced Enderby to gamble. Although refurbishing and provisioning whaling ships were the settlement's primary goals, Enderby envisioned a prosperous, self-contained colony complete with shops, churches, inns, and farms for sheep and produce. Huge boilers would render the blubber from countless whales, and contented workers would process tons of baleen. Full of confidence, he predicted self-sustaining profitability for the settlement within two years.

In August 1849 Enderby formed the Southern Whale Fishing Company. In order to persuade wealthy financial backers to invest in his enterprise, he described Auckland Island as covered with woods with "a very rich virgin soil" suitable for grazing sheep, horses, and cattle. Although he admitted that high winds and rain buffeted the islands, they were nevertheless "exceedingly healthy," he declared in his prospectus. As an added incentive, he assured subscribers that the land was "free from aboriginals." Young married couples with a wide range of skills and trades were targeted and recruited. In late summer 1849, three whaling ships sailed with Enderby and about three hundred colonists, together with assorted farm animals, to an island lying within latitudinal boundaries known as the "Furious Fifties" for its endless gales and high swells.

When the ships arrived in December, about seventy Maoris greeted the surprised settlers. Instead of luxuriant woods, the shores were lined with stunted and gnarled rata forests that gave a haunted look to the land. Deeper inland were swamps and peat bogs, instead of the imagined deep rich soils. But Charles Enderby and the colonists were determined to settle the land. Prefabricated buildings were unloaded and erected; cobblestone paths linked shops and homes; gardens were staked out and fenced, and sheep were turned loose to feed on scrub grass. On New Year's Day, 1850, the colony was formally named Hardwicke, in honor of the Earl of Hardwicke, one of the company's principal investors.

From January through March a sense of urgency permeated the colony as the settlers prepared for the first whaling season. Their optimism persisted even if the "healthy" climate included sustained gales with winds exceeding 80 MPH and daily drenchings of rain. Vegetable gardens were duly planted but had to be fenced every twelve to fourteen feet with rows of stakes in an effort to ward off the wind. The overall effect resembled sheep pens, with droopy vegetable greens in place of woolly ruminants.

Although whales were expected to bear calves in the island's many protected harbors and coves, none came during the first two years. Boredom and apathy settled over the colonists, and imbibing whiskey became the pastime of choice for many. A prison was built on tiny Shoe Island where inebriates and troublemakers could respectively sober up and cool down. By 1852 Hardwicke was in a shambles.

Then during the final weeks of Hardwicke's existence, the first—and only—whale was killed in the harbor and sank. Two days later the body surfaced. Every available boat was needed to tow the bloated carcass to a whaling ship six miles away. Once the whale was chained to the ship, the captain made the first cut. "The escape of foul air was so great and the stench so unbearable that I was sick for some time," noted R. E. Malone, assistant surgeon on the *Fantone*, "and the noise was like a rush of steam from a boiler of a steam engine." Within a month the entire colony abandoned their homes and sailed for Sydney. Charles Enderby returned to London, only to discover that Enderby Brothers, the illustrious whaling company founded by his father, was liquidating all assets.

The Beginning of the End

Nine years after the defeated Enderby colonists sailed for Sydney, America's formidable whaling industry collapsed during the Civil War. In 1861 the federal government purchased about forty older whaling ships and loaded them with boulders and smaller stones. When the fleet arrived at Charleston Harbor, South Carolina, the ships were sunk in an effort to obstruct the channel used by Confederate blockade runners. The plan failed miserably. The ships sank too deeply in the soft mud and Confederate ships sailed right over the "Great Stone Fleet," as it was subsequently named. Marauding Confederate ships also took a heavy toll. In 1865, the Confederate cruiser *Shenandoah* suddenly appeared in the Bering Straits, surprising New England whaling fleets. Within four days, twenty-five ships were burned and four others converted into Confederate transports.

The final blow came in 1871 when forty American whalers and fifteen hundred men headed toward the Chukchi Sea, north of the Bering Straits. From June until mid-August, the fleet worked its way through the floes, following the retreating bowheads deeper into the ice. On August 29 a gale drove the floes and fleet toward Wainwright Inlet, south of Barrow, Alaska. Within days the fleet was trapped. On September 14, two hundred small boats took to the narrow strip of water between land and pack ice. Within three days, other whaling ships in the vicinity had rescued all of the stranded men. Although not a single life was lost, American whaling, as an industry, was destroyed.

The void left by American whalers was quickly filled. British and Norwegian ships prowled the seas and killed whales with hand-thrown harpoons. Attacking whales in small open boats was a terrifying experience for first-time whalers, or "greenies." Sometimes the men simply froze, "temporarily incapable of either good or harm," wrote whaler Frank Bullen in *Cruise of the Cachalot*. Paralyzed with fear, the men often required "a not too gentle application of the tiller to their heads in order to keep them quiet." The remedy, administered by either the mate or harpooner, was effective, as Bullen and others learned during their first experience with a sperm whale in 1875.

Bullen and the now silent but wild-eyed men maneuvered their twelve-foot boat into position to wait for the whale to surface. With a blow that sounded like hissing steam from a boiler, the whale surfaced. It then flicked its flukes and hoisted the boat into the air. When it hit the water, the boat "collapsed like a derelict umbrella." The men grasped the remnant planks and hoped for a quick rescue.

The *Cachalot*'s captain, however, clearly saw his duty elsewhere and took the time to fasten a dead whale to the ship's side before launching a boat for the desperate men. Help arrived at last, and the shivering, blue-faced men were hauled on board. "We were fortunate to be rescued as soon as we were, since it is well known that whales are of much higher commercial value than men," noted Bullen.

Expecting a bit of commiseration for his ordeal, Bullen soon recognized his naiveté. "The skipper cursed us all . . . with a fluency and vigor that was, to put it mildly, discouraging. Moreover, we were informed that he 'wouldn't have no [adjective] skulking;' we must 'turn to' and do something after wasting the ship's time and property in such a [blanked] manner." In 1883, the twenty-six-year-old Bullen packed up his few belongings and walked away from the whaling life.

By the mid-1880s, many vessels were equipped with the latest whaling tool—the exploding harpoon. Invented by Norwegian Svend Foyn, a heavy iron harpoon was fitted with explosives and fired from a cannon. When the harpoon struck the whale, the charge detonated. Hunters now had the tool to pursue the rorquals, a group of baleen whales that included the blue, fin, minke, sei, and the slower-swimming humpback.

Before Foyn's invention, whalers ignored the rorquals because they were too swift to catch and sank when killed. "Finbacks were always pretty numerous, and, as if they knew how useless they were to us, came and played around like exaggerated porpoises," wrote Frank Bullen. Opinions quickly changed when late nineteenth century technology provided whalers with steam-powered ships and devices for injecting compressed air into a whale's carcass to keep it afloat. Now the final requiem for the rorquals was about to be performed on the seas surrounding Antarctica.

First Weddell Sea Whaling Attempts

By the late 1870s whalers had decimated the bowhead population. In response to the meager supply, the price of baleen skyrocketed; just one kill made the voyage profitable, but Scottish whalers who earned their livelihood by sailing into the Arctic seas were especially hard-pressed to find that solitary bowhead. In 1874 two Scottish brothers, David and John Gray, took matters into their own hands. Aware that Sir James Ross had seen many right whales in the Ross and Weddell Seas, the brothers methodically tallied every reference to whales in Ross's narrative, and then pushed their research back to the time of Cook. They published a pamphlet concluding that Antarctic whaling "would be attended with successful and profitable results." Although the brothers had hoped to attract investors, few responded and they abandoned their plan.

When the Grays' pamphlet was reprinted in 1891, a Dundee whaling company decided to gamble and sent four steam-powered whalers, the *Balaena*, *Diana*, *Active*, and *Polar Star*, to the Weddell Sea to hunt right whales. William S. Bruce, a naturalist from Edinburgh who would later lead his own expedition, was the surgeon on the *Balaena*. His friend, W. G. Burn Murdoch, was almost excluded from the voyage because of lack of space. Although the prospect of an Antarctic adventure excited Murdoch, others weren't quite so enthralled. On the day the ships were to sail, several last-minute recruits deserted and Murdoch claimed a bunk.

On September 6, 1892, the small fleet sailed from Dundee. The crew "threw off their shore togs and shore cares, had one last pull at the bottle, and were up on deck in a minute, drunk and glorious, ready to go to the world's end or beyond it," wrote Murdoch. "Fifty men—strangers an hour ago, brothers now—in the one spirit of whisky, devilment, and adventure."

The weather soon tempered their enthusiasm as the *Balaena* pitched, shuddered, and rolled in the high swells. "My diary...is one long wail at the wretched weather," noted Murdoch. Only those men who possessed ironclad constitutions and were untroubled with seasickness relished meals of "Dead Dog," a concoction of roasted biscuit mixed with salt beef and fat, or "Strike Me Blind," a thick slop of boiled rice and molasses. The only good days, food-wise, were Saturdays, or "plum-duff" days, when the cook dished out the popular pudding.

On October 25 the ships crossed the equator. As they approached the Falkland Islands, the quantity and variety of birds fascinated Murdoch. Tiny black storm petrels "look like flakes of soot driven about in a windy sky." But the many albatross that glided back and forth across the ship's foamy wake were his favorites. One day a wandering albatross hovered overhead and "slewed his head to one side and brought his left foot forward, a great pink, fleshy, webbed affair—and scratched his eye. It was very clever to do this on the wing, without changing his course," he noted.

The ships steamed into Port Stanley in early December. To honor the day, Bruce and Murdoch donned their least-mildewed trousers and morning coats, fastened geology hammers to their belts, and pulled on thick woolen caps. In their explorers' garb, the friends set out to discover the town and surrounding countryside. After a week, Murdoch concluded that Charles Darwin must have been "suffering from one of his frequent attacks of sea-sickness" when he described the islands as a "howling wilderness ... and unfit for man or beast." To Murdoch, the Falklands were "good and fair to see, and very like dear old Scotland."

The fleet sailed from Port Stanley on December 11. One week later the sea was littered with "large ragged pieces of ice, like the roots of huge teeth, rolling about in the swell," wrote Murdoch. When the ship reached broken pack ice, a gale struck the wooden-hulled whaler. "What a pandemonium of sounds," wrote Murdoch. "The wind howling and the timbers creaking and cracking as the ice pounds against our sides ... and we go crushing through the press head-first into the next block." While Murdoch enjoyed the adventure, the sailors threw their belongings into small sea chests and prepared for a quick scramble onto the ice if the ship were mortally wounded.

As the ships sailed toward Erebus and Terror Gulf on the eastern side of the Antarctic Peninsula (the area where Ross had seen many right whales) hundreds of fin and blue whales blew close to the ships, but no right whales were spotted. None of the ships were equipped to capture the swift rorquals and, desperate to fill their holds, the captains turned to the tried and true: hunting seals.

During the next six weeks, the crews killed many crabeater seals on the floes for oil and skins. Bruce and Murdoch did little scientific work, much to their bitter disappointment, "for our vessel was not yet filled with blubber." Shuttling the twosome to the nearby shore was out of the question since the crews were too busy sealing, so the naturalists had little choice but to pile into the boats and head to the nearest floe with its resident seals. The slaughter sickened Murdoch. "How mean and ugly we of the world of people feel in this lovely world of white beauty, making bullets sing through the cold, silent air, fouling the snow with blood and soot," he wrote.

Once killed, the seals were skinned and the hides, with blubber intact, were loaded on the boats. At the end of the day the men gathered on deck to finish their grisly work. First, they hung the sealskins, blubber side up, over waist-high upright boards. Then they detached the blubber with their blades, their arms sweeping back and forth across the hides. The only sounds that broke the silence were the swish of knives and the smack of blubber landing on the deck. As the days lengthened to weeks the work took its toll on the men. "Their faces are drawn and their eyes bloodshot," wrote Murdoch, "they are tired and . . . feel the cold more."

On January 24, 1893, Bruce and Murdoch shared dinner with Captain Carl Anton Larsen aboard the Norwegian whaler *Jason*. Unknown to the Dundee whalers at the time they sailed, the pamphlet that had launched the Scottish fleet had also spurred a Norwegian shipbuilder to commission the *Jason* to hunt for right whales in the same vicinity as the Dundee fleet. Larsen, like the Scots, had no success and turned to seals. But he also took the time to study the geology of the area—much to Bruce's envy. On Seymour Island, Larsen not only discovered fossils but also fifty clay balls arranged on small pillars, as though "having been made by man's hand." Bruce, intrigued with Larsen's find, studied the fossilized shells and petrified coniferous tree fragments with interest, concluding that they "indicate a warmer climate than now prevails in these high southern latitudes."

On February 18, the four Scottish whalers turned northward with their load of twelve thousand sealskins and oil. Although Bruce would return to the Weddell Sea in 1902 as leader of the *Scotia* expedition, Murdoch would never again sail to the high southern latitudes. As if sensing this, he wrote, "And so we turned from the mystery of the Antarctic, with all its white-bound secrets still unread, as if we had stood before ancient volumes that told of the past and the beginning of all things, and had not opened them to read. Now we go home to the world that is worn down with the feet of many people, to gnaw in our discontent the memory of what we could have done, but did not do."

Captain Larsen returned to Antarctica the following season with the *Jason* and two other ships. After sealing near Seymour Island in the Weddell Sea, Larsen headed southward and sighted the Foyn Coast, named after the Norwegian inventor of the harpoon gun. He also discovered Oscar II Land, and then skirted a massive ice barrier (later named the Larsen Ice Shelf) until tabular icebergs completely blocked his path. Meanwhile, the other two ships explored the western side of the Antarctic Peninsula as far south as Bellingshausen's Alexander Island. The ships rendezvoused at Joinville Island and headed for Tierra del Fuego for right and sperm whaling, with little success. Larsen would return to Antarctica in 1901 with the ill-fated Nordenskjöld expedition.

First Ross Sea Whaling Attempts

While Larsen was steaming toward the Weddell Sea in the fall of 1893, Henryk J. Bull sailed from Norway on September 20 toward Îles Kerguélen. Bull, a businessman who had immigrated to Australia from Norway, had tried to persuade Australian investors to finance his whaling voyage, but Australia's tottering economy collapsed in 1892. Undeterred, Bull traveled to Norway and met with Svend Foyn, who agreed to back the expedition. The *Antarctic* was equipped with eleven harpoon guns, an arsenal of explosives, eight whaleboats, and thirty-one men to search for right whales. Bull served as manager of the expedition.

During their three-month voyage from Norway, the crew saw many blue and fin whales, but not a single spout of a right or sperm whale. On December 19, the *Antarctic* arrived at Îles Kerguélen. Massive whale skeletons, cracked blubber cauldrons, splintered casks, weathered ships' timbers, and dilapidated huts lined the shores. While Bull remained on the ship, Captain Kristensen and the crew searched for food on shore. Suddenly, shot after shot rang out. "Two boats were lowered in feverish haste," wrote Bull, "and the first sea-elephants were brought on board in the moonlight."

Bull had never witnessed sealing, and the ruthless slaughter disgusted him. But profits had to be made—especially in light of the scarcity of right whales. In early February 1894, the *Antarctic* sailed for Melbourne, Australia, loaded with sixteen hundred elephant seal skins and ninety-five tons of blubber. "I could not see the Kerguelen peaks disappear without a feeling of attachment toward the rugged, desolate land... which had so bountifully rewarded our first effort," he wrote.

While Bull remained in Melbourne, the *Antarctic* sailed for Campbell Island in April to hunt whales. Many right whales were seen, but the crew managed to kill just one before running aground in Perseverance Harbor. The battered ship returned for extensive repairs, which wiped out the profits from sealing. Bull was also disappointed to learn that William S. Bruce, the naturalist who had sailed with the Dundee whaling expedition, would not arrive in time for the Ross Sea portion of the voyage. Carsten E. Borchgrevink, a Norwegian by birth but living in Melbourne, heard about the opening and applied for the position of scientist.

The *Antarctic* left Melbourne in September 1894 and stopped at Macquarie and Campbell Islands for sealing. Although most of the fur seal rookeries had been destroyed in the 1820s, a small band of seals had managed to escape the butchery and had increased the population. Still, the men found only a few seals after searching "under stones, out of caverns, behind breakers, and wherever Nature affords a precarious shelter," according to Bull. On November 1, the ship headed toward the Ross Sea but returned to New Zealand one week later with a damaged propeller. While the vessel was in drydock, nine sailors deserted. "I really do not believe that a single contented or happy being is found on board," wrote Bull.

The glum men steered south once more. The ship entered the broken pack ice near the Balleny Islands on December 8. Although many blue whales were spotted, the inept and undisciplined men fired the harpoon guns either too soon or too late. "Another whaling comedy terminated," wrote Bull on January 13, 1895. "Unless gun-drill is seriously undertaken even at this, the eleventh hour, I can see no chance of success, even if we penetrate the ice and find schools upon schools of Right whales." Later that day the *Antarctic* emerged from the pack ice into the Ross Sea and steamed toward Possession Island, near Cape Adare, where Ross had seen "numerous whales."

On January 19 a small party landed on Possession Island "to pay a visit to the natives—thousands of penguins." Borchgrevink discovered lichen in several sheltered nooks, proving that even in the rigorous Antarctic climate plant life could exist. Henryk Bull examined the penguin rookery, observing that the thin guano layer indicated only recent penguin colonization. "From this fact interesting inferences may be drawn regarding changes in the climate of Antarctica during recent times," he wrote.

The *Antarctic* continued southward but no shouts of "Blow!" rang out. "Still no whales," wrote a discouraged Bull three days later. "It appeared, therefore, commercially useless to continue our voyage South." The *Antarctic* turned northward with the hopes of procuring a cargo in warmer seas. As they neared Cape Adare at 1 A.M. on January 24, Bull, Borchgrevink, Captain Kristensen, and several seamen set off in a boat to land on the continent. When the boat approached the pebbly shore, a confused scramble—and subsequent squabble—occurred. New Zealander John Tunzelman claimed that he was the first man on the Antarctic continent because he had leaped over the bow to steady the boat for Captain Kristensen to climb out; however, Borchgrevink maintained that he had jumped over the side as soon as the order was shouted to stop rowing. Then Captain Kristensen insisted that he took the first steps on the continent. Bull, however, ignored the fuss and credited everyone. Although unknown to them at the time, the three men were not the first to land on the continent. That distinction belonged to sealer John Davis, who had stepped onto the Antarctic Peninsula in 1821.

To celebrate, the men erected a pole and hung a box painted with the colors of the Norwegian flag and inscribed with the ship's name and date. Borchgrevink collected samples of lichen, rocks, jellyfish, and mollusks. As the men rowed toward the ship, they were astonished to see the *Antarctic* head off in the opposite direction. With much shouting and arm waving, the men attracted the lookout's attention. When asked for an explanation, the man on watch told Bull that he had seen three men heading toward the shore some distance away, "which proves that penguins on the march can be easily mistaken through a telescope for human beings—at least, when Mr. M. H. is at the other end of it," noted Bull.

On March 11 the *Antarctic* anchored at Melbourne with 180 sealskins and oil from one sperm whale in its hold. As for the complete absence of right whales near Cape Adare, Henryk Bull correctly concluded that since the time of Ross's voyages, whalers had eliminated the species at the whales' breeding sites in warmer climates. However, fin and blue whales, he noted, "will form a source of great wealth to whoever sets about their capture."

Although Bull never returned to the Ross Sea, he later protested against the "criminally wasteful" business of sealing and called for international regulations to halt the "insane killing." He also concluded that "landing on Antarctica proper is not so difficult... and that a wintering-party have every chance of spending a safe and pleasant twelve months at Cape Adare." Carsten Borchgrevink also arrived at the same conclusion. He hurried to London to report his scientific findings to the Royal Geographical Society and to raise funds for a group of men to winter at Cape Adare.

South Georgia's Grytviken: The First of Many

The abundance of rorquals in the Weddell Sea was not lost on Carl A. Larsen. In February 1904 he convinced Argentine businessmen to invest in a whaling station to be built on South Georgia Island. Eight months later, a small fleet of ships and sixty Norwegians arrived at Cumberland Bay on the island's eastern side to build Grytviken, named after the abandoned try-pots that littered the shores. Prefabricated houses and buildings were assembled. Enormous meat, bone, and blubber cookers were erected near the slipway. Within five weeks the first humpback whale was killed, dragged up the ramp, and flensed. The carcass was then winched to the side and discarded.

The waters surrounding South Georgia teemed with humpback, blue, and fin whales. Larsen's speedy whale catchers did not have to venture far into the open sea to seek their prey. So successful was his station that the British granted leases to six more companies by 1912.

On November 23 of that same year, American ornithologist Robert Cushman Murphy arrived at South Georgia on the whaling ship *Daisy*. Sent by the American Museum of Natural History, he would soon collect and describe the birds and seals on an island that was "a small speck near the bottom of an unfamiliar map." As Captain Cleveland guided the ship into Cumberland Bay toward Grytviken to obtain the required sealing permit, Murphy wrote that "the whole shoreline of Cumberland Bay proved to be lined for miles with the bones of whales, mostly humpbacks. Spinal columns, loose vertebrae, ribs, and jaws were piled in heaps and bulwarks along the waterline and it was easy to count a hundred huge skulls within a stone's throw."

Even before Murphy's first view of Grytviken, the smell overpowered him. "The odor of very stale whale then increased as we entered [King Edward] cove," he wrote, "which might be likened to a great cauldron so filled with the rotting flesh and macerated bones of whales that they not only bestrew its bottom but also thickly encrust its rim to the farthest high-water mark. At the head of the cove, below a pointed mountain, we could see the whaling station, its belching smoke, several good-size steamers, and a raft of whale carcasses ... floating out to sea."

British magistrates controlled leasing, sealing permits, and port charges; and all vessels were required to register at the station. The sixty-nine-year-old Captain Cleveland could well remember the days when sealers were not burdened with laws and quotas. He grumbled, filled out the necessary paperwork, and then promptly ignored all sealing regulations. Since the *Daisy* was not powered by steam, Cleveland had little chance of competing with the Norwegian vessels for whales. Elephant seals were far easier prey.

During a luncheon with Carl Larsen, Captain Cleveland and Murphy learned that Larsen's stockholders were paid annual dividends from 40 to 130 percent on their initial investments. By the end of the meal, Murphy understood why Grytviken was so profitable. "I watched at least fifteen humpback whales being drawn up the slip and flensed with incredible dispatch. The Old Man [Captain Cleveland] is nothing short of goggle-eyed over the big-scale butchery of modern whaling."

Humpbacks, however, eluded Cleveland. While he turned his men loose on the shores for elephant sealing, Murphy sailed on the Norwegian whaler *Fortuna*. Many other steamers were within sight, and blasts from harpoon guns were almost continuous. Then came the shout—a blue whale was straight ahead. Engines were cut and the gunner ran to his platform. "Up swung the butt of the swivel gun. A flash and a deafening detonation split the frosty air," wrote Murphy. The iron harpoon with its tail of hemp line hit the whale's side. Three seconds later came the explosion. The wounded whale towed the vessel for several miles and then died. "Envision this magnificent blue whale, as shapely as a mackerel, spending his last ounce of strength and life in a hopeless contest against cool, unmoved, insensate man. Sheer beauty, symmetry, utter perfection of form and movement, were more impressive than even the whale's incomparable bulk, which dwarfed the hull of *Fortuna*." As the steamers killed the whales, "millions and millions of petrels and albatrosses, filling the air like snowflakes," hovered over the debris and settled on the water to feed. Overstuffed and bloated, the albatrosses and giant petrels were incapable of launching themselves into the air—even when the ship bore down on them. "Several of the huge birds were actually bumped out of the way by the bow of our steamer, but the experience only made them whirl around and look indignant."

That 1912 November afternoon was typical of the magnitude of a slaughter that continued until 1961, when the last blue whale was killed in South Georgia waters. Grytviken ceased operation four years later, not because of environmental campaigns but for the simple reason that whaling was unprofitable. Too few whales remained to justify the costs.

Today, bleached and wind-scoured whalebones rest on the deserted shores of South Georgia, the South Shetlands, and the South Orkneys. At Grytviken, the wooden ramps are gouged with parallel grooves from thousands of winches and heavy chains that dragged whale carcasses to the now dilapidated buildings that housed the meat, bone, and blubber cookers. Elephant seals lumber across the iron-plated slipway and heave their ponderous bodies over drainage pipes to wallow in the debris of a bygone era, a time when world demand fueled the rusty boilers.

Few rorquals glide through the waters surrounding South Georgia. The ramshackle relics stand today not as a memorial to whales or whalers, but as silent symbols of hope that future generations will again watch in astonishment, as William Bruce did in 1894, "hundreds of the Finner Whales blowing fountain-like spouts, and filling the air with their characteristic notes of booming resonance."

• 6 •

TRIUMPH AND TRAGEDY: THE SOUTH POLE EXPEDITIONS

On August 6, 1901, three years before the first whaling station was established at Grytviken, an obscure Royal Navy lieutenant sailed from England with ambitious plans to explore Antarctica's unknown interior. Robert Falcon Scott's *Discovery* expedition was the first to test the limits of human stamina, commitment, and courage on land. Ernest Shackleton, a member of Scott's three-man southern team, vowed to return to Antarctica after Scott ordered him home with the relief ship before the completion of this first trek. His epic 1909 South Pole attempt, and Peary's conquest of the North Pole that same year, galvanized Robert Scott and Norwegian Roald Amundsen to try to claim the last geographical prize on Earth.

In separate expeditions Scott and Amundsen trudged across the Ross Ice Shelf for hundreds of miles and endured its unrelenting monotony. Further south, they climbed glaciers and steep spillways of blue ice to reach the ten-thousand-foot Polar Plateau. Cold, in all its brutal guises, was their silent companion. It hovered inside tents, crept into reindeer sleeping bags, and marked faces, hands, and feet with hard white patches—pale emblems of the wilderness these men explored.

Wind and ice have long since erased their sledge tracks and footprints. The land now bears the relics of that era—abandoned huts, discarded sledges, the bones of dogs and ponies, the bodies of men. The long-silenced voices of those explorers who struggled for conquest and fame can be heard in their diaries, words that reveal each man's character. In the end, one man found glory in defeat; the other, defeat in victory.

The Discovery *Expedition: Unmasking the Unknown*

Sir Clements Markham, president of the Royal Geographical Society, was the formidable impetus behind the *Discovery* expedition. The supreme test of British naval manhood and heroism, he believed, lay in the icy wilderness of Antarctica. Although membership in the society included retired admirals and other naval personnel, the Royal Navy had not involved itself with Antarctic exploration since Sir James Ross's 1839–43 expedition. Markham set about to rectify that error in 1893.

After a six-year struggle, he finally gained enough financial support to launch the expedition. Markham next churned out pamphlets, letters, and newspaper editorials announcing the expedition's goals. In early June 1899, Robert Falcon Scott met with Markham; two days later he applied to be commander of the expedition.

Up to this point, Scott had shown no burning desire for exploration. Born in 1868 to a comfortable life near Devonport, England, he was tutored at home until he was thirteen years old. That year, his family sent him to school to prepare for the Royal Navy examinations. Scott graduated from the naval school in 1883 and later specialized as a torpedo gunner. Although competition for promotions among officers was stiff, several fortuitous encounters with Markham improved his prospects. Scott's application to lead the Antarctic expedition, together with Markham's political maneuvers, clinched his appointment and promotion in June 1900.

Scott now had less than a year to prepare for the voyage that would take him to the Ross Sea sector of Antarctica. At Markham's urging, he traveled to Norway and met briefly with legendary Arctic explorer

Fridtjof Nansen for advice about sledging and dog handling. Nansen, who had crossed Greenland in 1888, tried to convince Scott that dogs were indispensable on polar ice, but Scott continued to echo Markham's opinion. "In my mind no journey ever made with dogs can approach the height of that fine conception which is realized when a party of men go forth to face hardships, dangers, and difficulties with their own unaided efforts," Scott later wrote. "Surely in this case the conquest is more nobly and splendidly won." Nonetheless, Scott reluctantly agreed to take some sledge dogs as an experiment.

The next hurdle was the selection of men to participate in the expedition. Although the Royal Society recommended a small naval but large civilian contingency, Scott campaigned to staff the *Discovery* with navy and merchant marine officers who had the background to double as scientists. "I felt sure that their sense of discipline would be an immense acquisition," he wrote, "and I had grave doubts as to my own ability to deal with any other class of men." Scott compromised and took six civilians: two surgeons, a biologist, a geologist, a physicist, and the ship's cook.

ROILING ALONG TO ANTARCTICA

The *Discovery* sailed from England on August 6, 1901, after a rousing send-off. Less than one month later, the ship developed a leak, despite two thick layers of planking that lined its rounded hull. The 172-foot vessel also had a pronounced roll, even on a moderate sea. "The noise is indescribable—a roar of creaking and groaning timbers and banging doors, slamming cupboards, and the to and fro rattle of everything not immovably fixed in one's cabin," wrote artist, zoologist, and surgeon Dr. Edward Wilson.

Sketching with paints and pencils on a lurching ship was difficult. Wilson solved this problem by wedging his lanky body into his cupboard with his feet braced against the opposite chest of drawers during the forty-degree rolls. With his calm manner and gentle good humor, Wilson was soon christened Uncle Bill, the man who many turned to for advice.

Second Lieutenant Ernest Shackleton also became a favorite with sailors and officers. Born in Ireland in 1874, he joined the Mercantile

Marine when he was sixteen years old. With his gift for spinning a tale and quoting poetry "by the yard," he had the verbal deftness to hold his listeners spellbound. His natural knack for leadership—and practical jokes—endeared him to all, including those who would sail with him on the *Endurance* in 1914.

On November 28, the *Discovery* steamed into Port Lyttelton, New Zealand, for final preparations and remained there until its departure on December 21. On that day, crates of food and other supplies occupied every inch of space on the ship, including odd corners in the crew's quarters. A flock of forty-five terrified sheep thundered back and forth on the deck. "Amidst this constantly stampeding body stood the helmsman at the wheel," wrote Scott, "and what space remained was occupied by our twenty-three howling dogs in a wild state of excitement." The unflappable townspeople mingled on the deck and shouted their good wishes above the din. But the tethered dogs, frustrated by their inability to reach fresh meat on the hoof, turned to the next best thing—each other. The visitors paused in midsentence and helped to wrestle the dogs apart before cheerfully continuing with their farewells and good wishes.

The happy mood ended abruptly, however. As the ship pulled away from the port, a young seaman climbed above the crow's nest, clutched the weather vane, and waved to the cheering crowd. Suddenly, he lost his balance and screamed. "We turned to see a figure hurtling through the air, still grasping the wind vane," wrote Scott. The sailor's death was instantaneous. A dark shadow settled over the ship and the men as they ventured toward Antarctica.

On January 3, 1902, the *Discovery* crossed the Antarctic Circle. Five days later, the dark basalt columns of Cape Adare loomed beyond broken floes and icebergs. As the ship crept through heavy ice toward Carsten Borchgrevink's 1899 hut on Robertson Bay, Adélie penguins porpoised through the wide leads like hundreds of skipping stones on a quiet pond.

After a quick landing near Cape Adare, the *Discovery* continued southward. In the face of strong headwinds, the ship slowly steamed toward McMurdo Sound and Mount Erebus. The Victoria Land scenery mesmerized Shackleton. Early one morning, he woke Wilson to share the beauty of sunlit glaciers, ice floes, and unnamed mountains. The two men climbed to the crow's nest. "It was intensely cold," wrote Wilson. "I was in pajamas and a coat and slippers, but it held one spellbound." During the next three weeks, the ship crossed McMurdo Sound and skirted the Ross Ice Shelf to its eastern extremity where land was sighted. King Edward VII Land, as it was subsequently named, was the first Antarctic discovery of the twentieth century.

On February 3, the ship sailed into a bay within the Ross Ice Shelf. The next day, while Albert Armitage and five others harnessed the dogs to the sledges for a race, Scott ascended eight hundred feet in a hydrogen-filled balloon. Swaying in what Scott described as "a very inadequate basket," he gazed at the insect-size figures of Armitage's party, careening over the ice shelf's broad wavy surface. Shackleton soared next to take photographs, but Wilson remained firmly on the ground. "The whole ballooning business seems to me to be an exceedingly dangerous amusement," he noted in his diary. Scott named the bay Balloon Blithe.

Scott turned back toward McMurdo Sound and found a protected site, later named Hut Point, at the southern tip of Ross Island. There, the *Discovery* could be safely moored and frozen in for the long winter. Three prefabricated huts were erected on shore to store supplies and scientific equipment; the ship would serve as the men's living quarters. Scott now turned his attention to late-season sledging trips.

The first twelve-man party left on March 4 and headed to Cape Crozier. Disoriented by a blinding blizzard, the men blundered in the wrong direction, toward the edge of the Ross Ice Shelf. Two men, Vince and Hare, slipped. As the other men watched in stunned horror, Vince hurtled down the precipice and disappeared into the Ross Sea. They assumed Hare had met the same fate. Although no trace of Vince was ever found, Hare miraculously appeared at the ship two days later, exhausted and hungry but free from frostbite. Somehow he had stopped his slide in the nick of time and survived the frigid temperatures under a blanket of snow.

Another sledging party quickly returned to Hut Point when temperatures plunged to $^{-}42°$ F. Scott led the last group south on March 31, but they returned to the ship after sledging just nine miles in three days. "We were practically doomed for failure," Scott wrote. He blamed low temperatures, inadequate food, and inexperience with dogs as the reasons for defeat. "Our autumn sledging was at an end, and left me with much food for thought. In one way or another, each journey had been a failure; we had little or nothing to show for our labors."

WHITE NIGHTS

On April 23, 1902, the sun set for four months. Each day, work was done at a leisurely pace until 1 P.M. Meteorology readings were recorded twice each day; geology specimens were cleaned and polished; and magnetic observations were frequent. The men then devoted the rest of each day to reading, card games, debates, and sports. Shackleton edited *The South Polar Times*, a monthly collection of articles and drawings by officers and men. Wilson worked on his sketches, skinned the many birds he had collected, and wrote up his zoological notes.

Wilson and Shackleton took long afternoon hikes together, talking quietly at the top of nearby hills as the aurora australis billowed overhead like sheer green draperies in a breeze. Some days Wilson walked alone. "Everything was so still and dead and cold and unearthly... the silence one 'felt' as a thing that had been broken by nothing but wild nature's storms since the beginning of the world."

Scott worked on a new dog harness design and drafted a spring sledging schedule. Reading books about Arctic exploration and sledging techniques was a high priority but problematic. Many important books had been omitted because of time constraints. Fortunately, Wilson came to his rescue with his voluminous notes extracted from Fridtjof Nansen's *Farthest North*.

On June 12 Scott confided to Wilson his plans for a three-man southward march. "Our object is to get as far south in a straight line on the Barrier ice [Ross Ice Shelf] as we can, reach the Pole if possible," noted Wilson in his diary. When Scott asked Wilson's opinion about who the third man should be, Wilson suggested Shackleton. "So it was settled and we three are to go," wrote Wilson, delighted.

September and October were filled with hastily planned sledging trips that echoed the same problems Scott had faced the previous March: dog-handling frustrations, food shortages, inadequate clothing, and inexperience with heavy, ill-packed sledges. The single success during those two months was a brief excursion to Cape Crozier to erect a box containing directions to the expedition's winter quarters for the relief ship, *Morning*, due to arrive in February.

On October 30, a twelve-man advance party struggled southward, forced to maneuver the heavy sledges over bad ice with brute strength and curses. One of the handmade silk flags that decorated the sledges was embroidered with the words, "No dogs admitted." The men were proud to strap on the harnesses and haul the sledges without canine aid, at least at the start of the journey.

"SLEDGING IS A SURE TEST OF A MAN'S CHARACTER"

"We are off at last," wrote Scott on November 2. The nineteen dogs pulled the heavy sledges so well that Scott, Wilson, and Shackleton were forced to run to keep pace. Scott was surprised and delighted that he had underestimated the dogs' stamina. The next day they caught up with the man-hauling party who were "doing their best, but making very slow progress," noted Scott. On November 15 he loaded additional supplies on his sledges and then dismissed the support team. Unencumbered by the slower group, Scott now increased the pace. "Confident in ourselves, confident in our equipment, and confident in our dog team, we feel elated," he wrote.

His ebullient mood collapsed that same evening at their camp, barely three miles further south. "The day's work has cast a shadow on our highest aspirations . . . for the dogs have not pulled well today," he complained. The next morning matters were worse. The dogs strained at the harness, but the over-loaded sledges would not budge. Their only choice was to relay half their load at a time—which meant trudging three miles for every mile they gained southward. During the next few days, Scott mused on the reasons why the dogs had failed so miserably. He concluded that canine meals of Norwegian stockfish were the culprit—no one had checked the fish for spoilage before it was packed on the sledges. The dogs now suffered from bloody diarrhea.

The relay work was backbreaking for the men and heartbreaking for the dogs. Scott noted that "there is now no joyous clamor of welcome" in the morning from the dogs. Neither shouts of encouragement nor cracks of the whip increased the dogs' pace as they shambled over the ice with bowed heads and droopy tails. Day after day the men averaged just four miles before camping. "It is the dogs, and not we, who call the halt each night," Scott wrote.

On November 21, mountains appeared on the horizon toward the southwest, a majestic section of ten- to fourteen-thousand-foot peaks, later named the Transantarctic Mountains. Whatever pleasure the men derived from the scenery was overshadowed by hunger, for food and fuel had dwindled to the critical point. Instead of the usual hot meal at noon, they now chewed small chunks of frozen seal meat and dry biscuits. Each night before sleeping, the men tightened their belts to deaden the gnawing pains of hunger. Dreams of vivid feasts, with "sirloins of beef, [and] cauldrons full of steaming vegetables," turned into nightmares. "One spends all one's time shouting at waiters who won't bring a plate of anything," noted Wilson, "or else one finds the beef is only ashes when one gets it."

On December 9 the first dog died. Wilson cut up the carcass and distributed it to the others. "The dogs have had no hesitation in eating their comrade," wrote Scott. The next day a skua, the scavenger of the south, circled overhead. One by one, the weaker dogs were killed to keep the stronger ones on their feet for another day or two. "I act as butcher at night now," wrote Wilson.

One week later, the men reached a mile-wide chasm filled with a chaotic jumble of hundred-foot-high blocks of ice, the boundary between the Ross Ice Shelf and the continent. They stored two pairs of skis and stopped relaying their loads, "a blessed relief" after thirty-one days of almost unendurable hardship.

Their troubles were far from over. On Christmas Eve, Wilson examined Scott's and Shackleton's swollen gums. The signs of scurvy were unmistakable. Still, the next day the men put aside their worries and enjoyed a double ration of biscuit crumbs mixed with seal liver.

Then Shackleton, with a glint in his eye, fumbled with his clothing pack and withdrew a crumpled sprig of holly and a small tin of plum pudding. "We shall sleep well tonight—no dreams, no tightening of the belt," wrote Scott.

The excruciating pain of snow blindness robbed Wilson of sleep, however. Long hours of sketching in intense sunlight brought on attacks so painful that he trudged in the sledge harness, blindfolded. "Luckily, the surface was smooth and I only fell twice," he wrote. "I had the strangest thoughts or day dreams as I went along.... Sometimes I was in beech woods, sometimes in fir woods.... And the swish-swish of the ski was as though one's feet were brushing through dead leaves.... One could almost see them and smell them."

Wilson's eyes improved by December 30. On this day the men reached 82° 17' south latitude, about 350 miles from Hut Point. After they made camp, Scott ordered Shackleton to stay with the dogs while he and Wilson marched several more miles southward. Stunned that he was not allowed to share the glory of the "furthest-south" record, Shackleton never forgave Scott.

NO MARGIN FOR ERROR

The "furthest-south" record would have no meaning unless they arrived back to tell the tale. Now critically short of supplies, the return trip became a race against starvation and scurvy. Frustrated by an additional cut in their daily rations, Wilson wrote that "one *must* leave a margin for heavy surfaces, bad traveling, and weather, difficulty in picking up depots, and of course the possibility of one of us breaking down." His words were prophetic.

They rigged a sail for the sledge from the tent's canvas floor cloth and unhitched the sick dogs to follow at their own pace. During periods of dead calm, the men dragged the sledge over the icy surface while thoughts of food tortured them. "The food-bag is a mere trifle to lift," wrote Scott.

On January 12, 1903, Shackleton collapsed outside the tent, gasping for air and coughing up blood. Weak and very ill, he struggled to keep up with Scott and Wilson. The next day they reached Depot B and the supplies the support party had stored. Although they ate their fill, Scott confessed that they were all "a bit 'done.' " Wilson had one more task before he rested, however. The last two dogs were killed that night.

Shackleton's condition worsened. Scott's main goal now was to keep him on his feet. "In case he should break down and be unable to walk, I can think of absolutely no workable scheme," wrote Scott. "We cannot carry him," noted Wilson. Shackleton, operating on sheer willpower alone, carried the compass and skied one-half mile ahead of the other two, breaking trail and setting the course. A steady wind from the south filled the sledge's sail while Scott and Wilson floundered behind, sinking to their calves in snow. As they watched how easily Shackleton moved over the thin crust, both men regretted discarding their skis five weeks earlier. "In spite of our present disbelief in skis, one is bound to confess that if we get back safely Shackleton will owe much to the pair he is now using," wrote Scott.

On January 28, Shackleton spotted the lone black flag that marked Depot A. The three men ate their fill of raisins, sardines, biscuits, and chocolate. On February 3, lookouts spotted the three bedraggled men stumbling toward the ship. Out at the edge of the floe the relief ship, *Morning*, waited for the final breakup of the ice to free the *Discovery*.

Scott and Shackleton recovered within two weeks, but Wilson was still weak and exhausted by the end of February. The ice that held the ship in its grip showed no signs of weakening and Scott prepared for another Antarctic winter. Supplies were shifted from the *Morning* to the *Discovery*, and all hands were given the opportunity to return to England with the relief vessel. Eight men volunteered to return. When Scott ordered Shackleton home, second-in-command Armitage and others tried to intercede on his behalf, but Scott remained adamant. Wilson, on the other hand, chose to stay for another year.

On March 3 everyone on the *Discovery*, except the recuperating Wilson and two others, huddled at the edge of the pack ice to give the *Morning* a rousing send-off. Shackleton stood at the stern. When he waved a final farewell to his many friends, the men acknowledged the gesture with heartfelt cheers and shouts. The men watched the *Morning* retreat northward and then hiked the six miles back to their ship to face the second winter.

THE SECOND SEASON

Wilson slowly recovered his health. By April he was again painting and preparing bird skins. In the late evening when sleep eluded him, Wilson hunched near the acetylene lamp, rereading letters from friends and family he had received from the *Morning*.

By September, preparations for the spring and summer sledging trips were in full swing. Scott decided to explore the western region and follow Armitage's pioneering route up the Ferrar Glacier to the Polar Plateau, nine thousand feet above sea level. Armitage had successfully led this expedition at the same time Scott had journeyed south.

On October 26, 1903, Scott and his nine-man party marched west from Hut Point toward the mountains of Victoria Land. By mid-November, after a grueling climb up the Ferrar Glacier, they had reached the plateau. "Before us lay the unknown," wrote Scott. "What fascination lies in that word! Could anyone wonder that we determined to push on, be the outlook ever so comfortless?"

Scott sent all but two men back to the ship when several sickened from the nine-thousand-foot altitude. Petty Officer Edgar Evans and Leading Stoker William Lashly continued with him. By the end of November, the unknown plateau was little more than a "terrible limitless expanse of snow," wrote Scott. "And we, little human insects, have started to crawl over this awful desert, and are now bent on crawling back again."

About three weeks later, they discovered a small valley that narrowed into a seventeen-foot-wide passage, a gateway to the bizarre. They scrambled through and then stared in awe at a great U-shaped, snow-free valley with a mile-wide frozen lake. Stark mountains with colorful horizontal rock layers towered above the men. Boulders, sculpted into graceful shapes by wind-borne fine sand, decorated the valley floor. Meltwater streams trickled down from small retreating glaciers. Scott and his companions searched for life but found no lichen or moss—just a mummified Weddell seal carcass. "It is certainly a valley of the dead," wrote Scott, "and even the great glacier which once pushed through it has withered away." The three men picnicked next to a gurgling stream and then explored the region for a few more hours. They arrived at Hut Point on Christmas Eve with news of the first "dry" Antarctic valley seen by man.

On January 5, 1904, the *Morning* and a second ship, the *Terra Nova*, loomed in the distance. Scott was ordered to abandon the *Discovery* if it was still locked in ice at the end of six weeks. On February 14, the floe suddenly fractured and the ships were able to penetrate the widening channels. Two days later, explosives freed the *Discovery*.

Scott reached England in September. During the following year, he edited his diaries and wrote *The Voyage of the 'Discovery'*. The 1905 book was a popular and critical success. Ernest Shackleton, however, seethed in private because of Scott's portrayal of him as a weak man ordered home. He vowed he would return to Antarctica to finish what Scott had begun: to stand victorious at the South Pole.

The Nimrod *Expedition: Pushing the Limits*

Sir Clements Markham listened incredulously to Shackleton's bold and audacious plan. The man who had been ordered home by Scott wanted to use Hut Point as headquarters for his own expedition. He and his men would winter there, and the following spring one sledging party would head east to King Edward VII Land; another would claim the South Magnetic Pole in Victoria Land; and Shackleton would march south with Mongolian ponies "to reach the southern geographical Pole." Science, he assured Markham, would not be neglected—he would continue the biological, meteorological, and geological work begun by the *Discovery* expedition. In spite of these assurances the Royal Geographical Society "could not see its way to assist financially."

Shackleton persevered. His wealthy employer, William Beardmore, and other well-heeled sympathizers loaned Shackleton enough money to make his expedition a reality. Unhampered by the demands of time-consuming committees, Shackleton controlled every phase of the expedition. He traveled to Norway to purchase handcrafted sledges, reindeer sleeping bags, and wolfskin gloves. He sampled different brands of pemmican—small cans of dried beef (40 percent) mixed with lard (60 percent); bartered for crates of tea, vegetables, cheese, beans, sugar, and flour; and searched for a suitable ship. He settled on the *Nimrod*, a forty-year-old ship that was "very dilapidated and smelt strongly of seal-oil." But as he later found out, his initial reaction "hardly did justice to the plucky old ship."

Shackleton found nothing noble about using men to haul sledges. He planned to travel in style across the icy wilderness. Surefooted Mongolian ponies, trained sledge dogs, and an automobile would eliminate this backbreaking, soul-destroying task. Although he recognized the limitations of mechanical transportation, his fifteen-horsepower Arrol-Johnston car had a modified exhaust system that warmed the fuel line and used a specially developed nonfreezing motor oil. It also came with a good supply of spare parts—and its own mechanical engineer, Bernard Day. Shackleton imagined a quick jaunt southward over the Ross Ice Shelf, with the car dragging the heavy loads as far as possible. The ponies would then haul the sledges to victory. Although he had little faith in dogs, he would take them as insurance against failure if the ponies died and the car crashed. Man-hauling the sledges would be the last-ditch choice for achieving victory.

That spring he outlined the expedition's logistics in the *Geographical Journal*—and received a devastating blow that altered his plans. Scott wrote Shackleton a letter, requesting that the men not use his quarters at Hut Point since he, Scott, was planning his next expedition. Shackleton pondered his options and then acquiesced to Scott's demands. King Edward VII Land at the Ross Ice Shelf's eastern extremity, he decided, would be his expedition's winter home.

A STORMY PASSAGE TO A SAFE HAVEN

Plagued with financial woes, Shackleton did not travel to Lyttelton, New Zealand, on the *Nimrod*. Instead, he sailed on a faster steamer to Australia to raise funds for the expedition. In late December he arrived in Lyttelton, one month after the *Nimrod*, and now faced myriad last-minute details. He first selected ten of the hardiest Mongolian ponies from the fifteen that had arrived earlier from China. Next he chartered the *Koonya*, a steel-hulled steamer, to tow the *Nimrod* to the Antarctic Circle in order to save his coal supply.

On the day of departure, January 1, 1908, the ponies were tethered in ten stout stalls, their home for the next month. The fourteen-man wintering party crammed their bodies and paraphernalia into one narrow dank cabin, described by Shackleton as a "twentieth-century Black Hole." On the upper deck, bundles of baled pony maize, sacks of coal and potatoes, and crates of scientific equipment were wedged around pony stalls, nine howling dogs, and the boxed automobile lashed to the deck. By afternoon, the *Nimrod* was so overloaded that only three and a half feet of freeboard showed above the waterline.

Just hours before the *Nimrod* was scheduled to sail, a New Zealand sheep farmer, George Buckley, listened to Shackleton's men discuss their forthcoming expedition in their dingy, congested cabin. "There, in that uncomfortable place, the desire for the wind-whitened Southern Seas, and the still whiter wastes of the silent Antarctic grew stronger in his heart," wrote Shackleton. When Buckley could bear it no longer, he cornered Shackleton and begged to accompany the ship as far as the Antarctic Circle. He would then return to New Zealand on the chartered tow ship *Koonya*. Shackleton readily agreed, "for his heart was in our venture."

In just two hours, Buckley caught the train back to Christchurch, signed over his power of attorney to a friend, flung a toothbrush and a change of underwear into a valise, dashed back to Lyttelton, and elbowed his way through the throngs at the wharf. He clambered up the gangway with minutes to spare, wearing the only outer clothes he had brought—a summer suit.

At 4 P.M. the *Nimrod*'s dock lines were cast off. Booming guns and wailing sirens punctuated the cheers from thirty thousand townspeople who crowded the wharf. The heavy cables from the *Koonya* weighted down the *Nimrod*'s bow so much that the ship was towed "like a reluctant child being dragged to school." Although the sea was calm, water poured into the wash ports within an hour. Shackleton wondered if the ship could survive a storm. He didn't have long to wait to find out.

Gale-force winds with high swells struck that night and continued for the next ten days. Waves broke over the ship's sides and surged across the deck, knocking down men, ponies, and dogs. "We could hear the frightened whinnies of the animals, as they desperately struggled to keep their feet in the rolling stables," wrote Shackleton. During those terrible days, George Buckley lived in the stables. With each pronounced pitch of the ship, he spoke softly and heaved the wild-eyed ponies to their feet.

On January 7, the gale reached hurricane strength. Breaking waves wreaked havoc with supplies stored on deck. Three days later the storm finally abated. Shackleton felt new respect for his "plucky old ship." It had survived. For the first time since leaving New Zealand, everyone washed salt-caked faces and hair. "We had become practically pickled," noted Shackleton. Soggy clothes draped over every available inch of railing and billowed from the ropes of the rigging. However, their prized feather pillows, now pulpy lumps, were unsalvageable.

On January 14 the first iceberg was spotted. The next day, as George Buckley boarded a small whaleboat to be rowed to the *Koonya*, Shackleton and his men gave the sheep farmer three cheers of appreciation. The cables that had towed the *Nimrod* for 1,510 miles were then disconnected, and the *Koonya* turned toward New Zealand. Later that night, icebergs surrounded the ship. "A stillness, weird and uncanny, seemed to have fallen upon everything when we entered the silent water streets of this vast unpeopled white city," wrote Shackleton.

Eight days later the Ross Ice Shelf highlighted the horizon. Shackleton now headed east toward the inlet where Scott had ascended in a balloon in 1902. Since that time, enormous sections of the shelf had calved and the inlet had evolved into a wide bay, which Shackleton named the Bay of Whales. The unsettled ice shelf convinced Shackleton that erecting his winter hut anywhere in the vicinity would be foolhardy. His only choice was to head to Hut Point on Ross Island to winter—and break his promise to Scott. He weighed the consequences but had no choice. "It was with a heavy heart that I saw our bow swinging round to the west," he wrote.

On January 29 the ship entered McMurdo Sound, but pack ice extended for miles from Hut Point, blocking the ship. Ever the optimist, Shackleton waited two days for the ice to fracture and move out to sea with the currents. In the meantime, he tested his prized piece of equipment: the automobile. Bernard Day cranked the shaft and the car putt-putted to a start. It sputtered and rolled for a short distance and then jerked to a halt in soft snow. After a second dismal run, Shackleton's dreams of motorized transportation to the South Pole evaporated.

The barricade of ice didn't relent. The *Nimrod* followed the coast of Ross Island northward to Cape Royds, twenty miles from Hut Point. High volcanic hills sheltered the cove from the prevailing winds, several freshwater ponds lay nearby, and their meat supply—Adélie penguins—nested just beyond a low ridge. The site was perfect, Shackleton declared, and the men began unloading supplies at once. During the next three weeks, they erected the prefabricated hut, built a stable for the ponies, and hauled tons of provisions over the floes to shore. On February 23, Shackleton and his fourteen men gathered near the shore and watched the *Nimrod* steam northward toward New Zealand.

WINNING AND LOSING

The scientists were pleased with their location. Retreating glaciers had dumped a treasure trove of boulders that now waited for the crack of a geologist's hammer. A large penguin rookery offered food for the cook and behavioral studies for the zoologists. The ponds brimmed with microscopic life. And, fifteen miles to the northeast, volcanic Mount Erebus formed a dramatic backdrop. Most days, thick columns of steam rose thousands of feet in the air before leveling into long streamers—perfect indicators of upper-level wind currents.

Restless for exploration, six of the men strapped on the sledge harness and set out on March 5 for a grueling climb to the top of Mount Erebus. Five days later they huddled at the rim and peered into the mouth of the crater. Loud hissing sounds and thunderous booms emanated from the caldera, and "great globular masses of steam rushed upward," noted fifty-year-old professor, Edgeworth David, expedition leader. The descent took just one and a half days. The only casualty was one frostbitten big toe—which had to be amputated.

The men at Cape Royds fared better than the ponies. Less than one week later, four died suddenly. An autopsy on one revealed the reason: its stomach was filled with many pounds of sand. The ponies had not only relished a good roll on a nearby sandy beach but also had developed a taste for the volcanic sediments. Only Quan, Socks, Grisi, and Chinaman were left for Shackleton's southern journey. "They were watched and guarded with keen attention," he wrote.

The men settled into a routine of work and relaxation for the winter. Douglas Mawson trudged up the ridge to his meteorological shed twice each day, even during blizzards with hundred-mile-per-hour gusts. There he huddled over his hurricane lamp, protecting the small flame, and recorded his observations. James Murray discovered microscopic life in a melting ice chunk from the bottom of a pond. Intrigued with the organisms' apparent hardiness, he devised a hundred experiments to test their survival limits.

Everyone took a turn at nightly watch duty, to keep an eye on the stove and guard against fire. During this time the men darned woolen socks, baked elaborate cakes and breads, and listened to the mutterings of sleeping companions. Faithful renditions of snores and mumbles entertained the group each morning at the breakfast table. The men also wrote, typeset, and printed on a small hand press the 120-page *Aurora Australis*, the first book published in Antarctica.

The ponies, too, "entertained" the men—especially the mischievous Quan, who chewed through his neck rope to snack on the stacked bales of fodder behind him. A heavy chain that replaced the rope only made matters worse. Quan soon learned to toss his head and rattle it against the galvanized iron sides of the stable—but only when the men were sleeping. Desperate for uninterrupted rest, Shackleton strung a wire line along the length of the stable and tied Quan's head rope to it to limit the pony's access to the fodder. "Quan used to take this line between his teeth and pull back as far as possible and then let go." The resulting *twang* as the wire struck the iron sides never failed to rouse the men from their beds. Shackleton's annoyance passed when he saw the "intelligent look on the delinquent's face, rolling his eye round as though to say: 'Ha! Got the best of you again!' "

On August 12, ten days before the sun would rise above the horizon, Shackleton and two others hauled supplies to Hut Point, the start of his southern trek scheduled for late October. Inside the gloomy shelter, cases of tinned biscuits and meat still lined the walls. The men later climbed a nearby hill and looked toward the south. "We saw the Barrier [Ross Ice Shelf] stretched out before us—the long white road that we were shortly to tread," wrote Shackleton.

SHACKLETON'S LONG WHITE ROAD

By mid-October, after enduring temperatures that had plunged to -59° F, the men had established a single depot at 79° south latitude and returned to Cape Royds. For the next two weeks, Shackleton rested and enjoyed comfort and camaraderie before the start of his epic journey.

On October 28, 1908, as the men enjoyed a farewell dinner for the southern party, a shaft of evening sunlight suddenly illuminated the portrait of Queen Alexandra that hung on the wall. "Slowly it moved across and lit up the photograph of his Majesty the King," wrote Shackleton. "This seemed an omen of good luck." The next morning, everyone gathered around Grisi, Socks, Quan, and Chinaman, cheering and clasping the hands of Ernest Shackleton, Frank Wild, Eric Marshall, and Jameson Adams as they finished loading the sledges. At 10 A.M. the four men started on the sixteen-hundred-mile round-trip to "lay bare the mysteries of the Pole."

Even the weather favored the men—bright sunshine and a cloudless sky—but the auspicious beginning didn't last. Within an hour, Socks went lame. By lunchtime, Adams was limping from a kick on the shin from Grisi. On their third day out, Wild suffered the intense pain of snow blindness. Then, low thick cloud layers diffused the sunlight, dissolving all features into a shadowless white curtain. The men strained their eyes, trying to see mounds, small ridges, and dips in the murky gray expanse. Detecting crevasses was impossible until the ponies dropped to their bellies or the men plunged to their armpits, floundering over the "black yawning void" beneath their feet.

On November 7, still on the ice shelf, they camped unknowingly on a network of fine, crisscrossing fracture lines. Snug inside their tent, they cooked their dinner, wrote in diaries, and talked. Later that night Shackleton finished *Much Ado About Nothing*, while Adams tackled *Travels in France*. When they crawled out the next morning, bright sunlight illuminated the danger, much to the men's horror.

Even more dangerous were the undetectable crevasses that "had their coats on," thin crusts of snow that bridged the chasms. "Three feet more and it would have been all up for the southern party," noted Shackleton two days later. Chinaman, dragging a sledge packed with fuel and cooking gear, broke through a snow bridge and sank to his chest. The men barely managed to drag the pony and sledge to firmer ice. Worse still for the ponies were broad regions of deep snow. With every step they sank to their bellies, struggling against the weight of the heavy sledges. Rivulets of sweat froze into thick icy plates by evening. One by one, the ponies weakened.

So did the men. Shackleton cut food rations to stretch limited supplies. They notched their belts tighter and trudged southward on the ice shelf, each man alone with his thoughts. "We are but tiny specks in the immensity around us, crawling slowly and painfully across the white plain," wrote Shackleton. "Our imaginations take wings until a stumble in the snow, the sharp pangs of hunger, or the dull ache of physical weariness bring back our attention."

Chinaman collapsed and was shot on November 21. "We will use the meat to keep us out longer," Shackleton wrote. That night the men ate their fill and buried eighty pounds of pony meat. An upended discarded sledge with a small black flag, fluttering on a bamboo pole, marked the site. Five days later they passed Scott's "furthest-south" record. With each new mile south, the later-named Transantarctic Mountains grew on the horizon. "Mighty peaks they were, the eternal snows at their bases, and their rough-hewn forms rising high."

Wandering across the monotonous ice shelf was physically and mentally grueling. To keep alert during the gray days, Shackleton repeated endless lines of poetry to himself. On clear days he watched wispy clouds scud across the sky, pushed by high-altitude winds, yet no breeze stirred the grainy ice crystals at his feet. The stillness was uncanny. "We are truly at the world's end, and are bursting in on the birthplace of the clouds and the nesting home of the four winds, and one has a feeling that we mortals are being watched with a jealous eye by the forces of nature."

On November 28 Grisi was killed. Only Quan and Socks, both suffering from severe snow blindness, still dragged the sledges. Shackleton improvised shades for the ponies' eyes but they continued to shake from the pain. Four days later, Quan was shot and Shackleton mourned. "He was my favorite, in spite of all his annoying tricks." As the men lay in their sleeping bags, they listened to Socks whinny for his lost companion through the long, cold night.

They now crossed a chaotic jumble of huge blocks and pressure ridges of ice, an area that marked the boundary between the Ross Ice Shelf and the Transantarctic Mountains, about 440 miles from Cape Royds. On December 3, they discovered their "Highway to the South": the Beardmore Glacier. During the next four days, they worked their way over heavily crevassed terrain, webbed with snow bridges. The men faced the threat of oblivion with each step. Suddenly, Frank Wild dropped into space and felt "a violent blow on my shoulder and a fearful rush of something past me..." It was Socks, hurtling down into the darkness. Hanging onto the lip of the crevasse with his left arm, Wild yelled to the others. They pulled him out and then listened for signs of life from Socks. Only silence filled the abyss.

With their last pony gone, the men boiled Sock's maize and mixed it with their half-rations, hoping it would dull the searing hunger pains. As they trudged up the hundred-mile-long glacier, they crossed wide areas filled with rolling hills of glassy, wind-scoured blue and green ice. "Falls, bruises, cut shins, crevasses, razor-edged ice, and a heavy upward pull have made up the day's trials," wrote Shackleton on December 10. For the next five days, he recorded the agony of frostbitten fingers and bruised bodies. Symptoms of starvation shadowed each mile south.

"One more crevassed slope, and we will be on the plateau, please God," wrote Shackleton on December 16. Two days later they reached the head of the Beardmore Glacier, but ahead of them was an even more ominous region. A series of blue icefalls looked like storm-driven breaking swells frozen at the height of fury. The work was brutal. They relayed the loads and hauled the sledges up the steep slopes with alpine ropes. Ahead was always another icefall.

"If a great snow plain, rising every 7 miles in a steep ridge can be called a plateau, then we are on it at last," Shackleton wrote on December 27, exhausted. Frostbite blistered their thin, pinched faces. Blinding headaches, nausea, and nosebleeds—legacies of the ninety-five-hundred-foot altitude—further weakened them. "I cannot think of failure, yet," wrote Shackleton on January 2, "but I must look at the matter sensibly and consider the lives of those who are with me." Body temperatures fell to 94° F two days later, and the risk of hypothermia threatened all.

A shrieking blizzard with eighty- to ninety-mile-per-hour gusts struck on January 7 and continued the next day. "We simply lie here shivering," wrote Shackleton. During those agonizing hours, an occasional bare white foot protruded from a sleeping bag and was nursed back to life beneath its neighbor's shirt.

On January 9, Shackleton accepted the inevitable: they must turn back or die. That day the men left the tent and walked a few miles farther south. Then Shackleton called a halt and erected the Union Jack. After a few quick photographs and a hasty lunch, the men marched back to the tent. "We have shot our bolt," wrote Shackleton, "and the tale is latitude 88 degrees 23' South.... Whatever regrets may be, we have done our best." They were just ninety-seven nautical miles from the South Pole.

During the next ten days, a strong wind from the south filled the sledge sails. The men rushed across the plateau, averaging about twenty miles per day. On January 20 they reached the blue icefalls, the head of the Beardmore Glacier. Using axes for support, they scrambled down and crossed a network of crevasses, pausing only when they broke through the thin crusts of snow bridges. They raced against an enemy more horrifying than a quick death at the bottom of a crevasse: starvation.

Eight days later, they emerged onto the Ross Ice Shelf and dug out the stashed chunks of horse meat. "We are now safe, with six days' food and only 50 miles to the depot," noted Shackleton, happily. His mood did not last. The next day Wild developed dysentery, and Shackleton plunged into a hidden crevasse. "My harness jerked up under my heart, and gave me rather a shake up." His diary entries were stark, skeletal jottings during the next two weeks. Starvation coupled with dysentery had taken its toll. On February 13, they reached the "Chinaman depot" and clawed the snow for buried bits of meat and chunks of frozen horse blood. Six days later, the sight of Mount Erebus on Ross Island filled the men with hope. With superhuman effort they struggled against high winds. Drift ice blew into their noses and mouths as they labored for breath and stumbled toward the next depot.

"It is almost a farce to talk of getting to 'breakfast' now, and there is no call of 'Come on, boys; good hoosh.' No good hoosh is to be had," Shackleton wrote on February 20. They scraped the last bits of gristly meat from the pony's bones. Each time they cinched their belts to the next hole, they told each other everything would be fine once they reached Bluff Depot, the Promised Land for the four hollow-eyed, starving men. There, Shackleton hoped, would be the fresh supplies that he had instructed the men at Cape Royds to depot for him. "We must keep going.... Our food lies ahead, and death stalks us from behind," he wrote.

Daily rations were reduced to less than twenty ounces a day per man. Although no one joked about food, they discussed new culinary delights each man would cook if they survived this ordeal. "Wild's Roll was admitted to be the high-water mark of gastronomic luxury," Shackleton wrote. The imaginary creation consisted of well-seasoned minced meat, wrapped in thick slabs of bacon with plenty of fat, and covered with buttery pastry layers. Then the roll would be fried in lard to a toasty crispness and eaten while it was steaming hot.

On February 23, they stopped imagining meals. Wild saw a mirage of Bluff Depot, with black flags "waving and dancing" in the wind and a strange pinpoint of light flashed in the bright sun. "It was like a great cheerful eye twinkling at us," noted Shackleton. Four hours later the men found the source of the winking light: a tin canister of biscuits on top of a pyramid of food. The men ate plum pudding, eggs, cakes, and mutton, fresh from the *Nimrod*, which had arrived at Hut Point, about 130 miles away.

With renewed determination and a sledge loaded with food, they marched on the final leg of the journey. On February 27, Marshall collapsed with severe dysentery. While Adams stayed with Marshall, Wild and Shackleton struck out for Hut Point to reach the ship for help. For the next thirty hours, they marched with only three hours of rest and reached Hut Point late on February 28. There, the worst possible message awaited them: they had missed the *Nimrod* by two days. Wild and Shackleton wrapped up in roofing felt to keep warm, for they had abandoned the sledge and sleeping bags in the desperate race to the hut. For the rest of the night, they huddled together and discussed their limited options. The next morning they set fire to the small magnetic observation hut, hoping to attract the crew's attention if the *Nimrod* was close enough to see the flames. "All our fears vanished when in the distance we saw the ship." The crew had seen the burning hut; by late morning the two exhausted men boarded the *Nimrod*.

Shackleton did not rest. He guided the rescue party to Adams and Marshall, and by 1 A.M. on March 4 all were safe on board the *Nimrod*. As the ship sailed past their Cape Royds hut, the men gave three cheers. Only then did Shackleton sleep.

On March 23, 1909, Shackleton cabled London from New Zealand with news of the expedition's results, including the first ascent of Mount Erebus and the first successful trek to the South Magnetic Pole. His arrival in England was triumphant. When his wife, Emily, asked him why he had turned back from the South Pole, he replied that a live donkey was, perhaps, better than a dead lion. He delighted audiences across Europe with descriptions of penguin antics, life in the small Cape Royds hut, and the harrowing details of his South Pole journey. On December 14, 1909, Ernest Shackleton was knighted.

But Shackleton had not had his fill of Antarctica. In 1914, he would return with the ill-fated *Endurance* to attempt the first crossing of the continent from the Weddell Sea to the Ross Sea.

Robert Scott and Roald Amundsen: The Final Conquest

To Robert Scott, the motorized sledge was the new key to unlock Antarctica, but only a few experimental vehicles existed. He found a backer and helped to develop a prototype with a caterpillar track, the first to be designed specifically for snow. By the end of 1908, the motor sledge was ready for trials in Norway. The results were disappointing: the motor froze.

In March 1909, Scott heard the news of Shackleton's "furthest-south" record. When Shackleton returned to England in June, Scott attended a dinner party in honor of "the Boss"—a nickname Shackleton's men used. After dinner Scott announced his own intentions to return to Antarctica and claim the South Pole for Great Britain. He would follow Shackleton's pioneering route, he declared. The only obstacle barring his new expedition was money. Scott lacked Shackleton's financial savvy and charm which loosened philanthropic purse strings. With Markham no longer president of the Royal Geographical Society, the organization hesitated to fund a second expedition. The only new territory would be the ninety-seven nautical miles at the end—if Scott succeeded.

On September 10, Robert E. Peary announced to the world his conquest of the North Pole. Now, with only one important geographical "conquest" remaining, Scott intensified his fund-raising efforts. Three days later, he announced his Antarctic plans. Unknown to Scott at that time, news of Peary's victory had a profound effect on another man: Roald Amundsen.

ROALD AMUNDSEN: SECRETS AND SURPRISES

Born in 1872 near Oslo, Norway, Amundsen had planned to be an explorer since he was a boy. Books about Sir John Franklin, the British explorer who had died while searching for the Northwest Passage in 1847, had a profound effect on his imagination and heart. When he was fifteen years old, Amundsen focused all his energies to achieve that single goal. He skied during blizzards to increase his skill and stamina;

played football to develop muscles and leadership capabilities; and slept with his window open to harden his body against cold. At twenty-two, he worked as a deckhand on an Arctic sealer. Three years later, in 1897, he sailed to Antarctica on the *Belgica* and formed a deep friendship with Dr. Frederick Cook, the ship's physician. There, Amundsen strapped on a harness and pulled a loaded sledge—and never forgot the mind-numbing drudgery.

When Amundsen returned from Antarctica, he prepared to find the route through Arctic Canada from the Atlantic to the Pacific Ocean. Never one for serious scientific inquiry, he had, however, learned the value of dressing the expedition with research finery to raise funds. Still no scientists were on board when the *Gjoa* sailed at midnight—to escape creditors—on June 16, 1903. During the next two years, Amundsen filled many notebooks with observations of the Netsiliks—their culture, language, and dogsledging techniques. On August 26, 1905, near Point Barrow, Alaska, he sighted an American whaler and knew that the journey had ended. He had achieved his boyhood dream: he was the first man to navigate the Northwest Passage and survive.

In 1907, while Shackleton made final preparations for his *Nimrod* expedition, Roald Amundsen hauled boxes of lantern slides across America, lecturing to packed auditoriums. Although he had sailed for the Northwest Passage in obscurity and secrecy, he had returned in triumph. Amundsen finished the American tour, repaid his creditors, and focused on his next goal: the North Pole.

By early 1909, Amundsen had raised enough money to refit and update the *Fram*, a Norwegian polar ship, from steam to diesel propulsion. Every piece of equipment now had to be tested—or redesigned—for the rigors of the polar climate. Outerwear was patterned after the Netsiliks' reindeer anoraks; full sealskin suits were to be worn underneath. He oversaw the preparation of enough food to feed nineteen men for two years and initiated a search with the Danish authorities for the best Greenland sledge dogs.

On September 7 Amundsen's friend, Dr. Frederick Cook, declared that he had reached the North Pole. Three days later Robert E. Peary announced his claim. The news shattered Amundsen's dream; to him it mattered little who had reached the North Pole first, only that it had been done. Within days, Amundsen changed his plans from Arctic to Antarctic. He traveled to Copenhagen to congratulate Cook—and to double his order for sledge dogs. He confided his true destination to his brother, Leon, and swore him to secrecy. When Scott announced his Antarctic plans on September 13, Amundsen reasoned that if he revealed his true destination now, his financial supporters would withdraw their funds. The British press would interpret his intentions as a race to the South Pole and work the public into a patriotic frenzy, thus assuring Scott's success with generous donations of food and equipment. Norway, newly independent from Sweden in 1905, had no resources budgeted for exploration.

Amundsen kept his plans secret for one year. On August 9, 1910, the *Fram* departed from Bergen without bands, royalty, or crowds. The world and the nineteen-man crew believed the *Fram* would round Cape Horn, pick up the expedition scientists in San Francisco, and then continue north toward the Bering Strait.

When the ship reached the Madeira Islands, Amundsen requested all hands on deck. Since this was to be their last port of call for many months, most of the men were busy writing last-minute letters home. They gathered on deck, irritated at this unusual interruption. Amundsen finished tacking a map of Antarctica to the mainmast. Then he faced his men and spoke about his intention to sail southward, spend the winter on the continent, and then try to reach the South Pole the following spring. Second Mate Gjertsen recorded the crew's reaction in his diary: "Most stood there with mouths agape, staring at the Chief like so many question marks." The only sounds were the swish of waves against the bow, the creak of well-seasoned wood, and the squawks of gulls overhead. Amundsen outlined his plan—and the role each man would play. He told them where they were going to land, the hut they would build, how they would beat the English and win the South Pole for Norway. Since he had broken his agreement with them, they were now free to leave—he would pay their passage. No one chose the free ticket home.

Amundsen's brother, Leon, disembarked and posted the crew's letters. He also wired his brother's message to Scott, who would soon arrive in Melbourne, Australia: "Beg to inform you. *Fram* proceeding Antarctic. Amundsen."

ROBERT SCOTT: "IT IS A VERY, VERY TRYING TIME."

Once his plans were announced in September, Scott had just nine months to raise money, buy a ship, hire the crew, and secure supplies. Problems with the motor sledges prompted him to include Mongolian ponies and dogs as backups. At the last minute he purchased skis and convinced a champion Norwegian skier, Tryggve Gran, to join the expedition and teach the men to ski in Antarctica. Edward Wilson also agreed to return to Antarctica as Scott's chief scientist.

On June 1, 1910, the *Terra Nova* sailed down the Thames, accompanied by cheers so rousing that, according to Gran, "the air quivered on that blazing summer afternoon." The ship arrived in Melbourne, Australia, in early October. When Scott read Amundsen's cryptic message, he was not overly concerned and refused to comment publicly when questioned. Scott then sailed to New Zealand to pick up thirty-four dogs and nineteen ponies. Although Scott was "greatly pleased with the animals," Lawrence Oates, in charge of the ponies, was appalled. "Narrow chest.... Aged.... Windsucker.... Pigeon toes."

By the end of November Scott was ready to complete the last leg of the sea journey. The first day out at sea, gale winds of 55 MPH battered the ship. The *Terra Nova*, overloaded with thirty tons of coal and two-and-a-half tons of petrol, floundered in forty-five-foot waves. The ship's pitch "was so terrific that the poor dogs were almost hanging by their chains," noted Lieutenant Evans. Oates, meanwhile, cared for the ponies, "his strong, brown face illuminated by a swinging lamp, lifting the poor little ponies to their feet as the ship lurched..." wrote Evans. Two ponies died, despite Oates' constant attention.

With every lurch of the ship, cascades of water poured through seams in the deck planking and flooded the lower quarters. Worse was to come. First the main pump clogged, and then the hand pump failed. Water rose in the engine room and put the furnace fires out. The engines stopped. The *Terra Nova*, waterlogged and sluggish, listed as walls of water rolled over her. The engineers worked in water up to their necks and finally unclogged both pumps—but not before Scott ordered ten tons of coal dumped into the sea.

On December 9, Scott encountered broken pack ice much farther north than he had expected. Although he was following Shackleton's route, the heavy ice delayed him by three weeks. "Fortune has determined to put every difficulty in our path," he wrote. Scott's spirits were buoyed when the *Terra Nova* emerged from the pack ice on December 30; but that evening a blizzard blew from the south. "I begin to wonder if fortune will ever turn her wheel," he wrote.

ROALD AMUNDSEN: INCREASING THE ODDS

During the voyage from Norway to Antarctica, Amundsen also experienced his share of fierce storms. The nineteen men and ninety-seven dogs soon got their sea legs but none adjusted to *Fram*'s pronounced gyrations. "Perhaps it was worse for those who had to work in the galley," wrote Amundsen, "when for weeks together you cannot put down so much as a coffee cup without its immediately turning a somersault."

Although the dogs had the run of the ship, they were confined during bad weather. "In a storm, over 20 dogs can be pressed together... and when the ship lurches, the whole mass moves. And then there's a fight. Intelligent as dogs are, they cannot understand a lurch as anything but some devilment on the part of their neighbor who, naturally, needs a hiding," noted Hjalmar Johansen.

Amundsen divided the care of the dogs among the eight men who would winter with him. The dogs provided companionship, and each man became thoroughly familiar with the quirks of his canine group. Many evenings the dogs performed "howling concerts... suddenly and without warning," Amundsen noted. "The only amusing thing about the entertainment was its conclusion. They all stopped short at the same instant, just as a well-trained chorus obeys the baton of its conductor."

On January 2, 1911, the *Fram* crossed the Antarctic Circle and the next day entered the pack ice. Three days later, the ship emerged into the Ross Sea—a record passage. The ship's diesel engine had proved its worth. The *Fram* continued south to the Ross Ice Shelf. "The sea is still as a pond, and before one stands this Great Wall of China.... Far off, it is like a photograph that has just been developed on the plate."

Three days later, Amundsen anchored in the Bay of Whales where the ice sloped gently to the sea. Here, he would soon establish Framheim, his winter base. The next day, Amundsen prepared to drive the first team of dogs after the sledge was loaded with supplies. "With a flourish and a crack of the whip we set off," he wrote. "After moving forward for a few yards, the dogs all sat down, as though at a word of command, and stared at each other. The most undisguised astonishment could be read in their faces.... Instead of doing as they were told, they flew at each other in a furious scrimmage." Exasperated, Amundsen glanced toward the ship and the crew. "They were simply shrieking with laughter and loud shouts of the most infamous encouragement reached us."

The dogs soon learned "that a new era of toil had begun." For three weeks, forty-six dogs and five drivers moved tons of supplies two miles inland, the site of their home for the winter. The twenty-six by thirteen-foot prefabricated hut was then assembled and stocks of seal and penguin meat were stored. By January 28, Amundsen and eight expedition members moved in—just in time to receive a visit from the *Terra Nova*, under the command of Victor Campbell. Although neither the Norwegians nor the British discussed their plans, they enjoyed each other's company. The next day the *Terra Nova* sailed back to Scott's base at Cape Evans and the *Fram* headed for Buenos Aires.

There was no time to rest for Amundsen and the men. Before the onset of winter, depots had to be established at 80°, 81°, and 82° south in preparation for the South Pole journey. They shuttled three tons of supplies by dogsledge, built smaller depots with caches of food, and marked the route with black flags mounted on bamboo poles for flexibility and strength. They also buried drums of fuel, stored in galvanized containers with bronzed seams. On his Northwest Passage expedition, Amundsen had noticed the tendency of petroleum "to creep" and evaporate over short periods of time. These containers would prove airtight.

During the grueling weeks on the trail, Amundsen gained not only firsthand knowledge of the terrain but also noted clothing and equipment deficiencies, problems that would be corrected during the long winter. The men later returned to the first depot and encircled the cairn with six upended frozen seals for easy visibility. By March 23, all were back in the warm hut, feasting on Lindstrom's hot cakes and cloudberry preserves.

Winter and darkness descended over Framheim in April. During this time, the men excavated additional rooms underneath the snow: a carpenter's shop, a sewing room, and a sauna. There, under Amundsen's guidance, the men modified their equipment. Bjaaland redesigned four sledges, making them lighter and more efficient; Wisting sewed new tents from sailcloth to reduce weight; Hanssen lashed and relashed the modified sledges with rawhide, experimenting until just the right balance of elasticity and stability was found. Boots were enlarged to allow room for two pairs of reindeer socks rather than one.

Not every minute was spent working. For morale boosters, there were card games and dart contests. The dogs offered welcome diversions from gripes and irritations. Meals were simple but nutritious, with daily servings of seal steaks, penguin meat, whole wheat bread, and vegetables. So time passed pleasantly.

As the thermometer crept upward, Amundsen and his men fretted, plagued with thoughts of Scott and his motor sledges. The dogs were harnessed daily, but frustration deepened as blizzards delayed the start. Then on September 8, the dogs and men bursting with spirit, the cavalcade of sledges set off for the Pole. Three days later the temperature plummeted; two dogs froze to death. After reaching the first depot, they returned to Framheim to wait for signs of spring.

ROBERT SCOTT: DIMINISHING RETURNS

Scott's first choice for winter quarters, Cape Crozier, was blocked with sea ice, so he settled on Cape Evans at McMurdo Sound. On January 4, 1911, the men were in high spirits as they unloaded the ship. Motor sledges, dogs, and ponies dragged supplies from the *Terra Nova*, anchored at the edge of the ice, to the site of their future winter hut.

The ice deteriorated rapidly. During the next few days, the men worked from five in the morning until midnight. Soon exhausted, they suffered from snow blindness, sore faces and lips, and blistered feet. As the ice continued to weaken, the men worked even harder to finish unloading the ship. "A day of disaster," Scott recorded on January 8. One motor sledge, hoisted from the ship to the mushy ice floe, sank to the bottom of McMurdo Sound.

By January 17, the fifty-by-twenty-five-foot hut was finished and the wintering party moved in. Warm and cozy, the hut was insulated with quilted seaweed and had two stoves to provide heat. Crates of canned goods were stacked high, separating the crew's quarters from Scott and his officers.

Eager to begin his summer programs, Scott divided the men into five smaller groups. The southern party would lay supply depots along the polar route. A geological survey group would explore King Edward VII Land. The western party would investigate the region west of McMurdo Sound. The *Terra Nova* group would examine the Ross Ice Shelf. The last party would work to improve their living arrangements. As leader of such a wide diversity of tasks, Scott lost his focus in myriad details.

Unsure of the best means of transportation, the men experimented with motor sledges, dogs, ponies, and man-hauling. The motor sledges broke down often and were unreliable. The untrained and uncontrollable dogs inspired even less confidence. "I am losing all faith in them," wrote Scott. The ponies were also a disappointment. "I am not sure they are going to stand the cold." Scott concluded that the only reliable means of transportation was man-hauling the sledges.

On January 26, Scott set out with twelve men, eight ponies, and twenty-six dogs to establish depots along the route to the South Pole. Progress was slow. The ponies floundered in chest-high snow and grew weak in the blizzards. On February 17, Scott decided to push the march no farther. One Ton Depot was established—thirty miles short of his intended goal, 80° south latitude. "We shall have a good leg up for next year," he wrote.

While exploring the Ross Ice Shelf, the *Terra Nova* party found that Amundsen had established his base in the Bay of Whales, sixty miles closer to the Pole. This news rankled Scott. Although he felt that Amundsen had violated Great Britain's proprietary rights to the Ross Ice Shelf, he advised his men "to proceed exactly as though this had not happened." But he realized that Amundsen's plan represented "a very serious menace" and enabled him to start for the Pole "early in the season—an impossible condition with ponies." Scott expressed his concerns in his diary, but appeared unperturbed whenever Amundsen's name was mentioned.

Wintering at Cape Evans was pleasant and everyone settled into his own activities. Football scrimmages, science lectures three times a week, and lengthy evening discussions were highlights. Scott presented his South Pole plan on May 9, using Shackleton's figures, route, and timetable. They would start on November 3 and march across the Ross Ice Shelf to the Beardmore Glacier. Support parties with motor sledges, ponies, and dogs would then return to Cape Evans. His team and one support party would climb the glacier to the plateau. From there, one party would man-haul to the pole and arrive back at Cape Evans on March 27. The journey would take 144 days. Although Scott felt confident in the plan's feasibility, several members confided in their diaries that the plan left little margin for accidents, weather, and other contingencies.

On June 27, Henry "Birdie" Bowers, Apsley Cherry-Garrard, and Edward Wilson set out for Cape Crozier, sixty-five miles to the east of Cape Evans. Their goals were to test equipment and to bring back emperor penguin eggs for study. During their grueling trek, temperatures plunged to $^{-}77°$ F. When the haggard and exhausted men stumbled into the Cape Evans hut on August 1, their frozen clothes had to be cut from their bodies. Scott had "a sneaking feeling" that fur clothing could "outclass our more civilized garb." But no improvements were seriously considered.

ROALD AMUNDSEN: SKIING THE DISTANCE

In early October 1911, petrels soared over Framheim and the first seals hauled out on the ice at the Bay of Whales. On October 20, Amundsen,

Bjaaland, Hassel, Wisting, Hanssen, and fifty-five dogs departed for the South Pole, 870 miles away. The dogs were in high spirits as they towed the sledges across the broad undulations and snow-crusted crevasses on the Ross Ice Shelf, averaging twenty-five miles a day.

The first depot at 80° south, ringed with upright seal carcasses, was easy to spot. Inside the tent, the men relaxed in reindeer sleeping bags, pleased with their progress. Outside, a blizzard raged. The dogs, oblivious to the biting wind and blowing drift ice, gnawed on slabs of seal meat and blubber.

The men left the first depot on October 25 and Amundsen set the pace at seventeen miles a day. They had ample time to halt every five or six miles to build snow beacons along the route for the return trip. Buried inside each cairn were longitude and latitude directions to the preceding one, in case they had to steer the sledges in a whiteout, a common weather condition.

On November 1, after a day's rest at the 81° south depot, the men sledded into a thick ice fog and headed straight for an area crisscrossed with crevasses and snow bridges. Hanssen plunged into a narrow crevasse, landing on top of another snow bridge. He could not move without breaking through the thin snow crust. The dogs, however, crossed the three-foot-wide chasm safely and, without Hansen to stop them, enjoyed a snarling free-for-all. Amundsen leaped over the abyss and stopped the dogfight while Wisting rescued Hanssen. "We go with our lives in our hands each day," wrote Amundsen later. "But it is pleasant to hear nobody wants to turn back."

On November 7, they departed from the last depot, at 82° south. Ahead, the distant Queen Maud Mountains, a section of the Transantarctic chain, marked the boundary between the Ross Ice Shelf and the high Polar Plateau. As they approached the mountains, the ice shelf changed from flat to undulating. It seemed to the men that they were sledging over the high rolling swells of a frozen sea. Although Amundsen had expected pressure ridges and a honeycomb of chasms, they traveled with ease and reached the foot of the formidable Queen Maud Mountains, "each peak loftier and wilder than the last, rising to heights of 15,000 feet." The grim mountains barred the way south. To reach the plateau they would have to establish a winding route around towering blocks of ice, cross wide chasms, and ascend the later-named Axel Heiberg Glacier—a surface deeply rifted and scored with crevasses.

The men established a depot on November 16 and discussed strategy. They were about 350 miles from the pole and had forty-two dogs left. After much discussion, the final plan was to take enough supplies for 30 days and climb up the Axel Heiberg Glacier with all the dogs. At the top, twenty-four dogs would be shot to provide food for the remaining eighteen. Once at the pole, six more would be killed, leaving twelve dogs for the return to Framheim. It was a brutal plan, but Amundsen believed it necessary to ensure survival. All agreed to it.

The next morning they began the grueling ascent of the Axel Heiberg Glacier, scaling precipitous icy falls and plodding across snowfields fractured with deep crevasses. Men and dogs disappeared into chasms and had to be hauled to safety. The dogs clambered on their bellies and clawed their way up while the men pushed the sledges from behind. "The wildness of the landscape... cannot be described; chasm after chasm, crevasse after crevasse, with great blocks of ice scattered between, " wrote Amundsen near the summit. "Heigh ho, polar life is a grind," noted the irrepressible Bjaaland. In just four days men and dogs had traveled about forty-five miles and climbed to almost ten thousand feet, lugging a ton of supplies.

They made camp on November 21 and carried through with their plan. Twenty-four dogs were shot. "It was hard but had to be. We had agreed to shrink from nothing in order to reach our goal," wrote Amundsen. He remained in the tent and pumped up the Primus stove to high pressure, hoping to generate enough noise to drown the sounds of the massacre. But he heard every shot, "each time a trusty servant lost his life." The men named the camp "The Butcher's Shop."

The weather now took a turn for the worse. Gales roared around the tents and drift ice pelted their faces. For a week, they pushed forward, stopping when the blizzards were at their worst and continuing when the weather cleared for a time. Trudging on a surface of ice crystals was "wretched going... A sledge journey through the Sahara could not have offered a worse surface to move over." They trekked through Devil's Battlefield, an area that looked "as if a battle had been fought here, and the ammunition had been great blocks of ice." Then they came to Hell's Gate, a pressure ridge of ice about twenty feet high with a six-foot-wide fissure down the middle. Men, dogs, and sledges squeezed through the opening.

Blizzard after blizzard besieged them, and the ground seemed to boil with turbulent waves of drift ice. Bjaaland wrote on December 2, "We couldn't see in front of our nose tips, and our faces were white and hard as wax candles ... the hounds slid on the ice ... we forced our way 13 miles against the wind which burned like a flame, oh, oh what a life."

On December 4 they entered an area webbed with crevasses: Devil's Ballroom. About two feet beneath a deceptively thin layer of ice was a second surface of pulverized ice crystals, "the glacier's last farewell to us," wrote Amundsen. The first inkling of danger was the hollow echo of feet and paws. Suddenly, lead dogs broke through both layers and dangled at the end of harnesses over a great black chasm. The men pulled the wild-eyed dogs to the surface and then with yells of encouragement, men and dogs flew across the Ballroom.

They emerged onto firm snow—the featureless Polar Plateau. Now progress was rapid. On December 8 they passed Shackleton's furthest point. "Sir Ernest Shackleton's name will always be written in the annals of Antarctic exploration in letters of fire," wrote Amundsen. "Extra chocolate in honor of the occasion," noted Bjaaland.

The weather turned glorious. Only the swish of skis, the creak of sledges, and the hollow crunch of very cold snow disturbed the silence. Tension mounted and all craned their necks, scanning the surface for signs of Scott. The dogs, sensing the men's edginess, often stopped and faced south with noses high, sniffing the air and filling the men with foreboding.

The cry of "Halt!" broke the silence a little after 3 P.M. on December 14. The Norwegian flag was unfurled with ceremony, and "five weatherbeaten, frostbitten fists they were that grasped the pole, raised the waving flag in the air, and planted it as the first at the geographical South Pole," wrote Amundsen.

They remained at the Pole for three days. After a series of measurements, Amundsen decided to "box" the Pole due to the limitations of the navigational equipment. With the camp at one corner of the imaginary square, Bjaaland, Wisting, and Hassel skied ten miles in different directions, each carrying a wooden sledge runner. They each planted their runner in the ice, and attached a black flag with a note for Scott.

On the 17th the men broke camp. The tent was left behind with letters to the King of Norway and Scott. "And so, farewell, dear Pole," Amundsen wrote in his diary, "I don't think we'll meet again."

ROBERT SCOTT: UNFINISHED JOURNEY

In contrast to Amundsen's dependence on dogs, Scott couldn't decide which form of support transportation to use. The motor sledges were unreliable; the ponies were unsuited to Antarctic conditions; the dogs were irascible. Scott solved this dilemma by devising a plan that included all three means. The motor sledges would start first, followed a week later by Scott's group of nine men with the ponies. Cecil Meares and Demetri Gerof would bring up the rear with twenty-three dogs and two sledges. En route to the Beardmore Glacier, supply depots would be laid. Once they reached the glacier, two groups would man-haul the sledges up to the plateau. From there, Scott's polar party would march to victory.

On October 24, 1911, two motor sledges inched across the ice, each pulling one-and-one-half tons of supplies. One week later, Scott and nine men left Cape Evans with ten ponies. On November 3, the men found the first abandoned sledge; two days later, a black dot on the horizon confirmed Scott's worst suspicions. "So the dream of great help from the machines is at an end," he wrote. Ahead, the four-man motor party now dragged two overloaded sledges, laying a trail of depots along the route to the Beardmore Glacier.

A blizzard struck on November 7. While Scott's party waited for the weather to improve, Meares nonchalantly mushed the dogs into camp, undeterred by the squall that hampered the pony party. When the storm abated, he had to wait in camp to give Scott at least a half-day start with the ponies. Meares and Gerof then easily covered in two hours the same distance that took Scott the entire day.

On November 21, Scott's group caught up with the man-hauling motor party. From this point on, each day began with five separate starts: the man-haulers broke camp first, then the three pony teams, followed by the men with the dogs. The caravan stretched for miles across the ice shelf. The ponies weakened as they struggled to drag the heavy sledges through deep drifts, their flanks encrusted with thick shells of frozen sweat. Scott was determined that none should be shot until they were past "the spot at which Shackleton killed his first animal."

They reached that spot on November 24, and the first pony was shot. Overworked and starving from lack of fodder, four more were killed in early December. When the men reached the Beardmore Glacier on December 9, the remaining ponies were shot at a site named Shambles Camp.

Two days later Scott sent Meares and the dogs back to Cape Evans. Twelve men, including the exhausted motor party, harnessed themselves to three sledges and began climbing the Beardmore Glacier. A recent summer storm had turned the surface into a soupy slush. "It was all we could do to keep the sledge moving for short spells of perhaps 100 yards," wrote Birdie Bowers in his diary. "The starting was even worse than the pulling; it required from ten to fifteen desperate jerks on the harness to move it at all." At night their muscles cramped and cold, perspiration-soaked clothes kept them awake.

"We must push on all we can, for we are now 6 days behind Shackleton," Scott wrote on December 16. The next day he increased the pace, chasing the shadow of his nemesis every step of the way. Six days later, Scott sent one support party back to Cape Evans. "We have caught up with Shackleton's dates," noted Scott on December 30.

On January 3, 1912, the eight men reached the plateau. Scott now chose Evans, Oates, and Wilson to join him for the dash to the Pole, about 150 miles further south. At the last moment Scott included Bowers, although food and fuel had been allotted for just four men. His last orders to the three-man support party were to have Meares and the dogs meet the polar group at Mt. Hooper, south of One Ton Depot. The dogs' performance had, at last, impressed Scott. These instructions were given verbally—and contradicted written instructions left at Cape Evans.

Scott wrote "RECORD" in large letters in his diary on January 9. He had traveled "beyond the record of Shackleton's walk," the furthest south Shackleton had reached in 1909. Scott had, at last, beaten his rival.

On January 16, Bowers was first to spot a fluttering black flag tied to a sledge runner. Paw prints and ski tracks told the story. "The worst has happened," wrote Scott. "The Norwegians have forestalled us and are first at the Pole. . . . All the day dreams must go; it will be a wearisome return." They made camp but no one slept.

The next day they arrived at the South Pole. "Great God! This is an awful place and terrible enough for us to have labored to it without the reward of priority," Scott recorded in his diary. They trudged to the Norwegians' tent and found letters left by Amundsen. Some of the discarded equipment came in handy, especially for Bowers. "I was very glad too of a pair of their reindeer mitts, having lost my own dogskins some days back." Neatly piled fur garments lay on the floor. "It looks as though the Norwegian party expected colder weather on the summit than they got," Scott wrote. "It could scarcely be otherwise from Shackleton's account." The clothing was not touched.

Only Oates expressed appreciation for Amundsen's achievement in his diary. "I must say that man [Amundsen] must have had his head screwed on right . . . and they seem to have had a comfortable trip with their dog teams, very different from our wretched man-hauling."

The next day they began the long journey home. For the first three weeks they made excellent progress, averaging fourteen miles a day; but they felt the cold more and more, and finding each depot became a crisis. Evans' frostbitten fingers refused to heal, a deep scar on Oates' thigh split open, and Wilson suffered severe leg pains. Although the signs were unmistakable, Wilson made no mention of scurvy in his diary. Bowers, usually cheerful and optimistic, stopped writing in his diary on February 3.

They reached the upper Beardmore depot on February 7. The men faced a five-day march to the next depot with barely five days of rations. There was no margin for bad weather or other contingencies. Fortunately, the weather turned sunny and calm; but instead of pressing on, the men stopped for the afternoon and gathered thirty-five pounds of rocks at Mount Buckley to salvage the scientific part of the expedition.

They barely made it to the next food depot. "In future food must be worked so that we do not run so short if the weather fails us," wrote Scott. "We mustn't get into a hole like this again." Food wasn't the only problem. "It's an extraordinary thing about Evans," wrote Oates. "He's lost his guts and . . . quite worn out with the work, and how he's going to do the 400 odd miles we've still got to do, I don't know."

On February 17, at the bottom of the Beardmore Glacier, Evans unhitched himself from the sledge harness to tie his boots. "I cautioned him to come on as quickly as he could," wrote Scott. The three men continued on toward their next depot but stopped for lunch before reaching it. Evans still had not caught up. They skied back for him and, according to Oates, "We found him on his hands and knees in the snow in a most pitiful condition." While Oates remained with Evans, the other three returned for the sledge and dragged him back to the tent. Evans died that night at 12:30 A.M. The men struck camp at 1 A.M., leaving Evans's body, and arrived at a small food cache a short time later. They slept for five hours and reached Shambles Depot by midafternoon. The men dug out a horse carcass and enjoyed the first satisfying meal in a week. They loaded the sledge with horsemeat and followed the tracks of the returning support group to the next depot.

They reached it on February 24. "Found store in order except shortage oil—shall have to be very saving with fuel—otherwise have ten full days' provisions from tonight." Oates made another discovery that day. "Dug up Christopher's [pony's] head for food but it was rotten." This was the last entry in his diary.

The next day the men followed the last support party's deeply grooved sledge tracks for eleven miles. Meanwhile, the three-man support party, sick with scurvy, arrived at Hut Point more dead than alive. Scott and his men reached the next depot on March 1, but again discovered a fuel shortage. The leather washers and stoppers on the oilcans had disintegrated. Wilson stopped writing in his diary at this point. The next day, March 2, Oates "disclosed his feet, the toes showing very bad indeed," wrote Scott. The pain from gangrene had become unbearable. To pull on his boots in the morning was an ordeal, even after he had slit one boot down the front to make more room for his gangrenous foot.

On March 7 Scott wrote, "We hope against hope that the dogs have been to Mt. Hooper; then we might pull through. If there is a shortage of oil again we can have little hope." Late in the afternoon on March 9, they arrived at Mt. Hooper. "Cold comfort," wrote Scott. "Shortage on our allowance all round.... The dogs which would have been our salvation have evidently failed." In the confusion of contradictory orders, two men had waited six days at One Ton Depot with the dogs. They turned back to Cape Evans on March 10, unaware that Scott, Bowers, Wilson, and Oates were fighting for their lives sixty miles south at Mt. Hooper.

Their plight became desperate. After breakfast on March 11, Scott ordered Wilson to hand out the opium tablets. "Titus Oates is very near the end.... What we or he will do, God only knows... he is a brave fine fellow and understands the situation," wrote Scott. Oates struggled on for another five days. Unable to endure the pain of gangrene any longer, he crawled out of the tent into the arms of the blizzard. "He said, 'I am just going outside and may be some time,'" wrote Scott. The three men soon broke camp, leaving Oates's sleeping bag and his boots. His body was never found.

On March 21, eleven miles from One Ton Depot, the three men made their final camp when a severe blizzard prevented them from going any further. Scott now wrote farewell letters to friends and family. To the public he wrote, "Every detail of our food supplies, clothing, and depots made on the interior ice sheet and over that long stretch of 700 miles to the pole and back worked out to perfection... we should have got through in spite of the weather but for the sickening of a second companion, Captain Oates, and shortage of fuel in our depots for which I cannot account.... Had we lived I should have had a tale to tell of the hardihood, courage, and endurance of my companions which would have stirred the heart of every Englishman. These rough notes and our dead bodies must tell the tale."

He recorded his last message on March 29: "Every day we have been ready to start for our depot 11 miles away, but outside of the door of the tent it remains a whirling drift. I do not think we can hope for any better things now. We shall stick it out to the end, but we are getting weaker, of course, and the end cannot be far. It seems a pity but I do not think I can write any more. For God's sake look after our people."

Outside, beads of snow rasped against the canvas tent and drifted over the sledge. The long Antarctic night soon covered all. During the winter at Cape Evans, the search party waited for the sun to return.

On November 12, 1912, they found the frozen bodies of Robert Scott, Edward Wilson, and Henry Bowers inside their tent. After personal artifacts were collected, the search party collapsed the tent over the bodies and erected a cairn of snow to mark the site. Tryggve Gran's skis formed a cross on top. He strapped on Scott's pair of skis and completed his leader's journey to Cape Evans.

After Scott's death was announced, Amundsen's victory was belittled and bitter rhetoric filled newspapers and books. Although he continued his explorations in the Arctic, he never came to terms with the hostility. While attempting a rescue mission near Spitsbergen in 1928, his small plane disappeared into the cold polar sea without a trace.

· 7 ·

THE MAKING OF A TREATY

At the beginning of the twentieth century, maps of Antarctica were blank except for a few wavy lines representing partial coastlines, ice shelves, and landfalls. While the British forged inland toward the South Pole during the first decade, other important expeditions added new coasts and islands to Antarctic maps. One of the first to arrive at an unexplored area was German Erich Dagobert von Drygalski's expedition. On February 21, 1902, he sighted a stretch of coast on Antarctica's eastern side, naming it Wilhelm II Land. That same day his ship, the *Gauss*, was beset by ice and remained icebound for thirteen months.

Six weeks before the *Gauss* finally broke free, the Norwegian ship *Antarctic* succumbed to the crushing pressures of the Weddell Sea pack ice near the tip of the Antarctic Peninsula. Its loss stranded not only Captain Larsen and his crew but also the Swedish Antarctic Expedition team, who were forced to endure an unscheduled second year on Snow Hill Island.

Unaware of the *Antarctic*'s dire predicament in 1903, the Scottish National Antarctic Expedition, led by William S. Bruce, retreated from the Weddell Sea ice and wintered on Laurie Island in the South Orkneys.

There they built a meteorology station and maintained it until 1904. By the time his expedition had completed its goals, Bruce realized that important year-round weather data could be provided to science if the station were operated continuously. When he presented the station to the British government, officials rejected it on financial grounds. Bruce then offered it to the Argentines, who have since occupied it.

In the spring of 1903, news of the missing *Antarctic* prompted Frenchman Jean-Baptiste Charcot to mount a rescue mission. When he reached Buenos Aires in November, Charcot learned that all members of the Swedish expedition were safe. Still, he decided to continue southward to survey the western coast of the Antarctic Peninsula, since few maps of the region existed. After wintering on Booth Island, Charcot charted over six hundred miles of new coastlines and islands. Lured back in 1908, he surveyed about 1,250 miles along the Antarctic Peninsula and added many inlets, bays, and mountains to Antarctic maps.

In less than a decade, Antarctic exploration had acquired an international flavor. As charts slowly gave shape to the elusive continent on paper, one country took steps to protect its Antarctic interests.

A Claim-Staking Rush

On July 21, 1908, three weeks before Charcot's second expedition sailed, Great Britain claimed sovereignty over the Antarctic lands and islands discovered since 1775. To Great Britain's embarrassment, parts of South America were included in the hastily drawn document. These errors were corrected in 1917, when the British again claimed the South Orkneys, South Shetlands, South Georgia, South Sandwich Islands, and the Antarctic Peninsula. One reason for securing sovereignty in this region was to protect British interests in the flourishing Antarctic whaling industry. Great Britain controlled whaling stations on all the island groups that it now claimed.

Beginning in 1924, funds generated from lucrative whaling leases and licenses supported a series of British expeditions. Scientists studied the whales' feeding and migratory routes to learn how to control over-exploitation and thus preserve a highly profitable industry. In the same year, the British also reasserted their claim to the Ross Ice Shelf and surrounding lands, including King Edward VII Land and Victoria Land, as a British settlement. These lands were placed under New Zealand's control.

In 1924 France also claimed a small wedge of Antarctica, from 142° to 136° east longitude, the region Dumont d'Urville had discovered and named Adélie Land in 1840. Sir Douglas Mawson had explored a large tract of the same territory in 1913 and claimed it for Australia. The Australians were miffed, but made no serious challenge at this time.

America made no claims in 1924, but the navy produced an enthusiastic report about Antarctica's potential mineral and petroleum resources. Although the report also urged the government to assert American rights in Antarctica, no action was taken and the study was ignored. Secretary of State Charles Evans Hughes declared that "the discovery of lands unknown to civilization, even when coupled with the formal taking possession, does not support a valid claim of sovereignty unless the discovery is followed by actual settlement of the discovered country." Hughes' statement became America's unofficial policy toward Antarctic claims.

Within four years, Americans returned to Antarctica after a hiatus of almost ninety years. The charismatic man who organized the expedition was Richard E. Byrd.

"It is the effort to get there that counts"

Born in Virginia in 1888, Richard Byrd developed a love of flying during World War I while he trained as a navy pilot. He clearly saw the potential of the airplane, not only as a weapon of war but also as a tool to expand geographic horizons. On May 9, 1926, Byrd and copilot Floyd Bennett flew on a round trip from Norway's Spitsbergen to the North Pole. The only navigational tool onboard was a sun compass. Although Byrd expressed doubts in his diary about whether he and Bennett had actually reached the North Pole, he never revealed these concerns publicly during his life. Three days later and with much less hullabaloo, Roald Amundsen and two companions also flew over the North Pole and continued to Alaska—the first airplane journey across the Arctic Ocean.

Byrd returned to New York and enjoyed fame, if not fortune. At a reception for the young hero, Amundsen, the man who had reached the South Pole in 1911, asked Byrd what he planned to do next. "Fly over the South Pole," Byrd responded. Amundsen told Byrd there were just three prerequisites for success: a reliable airplane, a team of first-class dogs, and the best men to be found. Byrd took Amundsen's advice to heart.

Corporate America rallied behind Byrd with cash and equipment donations. In the political realm, news of the expedition's planned base at the Bay of Whales miffed Great Britain, especially since the United States had not sought formal permission to land Byrd's party on the Ross Ice Shelf. The United States ignored the fuss and refused to recognize Great Britain's, or any country's, proprietary rights to any sector of the continent.

Byrd, too, ignored the furor and involved himself in nearly every detail of the expedition, becoming "the cartographer, dietitian, purchasing agent, fund-raiser, haberdasher and jack of all trades" to get the job done. By the time Byrd sailed from America on August 25, 1928, with 650 tons of supplies, ninety-five dogs, and three airplanes, it was the best-organized and -provisioned private expedition to date.

After a stop in New Zealand, they reached the Bay of Whales on December 28. Unloading began at once and soon a prefabricated village, Little America, mushroomed eight miles inland on the Ross Ice Shelf. Accommodations for the forty-two-man winter-over party were spacious by Antarctic standards. Three main buildings and a dozen smaller ones, connected by snow tunnels, allowed the men mobility and a change of scenery. Three 70-foot radio towers assured an unbroken stream of two-way communication with the outside world.

On January 27, 1929, Byrd and two companions flew eastward and sighted fourteen unknown mountain peaks less than a hundred miles from Little America. "They lay in the shape of a crescent; their spurs and crags rising austerely out of the snow," wrote Byrd. "Here was something to put on the map: a fine new laboratory for geological research." He named his discovery the Rockefeller Mountains, after the expedition's major contributor.

In early March, geologist Laurence Gould and two others returned to the Rockefeller Mountains in the Fokker Universal monoplane, *Virginia*. For the next week the men surveyed the region and collected rock samples. Just as they finished packing their gear, the wind increased to over 100 MPH. The tethered plane bucked up and down as though flying through great turbulence. Gould tried to steady the plane and grabbed a rope attached to a wing. Moments later, he too was airborne, and his horizontal body flapped like a human flag. The wind quieted down for several hours, but one terrific gust lifted the plane from its moorings. As the men watched in horror, the Fokker flew backward for about a half mile and then crashed into a ridge of ice. The plane, including the radio, was demolished. Byrd mounted a search when his messages were not answered; within days the stranded men were rescued.

The men burrowed into their underground base and passed a comfortable winter as snow drifted over Little America. In the deep polar night only the ice-coated skeletal radio masts and a few chimneys marked the men's presence. "It is as quiet here as in a tomb," wrote Byrd. "Nothing stirs. The silence is so deep one feels one can reach out and take hold of it."

In October, teams of men and dogs established depots for Gould's trek to the Queen Maud Mountains along the same route that Byrd would fly to the South Pole. If he were forced to land, Byrd reasoned, the geological party would be close enough to rescue the plane party. On November 4, Gould's party left Little America with dog teams and began the fifteen-hundred-mile trek.

Two weeks later Byrd flew toward the Axel Heiberg Glacier, Roald Amundsen's treacherous pass to the Polar Plateau in 1911. Less than two hours into his flight, he spotted Gould's party, trudging through deep snow. "The men were in harness, pulling with the dogs; the dogs were up to their bellies in the snow," wrote Byrd. "If ever a conclusive contrast was struck between the new and the old methods of polar travel, it was at this moment." Byrd stored gasoline and food near the foot of the Axel Heiberg and then reconnoitered several other glaciers that offered lower-altitude passages to the Plateau.

Byrd returned to Little America and waited ten days for clear weather. His Ford Trimotor aircraft stayed packed with survival gear—tents, gasoline stoves, bulky food sacks, and rows of gasoline cans. "There was scarcely room to move," noted Byrd. Then, on November 28, 1929, Gould radioed that conditions were perfect toward the Queen Maud Mountains, cloudless with unlimited visibility. Bernt Balchen climbed into the pilot's seat, Richard Byrd sat next to him to navigate, Harold June operated the radio, and Ashley McKinley steadied his heavy aerial surveying camera near the window. At 3:29 P.M. the *Floyd Bennett*, named for his North Pole copilot, bounced down the ice runway and lifted into the air.

Although Byrd had intended to duplicate Amundsen's route to the South Pole by following the Axel Heiberg Glacier with its known altitude up to the Polar Plateau, he now decided on the wider Liv Glacier, "a niagric torrent doomed to rigidity, with frozen whirlpools and waterfalls." Its deeply scored surface resembled "the fluted surface of a washing board." The heavily laden plane ascended nine thousand feet before its speed diminished; the controls responded sluggishly in the thin air. Ahead was the narrow passage to the Plateau—two thousand feet higher. The *Floyd Bennett*'s nose bobbed up and down, and the whole plane shuddered and wobbled as they neared the narrow neck of the glacier. Weaving back and forth from one side of the glacier to the other, Balchen and Byrd searched for stable air currents. "We needed power," Byrd wrote, "and there was only one way in which to get it." They heaved three hundred pounds of food out the trapdoor and then watched the sack hit the surface of the glacier "with a soundless explosion."

With just a few hundred yards to spare, they rose over the hump to the Plateau. At 1:25 A.M. the men circled the South Pole. Byrd opened the trapdoor and dropped the American flag, weighted with a stone from Floyd Bennett's grave, onto the smooth sea of ice beneath their wings. "There was nothing now to mark that scene," wrote Byrd, "only a white desolation and solitude disturbed by the sound of our engines." June's Morse code message announcing the plane's South Pole arrival was picked up in New York and broadcast to the world. At 10:10 A.M. the four men arrived at Little America. The sixteen-hundred-mile journey had taken just eighteen hours and forty-one minutes.

Although his South Pole flight made headlines, Byrd considered his aerial surveys over the Ford Range, Marie Byrd Land, and Rockefeller Mountains to be his finest achievement. "An air of drama foreshadowed every mile of progress; north, south, east, and west—everything was untrodden and unknown. Here was the ice age in childhood. Here too was great beauty, in the way that things which are terrible can sometimes also be beautiful."

Another Round of Claim-Staking

Although the U.S. government had not responded to Byrd's inquiry about claiming new lands, he and second-in-command Gould quietly took possession of the territories for the United States on their own initiative. In June 1930, three months after Byrd's expedition returned to the United States, a Senate bill directed President Hoover to claim Antarctic regions that had been discovered or explored by American citizens. Although the bill was tabled, other initiatives followed. The measures focused on Wilkes Land, sighted in 1840 by Charles Wilkes, and Byrd's recent discoveries.

From 1929 to 1931 Norwegian whaling magnate Lars Christensen financed nine voyages to subantarctic islands and to East Antarctica. Captain Hjalmar Riiser-Larsen flew seaplanes over large tracts of Antarctica and discovered Queen Maud Land, as well as the Princess Martha and Princess Ragnhild Coasts. At this same time, Australian Sir Douglas Mawson claimed land in the same general vicinity. Supported by the British, New Zealand, and Australian governments, the primary objective for Mawson was "making landings to plant the flag," according to his written instructions. In January 1930 Mawson and Riiser-Larsen planted their respective flags along the same stretch of coast named Enderby Land, and the two men met on Mawson's ship to resolve their differences. Christensen interceded and instructed Riiser-

Larsen to explore lands west of 45° east longitude; Mawson agreed to claim land east of that demarcation.

In 1933 Great Britain formally claimed for Australia a large slice of Antarctic territory from 45° to 160° east longitude, excluding the narrow French wedge, Adélie Land. The United Kingdom now had carved out about two-thirds of the continent for itself. A few U.S. government officials balked at the Australian claim because it included Wilkes Land, but Hughes' dictum prevailed and the United States continued its nonassertive policy. In the meantime Rear Admiral Byrd organized his second expedition to Antarctica.

Alone for the Winter

American interest in Antarctica continued during the early 1930s, the result of the unprecedented media attention devoted to Byrd's first expedition. His well-documented struggle and triumph in an alien land inspired his countrymen. Weekly messages about the men's winter life beneath the snow, the discovery of Marie Byrd Land, and Byrd's South Pole flight gripped the public's imagination as much as the first moon landing years later.

Byrd's second venture was also privately funded. He raised money by selling exclusive story rights to newspapers and publishers. CBS won the radio broadcast rights and sent a professional announcer to accompany the expedition; General Foods Corporation sponsored the weekly radio shows (with commercials) that were broadcast from Little America's KFZ station.

When Byrd arrived on January 17, 1934, he found his first base buried beneath three to six feet of snow. The original base was salvaged and ten additional buildings, connected by a maze of tunnels, housed supplies and equipment. A generator provided electricity and power to beam live weekly programs from Little America II.

One important objective was to establish a weather station inland on the Ross Ice Shelf, to be manned during the winter. Byrd's men hauled supplies 123 miles inland from Little America II, and by the end of March, the nine-by-thirteen-foot hut had been entrenched in a pit on the ice shelf. Only the twelve-foot bamboo poles that supported the radio antennas and the anemometer were visible from a distance. Although Byrd had planned for three men to winter at Bolling Advance Weather Station, he decided at the last minute to stay by himself. On March 28, he watched the two Citroën tractors crawl northward "until only the vanishing exhalations of the vapor remained." Then he opened the trapdoor on the hut's roof and climbed down into the room that would be his home for the next six months.

Silence and solitude appealed to him at first. Here, his only responsibility was to climb out the trapdoor and record daily weather data from the anemometer. Books lined his shelves; his thick journal was filled with reflections about being alone on a sea of ice. "The silence of this place is as real and solid as sound," he wrote.

As winter's darkness descended over the hut, the burden of leadership lifted. He cooked simple meals on his oil-burning stove, listened to music on his gramophone, and tapped Morse code messages on his radio transmitter to Little America II. Byrd often stood outside his hut, reveling in a storm's approach. "First there is the wind, rising out of nowhere. Then the Barrier unwrenches itself from quietude . . . and begins to run like a making sea. . . . You become conscious of a general slithering movement on all sides. The air fills with tiny scraping and sliding and rustling sounds as the first loose crystals stir. In a little while they are moving as solidly as an incoming tide, which creams over the ankles, then surges to the waist, and finally is at the throat. I have walked in drift so thick as not to be able to see a foot ahead of me; yet, when I glanced up, I could see the stars shining through the thin layer just overhead."

The idyllic time didn't last. In late May the temperature plunged to -96° F. During a blizzard, the anemometer clogged with snow and stopped recording the wind's speed—which had reached gale force. "The Barrier shook from the concussions overhead; and the noise was as if the entire physical world were tearing itself to pieces," he wrote. Byrd climbed outside to clear the anemometer of snow while the blizzard raged around him. "It is more than just wind," he wrote, "it is a solid wall of snow moving at gale force, pounding like surf. . . . In the senseless explosion of sound you are reduced to a crawling thing on the margin of a disintegrating world; you can't see, you can't hear, you can hardly move." He reached the anemometer, but drift snow caked around it as soon as he cleared it. He then crawled back and pulled on the trapdoor. It didn't budge. "Reason fled. I clawed at the three-foot square of timber like a madman. I beat on it with my fists.... I lay flat on my belly until my hands were weak from cold and weariness...and said over and over again: You damn fool, you damn fool."

Suddenly, he remembered a shovel that he had stuck in the ice near the trapdoor a week earlier. In the blinding blizzard he groped for the tool with his feet and then used it to pry open the trapdoor. He stumbled down into his warm shelter, thinking, "How wonderful, how perfectly wonderful."

His relief didn't last long. On May 31, as he was sending a message to Little America II, Byrd heard the small generator in the tunnel sputter and choke. "Wait," he coded to the base and then headed toward the tunnel. The air was thick with black oily smoke and gaseous fumes. He passed out for twenty minutes and then crawled back to the radio transmitter. He signed off without a word of explanation. After turning off the generator he collapsed on his bed, too weak and nauseous to even light a candle. In his muddled mental state, he realized that the generator and his stove had slowly poisoned him with carbon monoxide. To signal Little America for help would put his men in jeopardy, Byrd reasoned. If a rescue attempt were mounted during midwinter there would be dire consequences and probable death. The sunrise was three months away. "I could not persuade myself that I had the strength to meet it." That night Byrd discovered just how alone he was.

The next day, which he referred to as "Black Friday," Byrd contemplated his own death and its consequences on those that he would leave behind. Weakened by the poisonous gas, he slowly wrote farewell letters to his wife and others. Like an endless benediction he repeated in his mind Robert Scott's last written words, "For God's sake, look after our people."

He moved in slow motion during the next few weeks, telling himself in simple sentences to get out of bed, to light the lantern, to change his clothes, to eat. Every action required deliberation and forethought, a step-by-step sequence that helped him to survive. He forced himself to send messages to Little America, to appear "normal" and thus alleviate any suspicions. To keep from freezing, he lit the stove for just a few hours a day and watched the flame burn red instead of blue, the hallmark of faulty combustion. Gradually, though, hope replaced despair.

His messages consisted of unintelligible fragments and nonsequiturs, which raised concerns at Little America. In mid-June, one of the scientists casually asked permission to start laying depots for the geological parties earlier than planned. The route would take the depot party to his hut. Byrd realized the significance of the request several days later. He replied that he would return to Little America with the party. No questions were asked, and Byrd didn't volunteer any reasons for his decision. The rescue party set out on July 20, but gales and low temperatures forced the men to turn back. Another attempt was made in early August but the tractor broke down. On August 8 the party set out and rolled over the ice shelf, averaging about 2 MPH. At midnight on August 10 Byrd watched the dim lights of the advancing group and burned flares to guide them to the station.

Several men stayed at the station to care for Byrd; another two months passed before he was strong enough to leave Bolling Advance Station. "I climbed the hatch and never looked back," he wrote. Behind him was the memory of a near-death experience; ahead, "the sheer beauty and miracle of being alive..."

Research teams with tractors, planes, dogs, and sleds fanned out eastward toward the Rockefeller Plateau, Edsel Ford Range, and Rockefeller Mountains; another party crossed the Ross Ice Shelf to the Queen Maud Mountains and ascended Robert Scott Glacier to the plateau. The various parties returned with a wealth of scientific data: 450,000 square miles were surveyed; fossilized tree trunks, leaves, and seams of coal were discovered; and meteorological information was meticulously recorded throughout the expedition. The depth and movement of the Ross Ice Shelf were measured for the first time. By early February 1935, the expedition was en route to the United States with boxes of scientific data. Left behind, scattered over vast areas of the continent, were brass cylinders with letters that proclaimed U.S. sovereignty.

Claims and Counterclaims

In 1938 President Franklin D. Roosevelt authorized a reexamination of the United States' "nonpolicy" toward Antarctica. Since American Lincoln Ellsworth was again heading to Antarctica, secret instructions were sent him via the U.S. consul in Cape Town, South Africa, his point of departure. He was advised to "assert claims" on all territory he would soon explore, map, and photograph—regardless of any claims previously made by other countries. Ellsworth was even given typed directions that described the proper procedure for asserting sovereignty. He took his task to heart. On January 11, 1939, Ellsworth and his pilot flew inland and claimed a three-hundred-mile-wide swath of East Antarctica, known today as American Highland, by dropping a brass canister with a document of U.S. sovereignty inside.

Three days later, Norway's King Haakon formally claimed the coastal area from 20° west to 45° east longitude as Norwegian territory. The timing of Norway's claim coincided with the launch of a German expedition heading toward Queen Maud Land on the *Schwabenland*.

As World War II loomed ominously over Europe, the German expedition, under the command of Alfred Ritscher, arrived at 4° 30' west longitude and anchored near the impenetrable ice on January 20. The purpose of the expedition was to claim territory so that future German whaling vessels could participate on equal footing with other nations. Pilots flew aerial reconnaissance missions and photographed the terrain from the coast to about 370 miles inland. Five-foot-long aluminum darts, engraved with the swastika, were dropped every fifteen to twenty miles along each route. Several landings were made and proclamations read to the ship's crew and one emperor penguin that was acquired along the way. According to one scientist, salutes and solemn words "did not make much of an impression on him." The *Schwabenland* arrived at Cape Town on March 1 and Hamburg, Germany, five weeks later.

Meanwhile, President Roosevelt had appointed Admiral Byrd to lead another Antarctic expedition from late 1939 to 1941. The government-sponsored venture, the United States Antarctic Service Expedition, established Little America III at the Bay of Whales and a base on

Stonington Island. The expedition abruptly ended for Byrd when Nazi raiders seized three Norwegian whaling ships near Queen Maud Land in March 1940. The rest of the expedition was terminated one year later.

Squabbles and Military Maneuvers

As World War II escalated in Europe, Argentina and Chile developed interests in Antarctica. In 1940 Chile declared sovereignty over a slice of the continent that included the Antarctic Peninsula, pointedly ignoring Great Britain's earlier claim. Two years later, Argentina asserted its rights to the peninsula and subantarctic islands, including the South Orkneys and the continuously occupied weather station turned over to them in 1905. Argentina followed up its proclamation with two expeditions in 1942–43.

Not to be outdone, Great Britain launched Operation Tabarin from 1943–45. This secret military maneuver established a base on Deception Island in the South Shetlands and a second station at Port Lockroy on Wiencke Island near the Antarctic Peninsula. The purpose of the operation was to keep Germany from using harbors in the South Shetland Islands or along the Peninsula to launch attacks against allied supply ships.

The feared Nazi invasion never materialized in Antarctica, for the Germans preferred subantarctic Îles Kerguélen's deeply penetrating inlets and hidden coves as cover for their operations. British attention turned to the Argentines, who had left plaques and cylinders with the usual notices of sovereignty. The crew of Britain's *Carnarvon Castle* collected the sovereignty tokens and sent them to their ambassador in Buenos Aires, who then presented them to the Argentine government. Within months, Argentina's *Primero de Mayo* arrived for survey work in the South Shetlands and the Antarctic Peninsula. British emblems were removed or painted over with renditions of Argentina's flag. The British retaliated and the Argentines responded. Huts were built, torn down, and rebuilt; emblems and slogans from both countries were painted, covered, and repainted on rocks, whale oil storage tanks, and buildings.

While Great Britain and Argentina spread graffiti, the first Chilean Antarctic expedition, led by Captain Federico Guesalaga Toro, arrived at the South Shetlands in 1946. One year later the Chileans established a base on Greenwich Island, further complicating British and Argentine claims.

American politicians for the most part ignored the three-country squabble over the South Shetland Islands and the Antarctic Peninsula. Postwar tensions between the USSR and the United States ushered in a new era of hype and hyperbole—the Cold War. American military leaders believed that, should tensions between these two former allies erupt into war, first-blood battles would be waged in the Arctic. Military maneuvers and practice operations conducted in a polar climate were viewed as essential ingredients for a healthy state of readiness.

During this period of heightened hostility, the U.S. Navy Antarctic Development Project—Operation Highjump—was started on August 16, 1946 and sailed in December 1946. In addition to training military personnel and testing equipment, one important objective was to consolidate and extend American sovereignty, if formal claims should ever be needed. Although the expedition's scientific aspects were emphasized to the public, only twenty-four men out of forty-seven hundred were civilian scientists. In early December 1946, thirteen ships carrying twenty-three planes and helicopters, assorted ground transportation, and construction supplies headed toward Antarctica.

The overall organization, under the technical command of Admiral Byrd, was divided into three groups. The central division headed for the Bay of Whales to build Little America IV; one contingent headed eastward around the continent; the other, westward. From these three positions, aircraft surveyed and photographed the interior and almost 75 percent of the coast of Antarctica. News correspondents focused on Byrd's second flight over the South Pole, where he dropped token flags of countries participating in the United Nations. Another attention-grabbing story was the discovery of an Antarctic Shangri-La, the ice-free Bunger Hills near the Shackleton Ice Shelf in East Antarctica. The expedition departed by early March 1947.

Although nearly fifty thousand aerial photographs were taken, accurate maps could not be drawn without reference ground points. Operation Windmill was launched the next season to survey thirty land features spotted during Operation Highjump. Weather and ice conditions prevented the men from achieving this goal, but seventeen reference positions were established by the time the expedition departed in February 1948. As a result, Antarctic maps now reflected reasonable geographic accuracy and detail. More importantly, at least in political circles, America had a solid basis to claim a piece of Antarctica.

The three-ring political circus continued on the South Shetland Islands and the Antarctic Peninsula. Any meaningful diplomatic dialogue degenerated into bitter rhetoric and accusations. However, the men dispatched to enforce their countries' claims solved sovereignty questions much more sensibly: with football games, soccer matches, and dart competitions. The winning country reigned over an island, base, or hut until the next rematch or the latest official directive arrived.

In 1950, eight countries—Great Britain, Chile, Argentina, Norway, France, Australia, New Zealand, and the United States—explored ways to settle territorial squabbles as a single international entity. Russia, whose history of Antarctic exploration extended back to Bellingshausen's 1819–1821 voyage, was excluded because the country had made no territorial claims—and as a response to the heightened Cold War paranoia that seeped into policies. The Soviet Union replied that it would not recognize any decision affecting territorial claims because its officials had not participated in the discussions. The formal memorandum also stated that since the Soviet Union made no claims on Antarctica, it would not observe any territorial boundaries—a stance later echoed by the United States.

An Eighteen-Month Year for Scientists

Independent of this unhappy state of affairs, a small group of scientists attended a dinner party in 1950 at the home of Dr. James Van Allen, an acclaimed geophysicist. One topic of conversation was the long delay before the next International Polar Year (IPY), an eighteen-month event devoted to Arctic research that occurred every fifty years, then scheduled for 1983–84. They noted that with all the technological advances in communications and transportation, there was no reason to wait thirty-three years. Their suggestion to hold the IPY earlier was presented to the International Council of Scientific Unions (ICSU), an organization that represented most scientific disciplines. Within months, the concept evolved into the eighteen-month International Geophysical Year (IGY), scheduled for 1957–1958. Two major themes dominated the worldwide scientific investigations: outer space and Antarctica.

In preparation for the IGY, the first Antarctic conference was held in Paris in 1955 for scientists from all participating nations. The chairman of the event, General Georges R. Laclavere, emphasized in his opening remarks that the meeting was about scientific investigations in Antarctica—territorial disputes would not be tolerated. United by their scientific disciplines, members from twelve nations ushered in a new era of cooperation that extended beyond arbitrary political boundaries.

Governmental officials from the participating countries embraced the IGY concept. Here, at last, was the means to establish a physical presence on Antarctica without political ballyhoo. Preparations began at once. Military personnel from the United States, the USSR, Great Britain, New Zealand, Norway, Australia, Argentina, Chile, Belgium, France, Japan, and South Africa established new bases or enhanced existing stations. The U.S. Navy launched Operation Deep Freeze in 1955, and within two years, seven U.S. bases were constructed, including McMurdo Station in the Ross Sea region and Amundsen-Scott Station at the South Pole. By July 1, 1957, the beginning of the IGY, over forty stations on the continent and twenty on subantarctic islands sheltered scientists and support personnel.

Scientific investigations during the eighteen-month period produced forty-eight hefty tomes on astrophysics, oceanography, glaciology, meteorology, and zoology. One of the most astonishing discoveries was the unsuspected thickness of the ice sheet. Teams from the United States and several other countries trekked hundreds of miles into the interior, periodically detonating bundles of TNT along their routes. By measuring the time interval between the explosion and the return of the shock waves that bounced off the underlying rock, the scientists were able to calculate the thickness of the ice. At first they doubted their results, but collaboration with teams from the Soviet Union, France, Australia, and Great Britain indicated that the ice sheet extended to a depth of two and a half miles in places. Their data also produced the first sound "pictures" of submerged mountains and deep troughs.

From Sea to Icy Sea

Although the Commonwealth Trans-Antarctic Expedition was not officially part of the IGY, it embodied the spirit of adventure for people worldwide. The plan was bold. Explorer Vivian Fuchs would cross the continent from the Weddell Sea to the Ross Sea via the South Pole, a journey of more than two thousand miles. Sir Edmund Hillary, knighted for his 1953 conquest of Mount Everest, would establish a series of depots for the last leg of Fuchs's trek.

On January 28, 1956, eight men and three hundred tons of supplies were unloaded at Vahsel Bay on Filchner Ice Shelf in the Weddell Sea region to establish Shackleton Base. While Fuchs returned to London to finalize preparations for his crossing, the land party endured harrowing hardships inside a large snowmobile crate during the winter. The following January, Fuchs returned and helped construct a hut and depot about 350 miles inland from Shackleton Base. Three men wintered at the new base, named South Ice. In mid-November 1957, Fuchs returned to Shackleton Base and began his continental crossing on November 24 with ten men, four Sno-cats, three weasels (a light, self-propelled track vehicle that travels well over ice and snow), and a muskeg tractor—a farm tractor modified for travel on snow and ice. He carried Robert Scott's pocket watch with him, wearing it around his neck on a leather braid.

While Fuchs was preparing for his assault from the Weddell Sea, Hillary arrived at McMurdo Station on Ross Island to select a site for New Zealand's research base. Scott Base was constructed at Pram Point, a few miles from McMurdo, during the 1956–57 season. He also established a depot at the foot of the Skelton Glacier, his route to the plateau. After wintering at the new base, Hillary and three men began their journey with three muskeg tractors and one weasel on October 14. Two weeks later, they reached the plateau without incident. During the next six weeks, food and fuel were flown in and Hillary established two well-stocked depots. On December 20, his tasks completed, Hillary decided not to return to Scott Base; instead he pushed on for the South Pole—over Fuchs' objections. He reached Amundsen-Scott South Pole Station on January 4, 1958, and then returned to Scott Base.

Meanwhile, Fuchs and his team faced horrendous terrain—appalling crevasses and wind-sculpted furrows and ice formations called sastrugi. He crossed land that was little more than a "chaotic mass of ice hummocks resembling a ploughed field." Time and again, his vehicles tottered on the lips of chasms. Fuchs chained the Sno-cats together so that if one plunged through a snow bridge, the others could drag it to safety. He and his men persevered and arrived at the Pole on January 20. After a five-day rest, Fuchs pushed onward and led his flag-decorated vehicles to Scott Base, completing his 99-day, 2,158-mile journey.

An Invitation to Write a Treaty

Even before the successful conclusion of the IGY, many researchers questioned why Antarctic science could not continue indefinitely. On May 3, 1958, United States president Dwight D. Eisenhower announced that letters had been delivered to the eleven other participating countries. He stated that Antarctica "should be used only for peaceful purposes" and not "as an object of political conflict." Mankind would benefit "if the countries which have a direct interest in Antarctica were to join together in the conclusion of a treaty."

Within one month the eleven nations had responded favorably. Representatives attended over sixty secret meetings to hammer out their differences and draft a preliminary treaty. They accomplished their mission on October 14, 1959, when the Antarctica Treaty was presented at a formal conference. Just six weeks later, Argentina, Australia, Belgium, Chile, France, Japan, New Zealand, Norway, South Africa, Great Britain, the United States, and the Soviet Union signed one of the most unique documents in human history. On June 23, 1961, the Antarctic Treaty went into full force.

The opening sentence in the preamble sets the tone. The twelve signatory governments recognize "that it is in the interest of all mankind that Antarctica shall continue forever to be used exclusively for peaceful purposes and shall not become the scene or object of international discord." The fourteen brief articles of the treaty emphasize freedom for scientific investigations and encourage the exchange of scientists and research data. Nuclear explosions, disposal of nuclear waste, and military aggression are banned; on-site inspections by member-nation committees are permitted if reasonable advance notice is given. The treaty addresses territorial disputes by freezing the status quo. No signatory country had to surrender its past claims; those countries that may have a basis for a claim—the United States and the Soviet Union, for example—did not have to renounce their right to make a claim. However, no new claims can be made while the treaty is in effect.

One important aspect of the treaty is concern for the Antarctic environment and its living resources. In 1964 the Agreed Measures for the Conservation of Antarctic Fauna and Flora were adopted, but these did not come into effect until 1978. The measures prohibit killing, wounding, or molesting wildlife—except by permit from the Scientific Committee on Antarctic Research (SCAR). The scientific community was granted the right to designate regions as Specially Protected Areas (SPAs), Sites of Special Scientific Interest (SSSIs), and Specially Managed Areas (SMAs), primarily as safeguards against tourism. No one can visit these areas without a special permit.

In 1978 the Convention for the Conservation of Antarctic Seals provided a system for regulating commercial sealing. Ross, fur, and elephant seals are fully protected. Four years later, in 1982, the Convention on the Conservation of Antarctic Marine Living Resources came into force as a response to unregulated harvesting of krill (a small shrimp-like crustacean) and uncontrolled fishing. The measure established a perimeter line around the continent, within which all living organisms were to be treated as part of a single ecosystem. In order to determine an appropriate quota for harvesting fish or krill, studies had to be undertaken to understand the organisms' interaction with their environment. Only then could quotas be set, fishing regions defined, and length of season controlled.

The Protocol on Environmental Protection, also called the Madrid Protocol, is now in effect and integrates existing measures into a powerful reconfirmation of the Antarctic Treaty. Antarctica is once again designated as "a natural reserve, devoted to peace and science." It establishes environmental principles that govern all activities, such as scientific research, new construction, and tourism. Environmental impact assessments and follow-up monitoring are required.

The Madrid Protocol prohibits mining (except for scientific research) for the next fifty years—unless the twenty-six signatory countries unanimously agree to an amendment modifying the Protocol. If an amendment is adopted but not implemented within three years, any country may "walk away" from the Protocol and be free to pursue mining or drilling activities. Although vast oil reserves and mineral resources may be present, none have been found to date.

A Continent Revealed

Science was not the primary focus for most Antarctic expeditions until the early 1900s. The search for new fur seal rookeries and the pursuit of whales had much higher priority than descriptions of birds or collections of meteorological data. Those scientists that did venture to Antarctica, usually doubling as shipboard surgeons, endured ever-present cold, faced near starvation, and battled scurvy to study, draw, and describe the last unexplored region on earth. Technological advances in transportation, communication, clothing, and shelter—the legacy of two world wars—enabled more scientists to work in reasonable comfort and safety. The Antarctic Treaty ignored arbitrary political boundaries and encouraged an exchange of information among researchers. Antarctica became their laboratory, dedicated to the peaceful pursuit of knowledge.

And now, scientists from many disciplines devote months—and sometimes years—to study the most pristine and complete ecosystem on earth. Their discoveries and those of their predecessors help unravel the mysteries of Antarctica's physical evolution, its unique ecosystem, and its future impact on mankind.

• PART II •

REGIONAL EXPLORATION: FOUR ANTARCTIC REGIONS AND SUBANTARCTIC ISLANDS

• 8 •

ANTARCTIC PENINSULA EXPLORATION

The Antarctic Peninsula is the most northerly section of the Antarctic landmass. Part of West Antarctica, it is an archipelago disguised as solid terrain by an icy mask that hides the diverse layers of bedrock. Like a thin crooked finger, the eight-hundred-mile-long peninsula beckons the adventurous toward a mountainous land skirted with deeply scored glaciers. Great walls of fractured ice obscure the coast and thwart human access to the peninsula's interior backbone, the north-south trending mountains that poke through the ice sheet crust.

On January 30, 1820, the mist dissolved as sunlight penetrated the thin low clouds. Edward Bransfield and the crew of the *Williams* stared with undisguised astonishment at "high mountains covered with snow." One year later, Russian explorer Thaddeus von Bellingshausen discovered two islands; in February that same year, American sealer John Davis waded ashore from a small boat near Hughes Bay to search for fur seals.

Eight years later, American Charles Wilkes led his four-ship armada southward from Tierra del Fuego to the Antarctic Peninsula. One ship, the tiny *Flying Fish*, battled gales and bergs to reach 70° 01' south, just short of Captain James Cook's 1774 record, near Thurston Island in the

Amundsen Sea. Thick fog swirled around the *Flying Fish*, and pale icebergs emerged from the gray mist "like tombs in some vast cemetery," wrote the ship's surgeon. This vessel and Wilkes' other three ships retreated to South America to avoid disastrous encounters with icebergs.

For the next thirty years, the Antarctic Peninsula held little interest for merchants or explorers. But by the 1870s, the fur seals had recovered a fraction of their former numbers after the devastating 1820–23 slaughter. American and British sealers searched rocky coves and shores on the South Shetlands, South Georgia, the Falkland Islands, and the Antarctic Peninsula for their prey.

In 1895 the International Geographical Congress declared Antarctic exploration a high priority and implored its members to fund expeditions to the far south by the end of the century. Only Belgium accepted the challenge. The man who led the expedition was Adrien Victor Joseph de Gerlache, a lieutenant in the Royal Belgian Navy. With him was an international group of scientists and crew, including Norwegian first mate Roald Amundsen, Romanian zoologist Emile Racovitza, Polish meteorologist and geologist Henryk Arctowski, and American physician Frederick A. Cook. On August 16, 1897, an old whaling ship, renamed *Belgica*, steamed away from Antwerp for what the men thought would be a nine- to twelve-month adventure. Little did they realize just how long their journey would last.

The First Long Antarctic Night: The Belgica *Expedition*

In January 1898 the 110-foot *Belgica* steamed down the Beagle Channel from Ushuaia, Argentina, and crossed the infamous Drake Passage, a 600-mile body of water often whipped to monstrous waves by gale-force winds. On January 20 the ship approached the Antarctic Peninsula. Although the austral summer was already far advanced, de Gerlache was determined to accomplish the expedition's research goals.

Two days later a terrible gale struck the *Belgica* near Sail Rock, about seven miles southwest of Deception Island in the South Shetland Islands. White-crested waves broke over the ship's sides and ribbons of foam streamed across the deck. Suddenly, the men heard a terrible scream. When they looked over the side, they saw Carl Wiencke struggling among the high waves. He had been swept overboard when he tried to clear loose coal from the scuppers. Although the men managed to tow Wiencke to the side of the ship with the log line, he lost his grip and disappeared beneath the waves, never to be seen again. Wiencke was the expedition's first death, but not the last.

The next day the *Belgica* crossed Bransfield Strait, headed toward the peninsula. De Gerlache then explored an unknown waterway separating the Antarctic Peninsula from north-south trending islands. This scenic, iceberg-choked passage was later named in his honor. The weather for the next few days was idyllic for Antarctica—clear and calm. Sunlight exposed in exquisite detail the fractured contours of glaciers, the soft sculpted shapes of weathered icebergs, and the frothy blows of whales. The men took advantage of these perfect days and made more than twenty landings during the next three weeks, including a weeklong sledging trek on Brabant Island.

As the ship continued southward, Dr. Frederick Cook hunched over his bulky camera equipment, the first in Antarctica, and exposed plate after plate of the rugged mountains that seemed to float on the still waters. The ever-changing panorama enthralled Cook. "It was a photographic day," he noted. "Not less than 300 were taken on this day." Sleep was impossible, and he wandered the deck at all hours. "There is a glitter in the sea, a sparkle on the ice, and a stillness in the atmosphere, which fascinates the soul but overpowers the mind."

On February 15 the *Belgica* crossed the Antarctic Circle. The next day de Gerlache entered deep into the broken pack ice, although it was very late in the season. "We were all anxious and uneasy," wrote Cook. The ragged-edged floes grated against the ship's sides, and specks of paint colored the ice, marking the route.

During the next two weeks, the ship zigzagged among the floes and attempted to penetrate further south. On February 28 the *Belgica* was at 70° south latitude in the Bellingshausen Sea. By late morning something ominous about the sky and sea promised "a night of unusual terrors." A few men swore they smelled mud and wet mossy rocks—land where there was none.

The wind rose to gale force, causing the masts to quiver and then sway independent of the rocking ship. The *Belgica* sought shelter from the tempestuous wind and roiling sea deep within the broken pack ice. Still, every swell hurled tons of ice against the ship's ribs with deafening blows and bone-jarring jolts. Howling wind and grinding ice silenced the men's desperate shouts as the ship moved deeper within the pack ice. Hours later the floes shifted and closed the leads behind the ship; however, tantalizing new channels had opened toward the south. De Gerlache didn't hesitate; he ordered the men to follow the fresh leads, full steam ahead.

By March 2 the ship could advance no further. Ahead, the pack ice was thick and seamless; behind, the leads had disappeared, sealing the *Belgica*'s fate. "We are now doomed to remain, and become the football of an unpromising fate. Henceforth we are to be kicked, pushed, squeezed, and ushered helplessly at the mercy of the pack," wrote Cook.

At first the men were cheerful. They sang, whistled, and squeezed out tunes on an old accordion; music boxes were regularly wound and filled the silence with merry melodies. One sailor practiced his trumpet on the deck, much to the amusement of penguins that came from great distances to listen to dance hall ditties. When the concert was over, the penguins waddled back to a fissure in the ice and disappeared into the black water. "Altogether we are making the dead world of ice about us ring with a boisterous noise," wrote Cook.

One day in March, meteorologist Arctowski set up his bulky equipment on the ice near a narrow lead. Engrossed in his work, he didn't notice a leopard seal evaluating him and his paraphernalia as a potential meal. Suddenly, the seal sprang from a lead onto the floe, opening its massive jaws with a snort. "Thinking that the creature contemplated an attack, Arctowski made warlike gestures, and uttered a volley of sulphureous Polish words," wrote Cook, "but the seal didn't mind that. It raised its head higher and displayed its teeth in the best possible manner." Arctowski quickly packed up his equipment and reboarded the *Belgica*.

Gradually, discontent gnawed at the men's optimism. Meals became unbearable. The cook racked his brain for new culinary combinations from his limited stock of canned beef and fish meatballs. His soups were full of "mystery" and the men were convinced that the evil-tasting beef had been "embalmed" before it was sealed in tin cans. Many doubted the beef was actually beef, but a mish-mash of slaughterhouse scraps spiced with rodents and roaches. "If these meat-packers could be found, they would become food for the giant petrels very quickly," noted Cook. Although plenty of penguins and seals were available for the galley, de Gerlache pronounced the meat unpalatable and refused to eat it. The men, of course, followed his example.

Alone in a cold, dead wilderness the men endured an endless cycle of raging blizzards that alternated with days of dense fog, but worse was the unchanging view everywhere that they looked. "It is a strange sensation to know that you are moving rapidly over an unknown sea, and yet see nothing to indicate a movement," wrote Cook. "We pass no fixed point, and can see no pieces of ice stir; everything is quiet. The entire horizon drifts with us. We are part of the endless frozen sea." The pack ice carried the *Belgica* about ten miles per day. When plotted on paper, the course looked like a child's scribbled drawing. At no time during their enforced journey on the unexplored Bellingshausen Sea did they see land.

DESCENT INTO MADNESS

On May 17 the sun disappeared below the horizon. "We are under the spell of the black Antarctic night, and, like the world which it darkens, we are cold, cheerless, and inactive," noted Cook. He forced himself to take daily ski treks and sometimes lingered beside small ponds that formed within the pack ice. In the moonlight, the water glistened with a silky sheen as lacy tendrils of new ice patterned the surface. Clusters of ice crystals bloomed like pale lotus flowers, floating serenely on frosted glass.

One night Cook shouldered his tripod-mounted camera and climbed down from the ship to the ice. He carried his equipment for a distance and then stopped to focus the lens on the hoarfrost-coated *Belgica*. Every rope of the three-masted rigging glowed in the moonlight. Cook exposed a photographic plate for ninety minutes and captured that ephemeral frozen landscape, with the ghost-like ship sailing across a white world.

A few times Lieutenant Emile Danco accompanied Cook on his outings, but by the end of May, the young man experienced severe shortness of breath and heart palpitations. On June 1 he was too weak to get out of bed. His condition further deteriorated and on June 5, 1898, Emile Danco died. The second engineer laid a bouquet of dried flowers on the shrouded body before sliding it through a deep hole in the ice. They didn't linger at the gravesite, for a bitter wind blew from the southwest and the air was filled with fine ice crystals that stung bare hands and heads.

Danco's death affected everyone deeply. A few men believed that the body floated directly beneath the ship; every mournful groan or creak was not the ice or ship but Danco. They looked at each other's pasty faces and wondered who would be next to slip through a hole in the ice. Paranoia lurked behind listless eyes and guarded facial expressions. A sideways glance, a whistled tune, or a remembered comment took on malicious significance. Arctowski studied the group and shook his head helplessly. "We are in a mad-house," he whispered to Cook.

The long Antarctic night also took its toll on the ship's mascot, a cat named Nansen. Once full of sass and spunk, the cat grew lethargic and short-tempered as the darkness deepened. "We have brought in a penguin to try to infuse new ambitions and a new friendship in the cat, but both the penguin and the cat were contented to take to opposite corners of the room," Cook noted. Nansen grew thin and unresponsive, until one morning he was found curled up in a corner, dead.

The death of the cat caused an unexpected problem: a proliferation of rats. With no feline to eliminate the surplus population, the rats multiplied with a vengeance. Within a short time, the rodents massed on deck and invaded cabins to search for crumbs and nesting materials. The men plugged their ears with bits of old cloth, but they still heard the rats scurry across the floorboards.

As the men grew more depressed and mentally unstable, Cook devised a plan of action, the "baking treatment." The patient stripped and stood in front of the stove's roaring fire for several hours each day for the next few weeks. The radiant heat and light against their pale flesh restored peace of mind and good humor. When the men complained of achy joints, bleeding gums, or swollen legs, Cook vanquished these symptoms of scurvy with "prescriptions" of fried penguin fillets or undercooked seal steaks, milk, and cranberry sauce. Although the cause of scurvy—vitamin C deficiency—had not yet been discovered, Cook based his fresh-food remedies on his previous experiences in Greenland. Recovery was rapid for the men who took their "medicine."

On July 22, 1898, Cook stared at the horizon and waited for the sun to once again push back the night. "Every man on board has long since chosen a favorite elevation from which to watch the coming sight," wrote Cook. "Some are in the crow's nest, others on the ropes and spars of the rigging." The more adventurous men climbed to the top of nearby icebergs and waited, their faces turned toward the coming light. At noon the rim of the sun peeked above the horizon. Cook watched the faces of the men nearest him. "Their eyes beamed with delight, but there was noticeable the accumulated suffering of the seventy dayless nights.... Their skin had a sickly, jaundiced color, green, and yellow, and muddy. We accused each other of appearing as if we had not washed for months." Roald Amundsen's chestnut brown hair had turned gray.

With the return of the sun, the men's mental attitude improved at first. They looked forward to daily marches around the ship during the brief daylight hours. When the novelty faded, the men grew sullen and morose once more. In October, despair turned to hope as more leads opened and lakes formed in the ice. The men readied the *Belgica* for the homeward journey and sang as they scrubbed floors and darned tattered clothes with renewed vigor. But their hopes were short-lived, for a ferocious storm shifted the pack ice and slammed the leads shut.

Throughout November and December this cycle of hope and despair was repeated each time the leads opened and then closed. During this terrible time, two men lost all sense of reality and had to be restrained. Others seemed ready to sink further into depression and paranoia. Cook had to cajole, threaten, or beg the men to take their daily jaunts around the ship. The well-trampled path was now called the "mad-house promenade."

On January 11, 1899, Cook submitted an escape plan to de Gerlache. He suggested cutting a two-thousand-foot canal through the ice to a small lake, or *polynya*, within the pack ice. The audacious plan rallied everyone, and hope replaced despondency. Officers and sailors worked shoulder to shoulder with pickaxes, hacksaws, and shovels for the next month. They ignored the pain of frostbitten toes and fingers inside their soaked penguin-skin boots and gloves. When they were within a hundred feet of freedom, the wind changed direction and the pack ice moved. Moments later, the canal suddenly closed. Cook feared that at least a few men would commit suicide by abandoning the *Belgica*. Fortunately, the next morning the ice shifted again and the lead snapped open. "We lost no time in steaming out," noted Cook.

For the next month the men battled the broken pack ice. Then on March 14, the *Belgica* emerged from the floes, thirteen months after entering the ice. Two weeks later the ship steamed into Punta Arenas. "A year hence, I am sure we shall all long to return again to this death-like sleep of the snowy southern wilderness," Cook wrote. But for now, the wind carried the fresh scent of flowers and grass, and for men of the *Belgica*, that was treasure enough.

Dr. Frederick Cook never returned to Antarctica; instead, he traveled to the far north and claimed to have reached the North Pole in 1908, nearly one year before Robert Peary. Most historians have dismissed Cook's account and awarded Peary the honor of being first. Roald Amundsen, however, believed his friend—even after Cook was sentenced to five years in jail for a bogus gold mine investment scheme. Without exception, members of the *Belgica* expedition praised Cook as the man who had saved their lives during that long Antarctic night with his "prescribed" fresh-meat meals, unfailing hope, and good humor.

Commander Adrien de Gerlache almost returned to Antarctica. In 1903 he joined Jean-Baptiste Charcot's expedition but changed his mind when he arrived at Buenos Aires. Years later, after traveling to Greenland and Spitzbergen, he sold his private yacht to Ernest Shackleton, who renamed it the *Endurance*.

A Touch of Class: Charcot's Expeditions

Dr. Jean-Baptiste Charcot was born in 1867, the son of an internationally renowned neurologist, Jean-Martin Charcot. Although he followed in his father's footsteps and studied medicine, his first loves were sailing and navigation. When his father died, Charcot inherited four hundred thousand gold francs and a painting by French artist Fragonard. Charcot used most of his inheritance to finance a research expedition to the northern polar regions in a 150-foot ship he had commissioned, the *Français*. When he learned Otto Nordenskjöld's men and the *Antarctic* had disappeared, Charcot changed his plans. If the Swedes had somehow survived the brutal Antarctic winter, then Charcot would rescue them and explore the largely unknown continent. He appealed to the public for contributions; the response was overwhelming.

On August 15, 1903, just two minutes after the *Français* had pulled away from the wharf at Le Havre, tragedy struck without warning. The towrope snapped and the heavy hawser swung loose, killing a young sailor. Charcot returned to port for twelve days. When the ship quietly sailed on August 27, Adrien de Gerlache, former commander of the *Belgica*, had joined the French expedition as an advisor.

The *Français* arrived at Buenos Aires on November 16. During the long sea voyage, de Gerlache announced that he had decided to return to Belgium to be with his new fiancée, rather than face the uncertainty of another perilous Antarctic journey. While in port, Charcot learned that the Argentines had rescued the Swedes just nine days earlier. Since it seemed senseless to explore the same region and possibly suffer the same hardships, Charcot decided to continue de Gerlache's work on the western side of the Antarctic Peninsula.

By February 3, 1904, the *Français* had reached the Antarctic Peninsula. Two days later the secondhand engine developed problems and the boiler pipes ruptured. The ship clunked and sputtered into Flandres Bay, where for the next eleven days the engineers hammered, soldered, and patched the old engine.

On February 19, Charcot discovered a deep sheltered inlet on Wiencke Island that he named Port Lockroy. The ship once more developed engine problems just as the weather turned violent. Tiny gale-driven snow granules stung skin and eyes "like fine needles, causing horrible pain," noted Charcot. The explorer persevered, however, and pushed southward to Booth Island. There, the *Français* anchored in an inlet for the winter.

The crew stretched a heavy chain across the inlet's narrow mouth to protect the *Français* from drifting icebergs, which could crush the ship within minutes. Next, the men erected a hut for magnetic observations and dug several deep cellars in the snow to store meat and other perishables. As soon as the ship was winterized, the men devoted most of their time to scientific pursuits, such as geology, zoology, meteorology, and astronomical studies. Meanwhile, Rozo, the cook from Buenos Aires, padded about the galley in his only pair of tattered carpet slippers and served delicious meals, including fresh-baked bread three times a week and croissants on Sundays.

During the winter months the men read and reread old newspapers, listened to lectures, discussed politics, and memorized poetry. For exercise, they strapped on skis and trekked to nearby islands. Although Charcot was not athletic, he tried to master this new sport. "I do not pretend to have small feet but I am not accustomed to their being about nine feet long, and I got into frightful tangles," he wrote, exasperated. "Now and again it seemed to me that I had definitely become one of the problems of the day."

Temperatures plunged in June and July. Most of the time the men remained bundled in heavy sweaters and fur-lined gloves. On the coldest days, they wrapped thick woolen mufflers around their necks and ears and thumped their feet on ice-slicked wooden floors. Hoarfrost grew on walls, ceilings, and inside drawers. Fingers ached from handling canisters and other metal objects.

Slowly, daylight hours lengthened. In late November, Charcot and four companions filled a small whaleboat with provisions for twenty days. The plan was to row to Petermann Island and continue to the Antarctic Peninsula, where they would survey a section of the coast. They maneuvered the boat through the broken pack ice without much difficulty and reached Petermann; however, the trek from there to the peninsula was grueling, for the leads were too narrow for the boat. The men had to drag their overloaded vessel through rotting ice that had the texture of thick slush. They sunk to their knees with each agonizing step. Snow blindness tormented them, with pain like "a handful of peppers in the eyes," noted Charcot. After five terrible days, they reached the mainland and surveyed the region for about a week before returning to the ship.

On Christmas Day the men decorated a cardboard tree with feathers, candles, and cutout tin trinkets. The next morning the ship pulled away from Booth Island and headed south to explore the Biscoe Islands and Alexander Island. On January 15, 1905, the ship struck a rock with such force that "the bow reared up almost vertically," Charcot reported. Frigid water poured in and, predictably, the engine sputtered and misfired, forcing the crew to man the pumps around the clock to keep the ship afloat. Charcot had no choice but to turn northward and hope that the pumps held out until the hole was sealed. Two weeks later the *Français* reached Port Lockroy for makeshift repairs to the gaping twenty-five-foot hole in the ship's false keel.

By mid-February, the *Français* had struggled past the South Shetlands, crossed the Drake Passage, and arrived in Tierra del Fuego. There, Charcot learned that his wife, Victor Hugo's granddaughter, was divorcing him for desertion. Charcot sold his ship to the Argentines, and he and his crew boarded another vessel home.

In France Charcot and his crew were welcomed as heroes, for they had charted over six hundred miles of mainland coast and offshore islands, endured an Antarctic winter, and returned with at least seventy-five crates filled with marine specimens, rocks, and scientific notes. During the next year, Charcot wrote a popular book about the expedition, *Le Français du Pole Sud*.

In 1907 Charcot's second wife, Marguerite, promised in writing never to oppose his expeditions. One year later her words were put to the test, for on August 15, 1908, Jean-Baptiste Charcot made final preparations to sail once more to Antarctica.

"NOTHING SEEMS CHANGED"

Nine men who had previously sailed with Charcot chose to accompany him on his second voyage in 1908. His ship, the *Pourquoi Pas?* ("Why Not?"), named after a favorite toy boat, was a 131-foot schooner that carried the most up-to-date equipment for polar travel. The ship had a 550-horsepower motor, a searchlight, telephones connected by a line that stretched from the crow's nest to the bridge, and a fifteen-hundred-book library. Slow-burning petroleum lamps were mounted in each room for use most of the time; electric lights, powered by an auxiliary eight-horsepower motor, were to be turned on just twice a week. A wine cellar was amply stocked with the finest labels and vintages.

Charcot arrived at Punta Arenas, Chile, on December 1, 1908, and sailed for the South Shetland Islands two weeks later. The ship anchored in a cove on Deception Island near a bustling Norwegian whaling station. Just above the tidal mark on the beach was "a regular hedge of whale skeletons," stripped of blubber and baleen. The sight and stench of so many carcasses sickened many men; the normally blue water of the cove was streaked with red.

On Christmas Day the *Pourquoi Pas?*, decked out with streamers, weighed anchor. With foghorn salutes to the whaling station, the ship sailed through the narrow passage between towering cliffs of rock and turned southeast toward the peninsula. The next day the ship reached Port Lockroy on Wiencke Island. Nothing seemed changed, much to Charcot's surprise. The letterbox on tiny Casabianca Island still stood upright in stark relief on a small rise; the glass vials that contained tightly rolled handwritten messages lay untouched. Three days later Charcot anchored in the same small cove on Booth Island where the *Français* had wintered in 1904. "My eyes are struck by the same familiar objects and the same buildings," he wrote, "my ears catch the same sounds from the rookeries of penguins and cormorants, which give forth the same powerful odor."

At midnight on New Year's Eve, in the profound silence of the rosy Antarctic twilight, "every bell on board, the foghorns and the phonographs gave forth their sounds in a deafening discord to welcome the New Year." Later, only the sharp reports of calving icebergs and the thunder of avalanches roaring down from steep snowfields broke the silence.

On January 3, 1909, Charcot anchored on the eastern side of Petermann Island. The next day, he and two other men set out in a small boat for Cape Tuxen to reconnoiter the surrounding area. Since they expected to return within a few hours, they carried just enough food for one meal. The men finished their work at 10 P.M. and stopped to eat just as snow began to fall. They searched for a lead to follow back to the ship but found none. Charcot guided the launch into narrow fissures that he hoped would open into leads, but each time the unbroken pack ice butted against the boat and squeezed shut the cracks. Even more ominous, the snow changed to sleet that soaked the men's clothes and clogged the boat's motor.

The men took turns hacking through the ice with a shovel, until it slipped from one man's cold white hands into the inky black water. "We laugh at the mishap and at the woebegone face of our good friend," wrote Charcot, "but our already feeble efforts now become almost useless." They battled the ice for a full day and then dragged themselves to shore. Exhausted, the men tried to sleep, but rest eluded them as the cold stiffened their saturated jackets and numbed their feet. Anxiety kept Charcot awake. Many times he walked to the top of the cliff and peered at the ice, hoping to see a fresh channel that would lead them to safety. Each weary trek ended in disappointment.

The next day, however, the floes did move apart. Within minutes the men had jumped into the boat and were picking their way among the broken floes when, suddenly, the motor sputtered and died. "In spite of all efforts, amiable encouragements, and harsh words, it is impossible to start it again," wrote Charcot. They abandoned the attempt at freedom and huddled together inside the boat for warmth.

A prolonged whistle jarred them out of their collective stupor. They climbed to the top of a rock and shouted at the *Pourquoi Pas?* Blessed with a loud voice, Charcot sucked in air, cupped his hands around his mouth, and yelled three times. Within seconds the ship replied with whistle blasts. A short time later the men cheered as the ship slowly but surely broke through the floes. Charcot named the rock where they were rescued Deliverance Point.

As the ship sailed beside the black cliffs of Cape Tuxen, the *Pourquoi Pas?* suffered a tremendous jolt and jerked to a stop, grounded on submerged rocks. The men peered over the side with looks of horror as large pieces of hull floated to the surface. Throughout the day the men shifted cargo from the forward part of the ship to the stern in an attempt to float the ship free. When the tide rolled in, Charcot revved up the engine. Then, with a violent series of jerks and agonized grinding of wood and iron against rock, the ship floated free. Charcot steamed the *Pourquoi Pas?* beyond the Antarctic Circle and entered a vast placid bay dotted with icebergs and broken floes, tenuously seamed with gray slush. Precipitous peaks and crumpled glaciers ringed the bay, and fractured walls of ice spilled into the still water. Charcot named it Marguerite Bay, for his second wife.

Continuing southward, Charcot saw very little animal life. Only a few crabeater seals lolled on the floes, and a pair of skuas scoured the ship's wake for galley scraps. "This is a forbidding country," noted Charcot, "and only at rare intervals does a whale break the silence with its heavy blowing." The ship retreated northward and anchored at Petermann Island on February 3 to prepare for winter.

During the next two weeks, the men erected four research huts and strung electrical wire between the small buildings and the ship. Now, for the first time, researchers had a good, steady light source by which to read the instruments and to calculate results. Great sheets of canvas were stretched over the upper deck, which created additional rooms for the two prized washing machines. Nearby were tubs of ice, ready to be melted over a seal-blubber fire on laundry days.

On February 23, everyone celebrated Mardi Gras with a picnic on Petermann Island. The men painted their noses red and strapped on tropical helmets and other outlandish hats. Bright red longjohns peeked out from trouser legs. To the rousing music of a bugle and clarinet, they stepped in time to the beat of a tin-box drum and paraded in front of unimpressed penguins. Even the cook celebrated the day by pretending to be a famous chef in a world-class hotel. Dinner was a feast of pancakes, and fine wines filled the men's cups. After the meal the men sang rousing songs that drifted across the snowfields to the hills beyond the harbor.

Winter activities included courses in geography, foreign languages, and navigation. There were also meetings of the newly formed Antarctic Sporting Club, which awarded commemorative tin-can medals for skiing excellence. The men who had played the clarinet, bugle, and tin drum during Mardi Gras organized a musical society, complete with Sunday afternoon recitals. Even the chief engineer, who strummed just one song over and over on his mandolin, joined to increase his limited repertoire.

The ship's lone rodent had far fewer distractions than the men during the long winter. After its mate had thrown itself down one of the scuppers, the rat disappeared for two months. Finally, a partially eaten stuffed bird provided ample evidence that it was still alive. "How this poor solitary rat must be bored, and how much he must regret his choice of a ship!" noted Charcot. Choice culinary tidbits were dropped in dark corners and behind crates to sustain the lonely rat.

As winter progressed, Charcot's usually robust health deteriorated. He suffered from shortness of breath, swollen legs, and heart palpitations with the least exertion. By the end of August, symptoms of scurvy were unmistakable in Charcot and other members. He ordered the removal of all preserved food from the menu and substituted penguin, seal, garlic, sauerkraut, and jams. Several weeks later, as the men's health continued to improve, flocks of cormorants arrived to herald the Antarctic spring.

In late October the penguins returned to the island, and the first penguin eggs were gathered for an omelet feast in mid-November. Two weeks later, the *Pourquoi Pas?* sailed from Petermann Island, home for the last nine months. They reached Deception Island the next day and stopped to check the ship's hull, damaged by the black cliffs of Cape Tuxen ten months earlier. A diver examined the *Pourquoi Pas?* and rendered his verdict: the slightest jar could send the ship "to the bottom." Charcot swore the man to secrecy and decided to continue the expedition. On January 6, 1910, the *Pourquoi Pas?* turned southward once more.

Five days later the men sighted Alexander Island. Up in the crow's nest, Charcot peered through his field glasses and studied the horizon. "I seem to see something strange in the southeast," he wrote. To everyone's astonishment, and without explanation, Charcot ordered the men to steer in that direction. Later that afternoon, all of Charcot's doubts evaporated. "Those are not icebergs which lift their pointed summits to the sky; it is a land, a new land, a land to be seen clearly with the naked eye, a land which belongs to us!" he wrote. He named his discovery Charcot Land, in memory of his father. Charcot continued to explore and chart the coastline to about 124° west longitude. On January 22, Charcot turned the ship northward.

In early June the *Pourquoi Pas?* sailed up the Seine toward Rouen after an arduous but rewarding expedition. The villages along the route were decked out with flags of welcome for the Antarctic explorers, who had charted 1,250 miles of coastline and discovered new land. The scientific results filled twenty-eight volumes, each illustrated with some of the three thousand photographs taken during the two years.

The polar regions continued to fascinate the French explorer. One night in September 1936, Charcot, once again commanding the *Pourquoi Pas?*, battled a terrible gale off the Icelandic coast. When the ship wrecked on the rocks of Akranes in the Bay of Reykjavik, Charcot was last seen on the bridge, calmly directing the evacuation of his crew. Only one sailor from the forty-four-man crew survived.

Flying High: The First Flight over the Antarctic Peninsula

Working the family farm did not appeal to Australian Hubert Wilkins. An adventurer at heart, he wandered throughout Europe and America instead of planting crops. In 1912 Wilkins worked as a photographer for the Turkish government during the Balkan War and later joined Ernest Shackleton's 1921 *Quest* expedition. After exploring central Australia for the British Natural History Museum, Wilkins headed to the Arctic. For the next three years, he planned and executed long flights over the Alaskan and Canadian tundra. Within months of his return to Australia, Wilkins headed south under newspaper scion William Randolph Hearst's sponsorship.

On November 6, 1928, he arrived at Deception Island with two Lockheed Vega monoplanes and an Austin car, which had two tires strapped together with heavy chains for each wheel. Several test runs of the planes were made before Wilkins encountered a problem that he had not anticipated: competition for airspace with hundreds of albatrosses and giant southern petrels. The birds were attracted to Deception Island because of whale carcasses decaying in the water and on shore near the station.

Ten days later, Wilkins flew the small aircraft for a very short journey from Port Foster toward the peninsula. Although he hoped that the harbor ice would serve as a runway, it never thickened enough to support the plane. Wilkins and a few men that could be spared from the whaling station had to extend the runway on land. By December 20, 1928, he was ready once more to explore Antarctica from the air.

Wilkins and his copilot flew over Bransfield Strait. Ahead of them was the Trinity Peninsula, with six-thousand-foot mountains rising from the plateau of ice. Aerial mapping was easy when compared to sledging. Forty miles took a mere twenty minutes to survey instead of months on the ground. The two men continued southward above the peninsula for about thirteen hundred miles before they turned back for Deception Island, loaded with sketch maps and photographs.

One more five-hundred-mile flight was taken before Wilkins returned to New York for the winter. Back again in September, he searched for a more suitable landing strip. Smooth, flat ice was found in Beascochea Bay on the Antarctic Peninsula, but after unloading the plane and car, Wilkins had to abort his plans when the temperature climbed to a record 54° Farenheit. Just as his equipment began to sink through the slushy ice, Wilkins and the crew hoisted the plane and car to safety onboard ship. After a series of short flights, Wilkins sailed for home in February 1930. Three years later, he served as technical adviser to an American millionaire's astounding flight.

Lincoln Ellsworth's Excellent Adventures

One of American Lincoln Ellsworth's passions was collecting memorabilia about his hero Wyatt Earp, Marshal of Tombstone, Arizona, and survivor of the OK Corral shoot-out. Exploration and adventure were his other consuming interests. In 1926 the independently wealthy Ellsworth flew with Roald Amundsen and Umberto Nobile over the North Pole, just days after Richard Byrd's questionable conquest. Like Byrd, Ellsworth would look southward to exploit the commercial and scientific potential of the airplane.

Ellsworth purchased a small Norwegian fishing vessel, renamed *Wyatt Earp*, and sheathed the hull in oak and plates of armor. Just before sailing, he packed Wyatt Earp's gun holster into his suitcase and slipped the gunslinger's wedding ring on his finger. Ellsworth then headed to the

Bay of Whales in the Ross Sea region and arrived in January 1934. His Northrop Gamma monoplane, the *Polar Star*, was unloaded on the ice shelf and taken for a brief test flight.

By the next morning, preparations were completed for one of the most ambitious journeys ever attempted. Ellsworth and his pilot, Bernt Balchen, hoped to fly across the continent from the Ross Sea to the Weddell Sea. Suddenly, as they prepared to leave, at 7:30 A.M., the ice shelf began to vibrate with a series of ear-splitting crashes. Terror paralyzed the men as a huge portion of the ice shelf disintegrated into floes and bergs. One of the mechanics stood on a section of what had been the runway; another floe drifted away with a load of supplies. The *Wyatt Earp*'s crew rescued the stranded men and then searched for the *Polar Star*. They found the plane barely floating, its wings straddling two ice floes. The men hoisted the plane onboard and tallied the damage. Ellsworth stared with dismay at the demolished landing gear and knew he would have to abandon his plans.

After returning to San Francisco, Ellsworth changed tactics and decided to begin the flight from the tip of the Antarctic Peninsula and end it at Little America on the Ross Ice Shelf. He arrived at Deception Island on October 14, 1934, and within days the airplane was ready for a test flight. The plane's motor never roared to life, however, for a connecting rod had cracked. Ellsworth sent the *Wyatt Earp* to South America to pick up a new rod he had ordered from New York, and by the middle of November the ship had returned with the critical part.

Once again, Ellsworth was ready to fly; unfortunately, the weather didn't cooperate and he was forced to wait until January 3, 1935, hoping for better weather. On that day the exasperated Ellsworth gave the order to pack up the plane, but just as they were about to leave, the skies cleared. Ellsworth and Balchen scrambled into the *Polar Star*'s cockpit for the start of their epic transcontinental flight. Less than an hour later, Balchen saw dark clouds on the horizon and turned north—to Ellsworth's extreme irritation.

Ellsworth hoped his luck would improve with his third attempt, on November 21, 1935. He and new pilot Herbert Hollick-Kenyon took off from Dundee Island, just north of the Antarctic Peninsula, in clear weather. A faulty fuel gauge forced their return. Two days later the twosome was again in the air, flying south toward the Bay of Whales.

Conditions were perfect as pilot Hollick-Kenyon and the fifty-five-year-old Ellsworth flew over mountain ranges and glaciers never before seen by human eyes. Ellsworth stared in awe at the later-named Eternity and Sentinel Ranges, the peaks as fresh and sharp as newly honed arrowheads. The men photographed every new vista that opened before them on their flight path. Then, like a curtain drawing across a magnificent stage, mist suddenly obscured their view. The men were forced to land after flying about eighteen hundred miles in fourteen hours.

The last five hundred miles of their epic flight to Little America took thirteen days. Blizzards and poor visibility forced the pair to camp four times and wait for better conditions. When the weather finally cleared, the men tossed their paraphernalia into the plane and continued the journey over the western Antarctic ice sheet until their fuel ran out. Hollick-Kenyon landed the sputtering *Polar Star* on the Ross Ice Shelf, about sixteen miles from the unoccupied but well-stocked Little America base.

The men's navigational skills proved to be much better in the air than on the ground, however. Disoriented during whiteouts, the pair wandered for eight days searching for Little America. Finally, on December 15, the men climbed down through a skylight into one of the recently abandoned buildings. They had at last completed their epic twenty-three-hundred-mile journey. They remained there for the next month in relative comfort until they were rescued, first by the Australians and then, four days later, by the *Wyatt Earp*'s crew.

Ellsworth returned to Antarctica one last time. On January 11, 1939, he and J. H. Lymburner flew inland from the Ingrid Christensen Coast and discovered the American Highlands. During the flight, Ellsworth dropped copper canisters that contained declarations of American sovereignty over the 430,000-square-mile area. In the back of the plane, stowed deep inside one of his packs, Ellsworth still carried his Wyatt Earp mementos.

• 9 •

WEDDELL SEA EXPLORATION

The Weddell Sea region extends from the eastern edge of the Antarctic Peninsula southward to the Ronne and Filchner Ice Shelves, and east through Princess Martha Coast. For history buffs, this region is second in significance only to the Ross Sea sector for celebrated—and doomed—expeditions. In 1823 Scottish sealer James Weddell discovered a broad, ice-free area at 34° west longitude and sailed to 74° 15' south latitude in the sea later named for him. Dumont d'Urville tried to repeat Weddell's record voyage in 1838, but the pack ice did not relent that year, or the next for American Charles Wilkes. In 1843 James Ross penetrated the Weddell Sea as far south as 71° 30' before turning northward.

Between 1902 and 1917 three major expeditions, led respectively by Otto Nordenskjöld, William S. Bruce, and Ernest Shackleton, ventured into the Weddell Sea pack ice. Each is a tale of survival against the worst seasonal foe on Earth: the Antarctic winter. Some of the men were stranded and forced to build rough shelters that would withstand howling gales and bitter cold. Many faced the threat of starvation and the fear of uncertainty, yet no one forgot his humanity, even when little remained except hope.

Against All Odds: The Swedish Antarctic Expedition

Polar exploration was a family affair for Swedish geologist Nils Otto Gustaf Nordenskjöld. His uncle, Baron Nils Adolf Erik Nordenskjöld, had discovered the Northeast Passage when he sailed the *Vega* along the northern coast of Europe and Siberia from 1878 to 1879. Otto Nordenskjöld followed in his uncle's footsteps by exploring Tierra del Fuego, Alaska, and east Greenland in the 1890s. An ambitious man, he saw an opportunity to apply his Arctic knowledge to Antarctica's little-explored Weddell Sea region. He invested his own funds and solicited public and private donations to finance the Swedish Antarctic Expedition.

On October 16, 1901, Nordenskjöld sailed from Göteborg, aboard the *Antarctic*, the same whaler that had carried Henryk Bull's 1894–95 expedition to the Ross Sea. Carl Anton Larsen, who led a whaling expedition to the eastern side of the Antarctic Peninsula in 1892–94, captained the ship for the seven scientists and twenty-two crew members. The eighth scientist, Dr. Gunnar Andersson, planned to join the expedition when the *Antarctic* returned to the Falkland Islands after dropping off Nordenskjöld and his wintering-over party. The following spring the ship was to fetch Nordenskjöld and continue exploring the eastern side of the Antarctic Peninsula. That was the plan on paper, but Nordenskjöld had underestimated the power of the Weddell Sea pack ice.

THE SNOW HILL MEN

The *Antarctic* skirted the South Shetland Islands in mid-January 1902 and crossed Bransfield Strait, sailing southward along the peninsula's western coast to the Antarctic Sound between Joinville Island and the northern tip of the peninsula. The men landed on Paulet Island and watched Adélie penguins "chatting" on ice floes. They then crossed Erebus and Terror Gulf to Seymour Island, known for the extraordinary fossils Larsen had collected in 1893. The men landed near Penguin Point and stashed boxes of sugar, margarine, salt, matches, petroleum, shoe soles, and dried vegetables.

They continued southward until a hundred-foot wall of ice blocked further passage and forced the *Antarctic* to retreat through broken pack ice. Navigating a wooden-hulled ship through a maze of floes was difficult and dangerous. Captain Larsen wrapped an old brown comforter around his shoulders, climbed to the *Antarctic*'s crow's nest, and attacked the problem with the mind of a dedicated billiard player. Scanning the leads and floes, he shouted his orders for where the bow of the ship should strike. The shock of wood ramming ice sent shivers from the hull to the top of the mainmast and set the crow's nest swinging. "Not only was that floe pushed aside in the right direction to a place where there was room for it," noted Otto Nordenskjöld, "but, in consequence of the recoil, the adjacent floe came into movement too, and made way for the vessel."

On February 9, they once again passed ice-free Seymour Island. Although there were plenty of landing places, Nordenskjöld hoped to find an even better island on which to winter. He chose Snow Hill Island when the men discovered magnificent fossils in the rock piles at the bottom of the cliffs. He and five companions unloaded crates of food and equipment, sections of the prefabricated hut, and sledge dogs eager for a romp on solid ground. As the *Antarctic* steamed away to the Falkland Islands to winter, the six men stood on shore, silent and overwhelmed by the vast landscape.

Hard work suppressed fear and self-doubt. They quickly erected the research hut and the cozy twenty-one-by-thirteen-foot house. By the first week in March, pictures hung on the bare walls, an embroidered cloth covered the table, and red checked curtains tied back with gold tassels framed the windows. Temperatures inside the house averaged about 60° F, and Nordenskjöld padded across the wooden floor in coarse goat-hair socks.

When the sea ice thickened, the men took brief camping trips to nearby islands. Without the protection of a sturdy house, they quickly learned to read ominous weather warnings: warm southern winds, a rapidly falling barometer, a strange light in the southwest. Just before the most ferocious storms, the air became saturated with electricity. Hair crackled and fingertips tingled; even the equipment sparked and glowed with an eerie green light, like St. Elmo's fire. "We had scarcely entered the tent ere the storm came on," noted Nordenskjöld. That particular storm turned into a fierce tempest with hurricane-force winds, the first of the brutal winter.

Wind terrified the men, especially during the night. The walls and door of the house bowed inward against its sustained fury, and the windows rattled incessantly. Water in the washbasin splashed on the floor. To Nordenskjöld, the house behaved like a ramshackle sleeping car on a train, hurtling through the night on a poorly constructed track. Outside, the fine snow swept along the ground like a foamy tide without ebb.

When spring arrived in early October 1902, Nordenskjöld and two other men loaded two sledges and harnessed five dogs for a trek toward the eastern coast of the Antarctic Peninsula, or Oscar II Land. At first the journey over the sea ice was easy, but as they neared land the ice became torturous. They passed over a network of crevasses connected by snow bridges. "Before I could properly grasp the situation, I sank to the armpits," noted Nordenskjöld. He scrambled out just in time. The fragile snow bridge collapsed, revealing a crevasse large enough to swallow men, dogs, and sledges without a trace. They reached the edge of the Antarctic Peninsula and pitched the tent at the foot of the weathered rocky headland, "a mass of mighty blocks" fractured by frost from the cliffs. The next day they started back to Snow Hill and arrived on November 3, after trekking four hundred miles.

Throughout November several men at a time climbed a nearby hill and gazed at the unbroken pack ice, hoping for a ferocious storm to disperse it. As the days passed and the ice remained unchanged, the men skied to the nearby islands to help pass the time. On Seymour Island in early December, Nordenskjöld found the scattered fossilized remains of a six-foot penguin and of plants such as firs, ferns, and southern beech. "Could it have been a dream which led me to choose just these tracts for my field of labor?" he wrote. The fossils linked Antarctica to South America, Africa, and Australia during the same geologic time period. Therefore, he reasoned, Antarctica must have been part of a great southern continent.

As January faded into February with no sign of the ship, Nordenskjöld forced the men to think about the unthinkable: how they would survive for another winter. On February 19 the temperature plunged, and a web of new ice seamed the floes together for as far as the eye could see. The men lost heart and hope as they prepared to slaughter penguins for enough meat to last another year. "Much as I had longed for the *Antarctic* during the course of this summer, I never did so more than during these few days . . . that the vessel would come for the penguins' sake, before we had slain all those we needed for our winter supply."

STRANDED! THE HOPE BAY MEN

After settling Nordenskjöld and his men at Snow Hill in January 1902, Larsen captained the *Antarctic* to the Falkland Islands for supplies, and to pick up Dr. Gunnar Andersson. On April 11, as the ship pulled away from the harbor, "the girls of Port Stanley waved a last farewell," wrote Andersson. At South Georgia the men mapped the island's deep bays and fjords until mid-June, at which time they sailed for Tierra del Fuego. In early November, Captain Larsen headed south toward the Antarctic Peninsula.

Pack ice, much farther north than usual, slowed the ship. Throughout most of November Larsen rammed the floes and squeezed the *Antarctic* through ever-narrowing leads toward the South Shetland Islands. After a brief visit at Deception Island, the men sailed across Bransfield Strait and charted sectors of the Antarctic Peninsula. Maps completed, Larsen now guided the ship toward Snow Hill Island in early December to pick up Nordenskjöld. Andersson prepared a welcome for Nordenskjöld by decorating the expedition leader's cabin with fresh beech branches from Tierra del Fuego.

The route around the tip of the peninsula toward Snow Hill Island was congested with floes and bergs, unyielding and immune to Larsen's ramming tactics. By the end of December, he realized that the only way to reach Snow Hill was to send a few men on foot from Hope Bay, at the tip of the Antarctic Peninsula, to the island. On December 29, Gunnar Andersson, Samuel Duse, and Toralf Grunden rowed to shore in the shadow of Mount Bransfield to begin their two-hundred-mile trek to Snow Hill Island.

After storing extra food and supplies, the trio climbed to the top of a high slope and watched the *Antarctic* pick its way through the floes. The next day they traveled along the eastern edge of the Peninsula. Ahead, expanding sea ice in one bay had captured numerous icebergs within its seamless white mantle. Here was "a gigantic snow-clad city with houses, and palaces in thousands, and in hundreds of changing, irregular forms—towers and spires, and all the wonders of the world," wrote Andersson. The men named the spectacular scene Bay of the Thousand Icebergs.

They tackled the iceberg labyrinth and skied toward Vega Island. After an exhausting fifteen hours wading through slushy ice, the men reached solid ground. Shivering with cold, they pitched the tent, cooked dinner over a cheerful Primus stove, and crawled into stiff reindeer sleeping bags. "Fatigue soon rendered us insensible to cold and moisture," noted Andersson.

The next day, January 3, 1903, "stockings, mittens, and nightcaps" dangled from an Antarctic-style clothesline: a rope stretched between two skis. Two days later, the men climbed to the top of Vega Island and stared at the wide stretch of open water that blocked their way to Snow Hill but would allow the *Antarctica* easy passage to Nordenskjöld's base. The trio headed back to Hope Bay to wait for the ship, according to prior arrangements. Blizzards slowed them down, but on January 13 they camped "amid the screaming crowd of penguins" at Hope Bay. They settled into a routine of work, meals, and sleep while they waited for the ship. As the days turned into weeks, the three men faced the inevitable: something terrible must have happened to the *Antarctic*. Now, they would have to endure the brutal winter alone.

They constructed a stone hut with walls as high as a man; the sledge and a tarp formed the roof. Winter's meat supply nested right outside the canvas-flap door, for thousands of Adélie penguins surrounded the hut. Inside, the men pitched their tent for more protection against the cold Antarctic winds and carpeted the ground with penguin skins. They swept away grime with a giant petrel wing.

Throughout the long winter, the stranded men rotated all chores, including taking turns as cook every third day. Whenever Grunden prepared the meals, the other two men were treated to the English and German songs he had sung as a child on rowdy Australian streets for the pennies passersby tossed at his feet. Now, as seal blubber crackled in a frying pan he had made from a large tin canister, Grunden carefully cooked penguin steaks and sang about life on the high seas.

Occasional meals of fish, caught with a modified brass-buckle hook, supplemented their monotonous diet of penguin soup or canned herring. Porridge was served twice a week and was "a great luxury in the midst of our chronic hunger for carbohydrates." They boiled moldy barley with a bit of sea water for flavor, and added browned seal-fat squares to the pot as a special treat. Andersson declared that no porridge in Sweden "ever tasted so well as did this wretched dish in our stone hut."

Evenings were the best time. As the blubber lamp cast flickering shadows on the rough stones, the men shared memories about Swedish nights when "winter darkness lay heavy o'er the land." They entertained each other by retelling Dickens' classic stories and acting out the parts. The men had no books but read and reread the labels on cans of food, the words soon smudged by sooty fingers.

During those long dreary days, the men learned that the power of friendship "subdued the dark night of isolation and extreme distress." Little courtesies, such as thanking the cook for each meal, helped the men remember their humanity. Andersson was, perhaps, the most optimistic. He cheerfully noted that now "we could cut our toenails without removing our boots" because of gaping holes.

Early September brought longer days and rivulets of melted snow that dripped through the frayed canvas roof. Unable to endure inactivity any longer, the men prepared to ski to Snow Hill. To protect their eyes from glare, they carved snow goggles with narrow slits from the staves of oak barrels. They packed enough penguin steaks to last several weeks and repaired their boots with seal and penguin skins.

On September 29, 1903, they printed a message on a wooden plank and inserted a rough map of their route into a flask. Then, the three men with soot-blackened faces and grease-coated hair crawled out of the "Crystal Palace" and began their last great journey together.

STRANDED II! THE PAULET ISLAND PARTY

As Andersson, Duse, and Grunden were rowing toward Hope Bay on December 29, 1902, botanist Carl Skottsberg watched from the *Antarctic*. He felt no misgivings and believed that within days Nordenskjöld's group at Snow Hill Island and the Hope Bay trio would be back on board. Captain Larsen continued to push toward Snow Hill, but the ice did not relent.

On January 10, 1903, Skottsberg and several others had just sat down for a card game when "the ship began to tremble like an aspen leaf, and a violent crash sent us all up on deck to see what the matter was." The pack ice had shifted, and now the floes' horrendous pressure lifted the ship four feet out of the water. The men thought they were safe as long as the vessel remained above the crushing pressure. However, during that night, after another series of bone-rattling crashes, the *Antarctic* settled back into the sea. Water poured through a hole in the stern, and the men just barely kept pace with the pumps. Larsen still tried to maneuver the ship through the pack ice, but the *Antarctic* remained its prisoner.

By February 12, not even six pumps could contend with the rising water. Larsen ordered the men to abandon ship as the pack ice closed in once more. Barrels of flour, sacks of potatoes, and large tin canisters of preserves were heaped upon the ice beside mattresses, tools, and planks of wood. "Even the ship's cat was carried down in a state of terror," wrote Skottsberg. With just one bundle of pressed plants under his arm, he, too, climbed down to the ice.

The men huddled on the floe and stared at the doomed ship as the sound of the pumps grew fainter. "She is breathing her last," wrote Skottsberg. "Now the name disappears from sight. Now the water is up to the rail, and bits of ice rush in over her deck. That sound I can never forget, however long I may live.... The crow's nest rattles against the ice-edge, and the streamer with the name *Antarctic* disappears in the waves.... She is gone!"

The *Antarctic* sank about twenty-five miles from tiny Paulet Island. During the next two weeks, the men dragged their goods from floe to floe with a small whaleboat and hacked paths through high-pressure ridges. Leads opened and closed at the whim of wind, pressure, and current. Large floes splintered into smaller pieces, separating men from goods and from each other. Mattresses, skis, a canoe filled with woolen clothes, and most of their salt supply drifted beyond reach.

On February 28 the men rowed six hours before they finally stumbled ashore on Paulet Island. "What joy, what happiness there was in once more treading firm earth, after 16 days' incessant strife," noted Skottsberg. Their refuge was a small extinct volcanic island with a three-mile circumference. A lake in the center with yellowish-green water lapped the base of the surrounding scree-covered hills.

The men decided to build a hut at the bottom of a slope because the location provided the perfect building material: flat basalt stones with an even thickness. Within days the double-walled hut rose from a sturdy foundation. Pebbles filled the space between the walls and penguin guano cemented gaps between the stones. Skottsberg referred to the hut's unique design and building materials as "Paulet Island architecture." When finished, the twenty-man hut measured twenty-two by thirty-four feet. The men spread their sleeping bags over a layer of pebbles. To soften the beds, several men stuffed penguin skins under their bags. "If anyone happened to move the bundle," wrote Skottsberg, "a terrible odor at once filled the room."

As the winter deepened, the men's lives centered on eating and sleeping. During the brief daylight hours, they hunted seals to supplement meals of stewed, fried, or steamed penguin. Adélie soup was the least-appetizing food the cook ladled into tin mugs. When the iron kettle was set before the men, the thin brownish-yellow soup didn't smell any better than it looked. "Still, there are many who cast a look of regret toward the empty kettle as it is carried out," noted Skottsberg.

During the evenings, someone told a story or read aloud from one of the few books they had salvaged from the ship. Humor and a "hang-gallows wit," a guise for their desperate situation, flourished inside "Paulet Cottage." Dreams involved food and rescue. "Why, we could dream through a whole dinner, from the soup to the dessert, and awaken to be cruelly disappointed," lamented Skottsberg. Still dreaming, the happy men boarded relief ships as rescuers patted their grime-covered backs.

In their dreams they survived their ordeal and boarded relief ships, all homeward bound, but reality was another matter. "Death was the one guest who could reach us," wrote Skottsberg. On June 7, 1903, seaman Ole Wennersgaard died of probable heart disease after weeks of moaning softly throughout the nights. His sleeping bag served as his coffin, a deep snowdrift his grave.

Midwinter Eve, June 23, brought a rice-porridge dinner and fine weather with outside temperatures rising to 31° F. The men spilled out of the hut and spent hours wandering around without the misery of cold feet. "Even the cat ran about as if she was mad," noted Skottsberg, "and enjoyed life to the utmost."

July faded into August, and the sun rose higher in the sky with each day. Spring storms carried a hint of better times to come during September; in October, the penguins and seals respectively announced their return with full-throated kazoolike noises and sanguine indifference. One day, a few weeks later, Skottsberg heard a wild commotion. When he peeked out of his bag, he saw a man standing in the middle of the room, grinning. "His hat is full of eggs; large, white, round penguin–eggs! How we shout and laugh all together; we possess a poultry yard worth having!"

The once-solid pack ice soon broke up into a maze of leads and floes. On October 31, Larsen and five others dragged the small boat beyond the breaking waves and climbed aboard. The men on shore cheered and waved until the boat, headed toward Hope Bay, disappeared among the floes.

STRANGE ENCOUNTERS AT CAPE WELL-MET

At Snow Hill, Otto Nordenskjöld was unaware that three men had been stranded at Hope Bay, or that Larsen and the *Antarctic* crew were forced to winter on Paulet Island. When the leads disappeared into a seamless white sheet, the Snow Hill men prepared for a second winter.

Although they still had coal, the men preferred to burn seal blubber because it generated more heat and less smell. The biggest problem was not the cold but the damp, for vapor condensed on the walls and froze. Mold and mildew grew on diary pages, crept into mattresses, and ruined wool rugs. The once-crisp red checked curtains now hung limp and gray. When the last candle flickered out, Nordenskjöld hung a blubber lamp that lit the room too well. Its clear, bright light illuminated dank walls covered with sticky cardboard. Dirty clothes hung from ropes and fixtures; everything made from iron was coated with rust. Sometime during the winter, a small pea had rolled from a plate and lodged among the damp blankets, where it had taken root. The seedling grew a long stalk with pale, anemic yellow leaves, untouched by the sun.

At the end of September, Nordenskjöld and Ole Jonassen packed the sledge and started on a trek to Vega Island. Once there, the men planned to climb a prominent headland to check ice conditions in Erebus and Terror Gulf. As they approached the island on October 12, 1903, they spotted strange shadows and movement. "It was as though a premonitory feeling told me that something important and remarkable awaited us there," Nordenskjöld later wrote. At first he convinced himself that the three black specks in the distance were penguins on the march, but when Nordenskjöld peered through his binoculars, his hands trembled as the three blurred black spots snapped into crisp focus. They were not penguins but "men with black clothes, black faces and high black caps, their eyes hidden by peculiar wooden frames." So stunned were both groups that they resorted to stiff formal greetings in English until the Hope Bay men said their names: Duse, Andersson, and Grunden.

The Hope Bay trio had walked away from their stone hut on September 29. The next day a ferocious storm had engulfed the men, forcing them to remain inside the tent. The reindeer sleeping bags, once soggy with the spring thaw, froze and looked like fur-covered coffins.

The storm didn't subside until October 2. For the next few days, Andersson, Duse, and Grunden toiled across the ice until they were close to Vega Island. Another violent sleet storm soaked them to the skin and streaked the grime on their faces. Clothes stiffened on their bodies like plates of armor, and their penguin-skin boots weighed down their feet like blocks of wet cement. After three grueling days, the trio reached Vega Island and nursed frostbitten toes back to life. On October 11, they collected rock samples and dried their sleeping bags in the warm sunshine. The next day, they stopped to watch two strange seals moving upright across the ice. "A delirious eagerness seizes us," wrote Andersson. "A field-glass is pulled out. 'It's men! It's men!' we shout."

Everyone talked at once as the reunited men headed back to Snow Hill and continued to share stories and answer questions for the next two weeks. One question in particular haunted the men: What had happened to Captain Larsen and the *Antarctic*?

MORE PENGUINS ON THE MARCH?

Unknown to any of the Snow Hill men, Captain Carl Larsen and his five-man crew rowed away from Paulet Island toward Hope Bay on October 31. When the wind freshened, Larsen set sail and they flew "at a smacking pace" among the floes. That night Larsen and the men camped on the ice, but wind and a high sea made for a miserable time. Twice the floe nearly upended, almost sending the men to their deaths. The next night, as they rowed in the bright moonlight, the icebergs glistened like a white-flowered garden at midnight.

Although progress was slow, they reached Hope Bay on November 4, just as a violent storm struck. They took refuge in their sleeping bags, but "the shriek of the penguins, the howling of the storm, and the thunder of the sea" kept sleep at bay. Later, they found the stone hut and the map inside the flask that showed the route the three men had taken toward Snow Hill.

On November 7 at 4 A.M., the men pushed the boat into the sea and began the last leg of their journey. They rowed until 2 A.M. the next day and rested on the edge of thick pack ice. Since they could go no further in the boat, the last fifteen miles had to be covered on foot. "It was a toilsome and troublesome march," noted Larsen, "for the snow was so loose that every now and then we sank into it up to the knee."

Meanwhile, the *Uruguay*, an Argentine rescue ship, plowed through the ice near Joinville Island toward Snow Hill. Before sailing from Tierra del Fuego, Larsen had informed the Argentines that Nordenskjöld and his men were wintering at Snow Hill; since nothing had been heard from the *Antarctic*, Lieutenant Julian Irizar organized his country's rescue efforts.

On November 8, 1903, the Snow Hill men stared at four black objects moving across the ice. Within minutes the specks grew into humans as the men rushed down the hill to greet the newcomers from the *Uruguay*. Nordenskjöld's joy was dampened only when he learned that Lieutenant Irizar had heard nothing about Larsen and the crew of the *Antarctic*.

Later that same evening, Nordenskjöld sat at his small desk and finished his final report while the other men hurriedly packed a few tattered clothes and notebooks. Suddenly, the dogs created a ruckus, barking first and then howling. Annoyed, the men walked outside to see what was bedeviling the dogs. Irritation quickly changed to puzzlement and then to joy as a group of men approached the hut. It was Larsen and his boat crew, tired but happy after their fifteen-mile trek over the ice floes. "No pen can describe the boundless joy of this first moment," wrote Nordenskjöld.

Two days later at Paulet Island, Skottsberg was restless inside the stone hut. He crawled out of his sleeping bag a few minutes before 4 A.M. and went outside. He studied the sea and then searched the horizon. Finding no reason for his uneasiness, Skottsberg tried to sleep. Suddenly, a horn blast broke the silence. When the sound was repeated, he jumped out of his bag and thumped the other sleepers awake. " 'Can't you hear that it is the boat—the boat—THE BOAT!' Arms wave wildly in the air; the shouts are so deafening that the penguins awake and join in the cries; the cat, quite out of her wits, runs round and round the walls of the room; everybody tries to be the first out of doors, and in a minute we are all out on the hillside, half-dressed and grisly to behold... we shall see home again."

The men from the *Antarctic* boarded the *Uruguay* in haste. As the ship slowly sailed away from the island, Skottsberg wrote, "I cannot take my eyes away from it. I can hardly grieve that our prison doors have been opened, and yet it is with a sense of sadness and of regret that I see Paulet Island disappear behind the dazzling inland ice—disappear maybe forever. Has it not been my home?"

Otto Nordenskjöld received fame, but not fortune, when his two-volume summary of the expedition was published. He remained in debt for the rest of his life. Although he never returned to Antarctica, Nordenskjöld led an expedition to Greenland in 1909. He taught geography at Göteborg until his death in 1928.

Captain Carl A. Larsen returned to South Georgia and established Grytviken, the first shore-based whaling station, in 1904. Twenty years later, he took the *Sir James Clark Ross* to the Ross Sea. As the ship approached the edge of the pack ice, Larsen suddenly collapsed and died.

Bravehearts and Bagpipes: The Scotia *Expedition*

William S. Bruce shunned publicity. An intensely private and patriotic man, he was born in London in 1867 but considered Scotland his home and country after attending Edinburgh University. Several years after graduation, Bruce sailed with the 1892–93 Dundee Whaling Expedition, serving as surgeon and naturalist. Although in 1900 Robert Scott offered Bruce a place as naturalist with the 1901–03 *Discovery* expedition, the Scotsman decided to lead his own venture to obtain oceanographic material in the Weddell Sea and to map the South Orkney Islands during the winter.

Bruce tried to secure funding from the British government and London financiers but was turned away in favor of Scott, who was endorsed by the influential Sir Clements Markham. His luck changed, however, when the wealthy Coats brothers in Edinburgh donated a substantial sum. From that point on, Bruce sought only Scottish support.

In 1901 he purchased an elegant Norwegian whaling boat, which he renamed *Scotia* in honor of his adopted homeland. Captain Thomas Robertson, who had commanded one of the Dundee whaling expedition's ships, volunteered to return to the Weddell Sea. A twenty-five-man crew from Peterhead, Aberdeen, Lerwick, Dundee, and other Scottish towns signed up for the two-year expedition.

On November 2, 1902, Gilbert Kerr, dressed in full Scottish regalia, played *Auld Lang Syne* on the bagpipes to the small crowd on the wharf. With Scotland "emblazoned on its flag," the *Scotia* sailed into the Firth of Clyde. The piper played until the green hills dissolved into a soft mist.

The men arrived in Port Stanley, Falkland Islands, on January 6, 1903, and departed three weeks later. As the *Scotia* passed the Cape Pembroke Lighthouse, the lamp flashed good wishes for a safe journey to Bruce and his crew. Two days later a terrible storm with hurricane-strength winds buffeted the little ship. In the heavy seas the *Scotia* rolled more than forty degrees from the horizontal, accompanied by the clatter of crockery in the galley and shattered glass bottles in the small laboratory.

Saddle Island in the South Orkneys loomed into view on February 4. Bruce and the crew were the first men to visit since 1838, when Dumont d'Urville sent a small boat and several men to gather penguins for the galley kettles. Like d'Urville, Bruce battled pack ice, moving through a maze of floes, many with resident crabeater seals and chinstrap penguins.

On February 18, the *Scotia* crossed the Antarctic Circle and reached 70° 25' south four days later. Suddenly, without warning, the temperature plunged and the Weddell Sea congealed, glistening like black silk. For six days the men fought the freezing sea, and then turned northward. The South Orkneys loomed on the horizon in March. "The whole prospect was gloomy and forbidding in the extreme," noted J. H. Harvey Pirie, "the land being enveloped in a dense pall of cloud." Four days later, the *Scotia* dropped anchor in a large but protected bay on the south side of Laurie Island. During the first week in April, a violent storm pushed pack ice into the bay, imprisoning ship and men for the winter.

The men referred to the extensive shore at the head of Scotia Bay as the Beach. They constructed a small stone building, named Omond House, to store their meteorological equipment, and built tiny Copeland Observatory with scraps of wood painted cherry red. On April 19, Bruce tried to operate the first movie camera in Antarctica by shooting footage of the sheathbills eating kitchen scraps, but the film jammed. These opportunistic birds, however, quickly made friends with the cook and remained near the ship throughout winter and "prospered exceedingly," noted Bruce.

The men kept busy, too, cleaning seal and bird skeletons, inspecting fishing lines, and collecting rocks from the glacial moraines. During a severe cold snap that stopped all work, they watched "vast droves of penguins heading toward the north," wrote naturalist R. N. Rudmose Brown. Winter was fast approaching.

During the bitter cold months, work parties surveyed small islands, camping out for several days at a time. The work was grueling, but knowing that hot cups of tea awaited them at day's end quickened their steps and lightened their tasks. The memories of tea inside the tent were for Pirie most vivid, "how both hands having clasped the cup so as not to lose any heat, the warm glow gradually spread and spread, till at last even the toes felt warm ere the cup was drained. Truly it was a cup that cheered." After their dinner the men worked by candlelight, plotting the results on paper.

One expedition member, chief engineer Allan George Ramsay, could not participate in surveying trips because he suffered from severe heart disease. On August 6, he died with Bruce by his side. Two days later a small procession of men slowly pulled the coffin-laden sledge over the ice to Laurie Island. There, in the shadow of the hill that was named for him, Ramsay was interred. Piper Kerr played "The Flowers of the Forest" on the bagpipes, and the men did not return to the *Scotia* until the last mournful notes were lost among the snow-covered hills. Later, they gathered bundles of bright green moss to cover the grave.

Two weeks later Bruce took a long walk and noted with surprise vast numbers of Adélie penguins marching over the ice toward the islands. Cormorants, also called shags, cavorted above the widening leads before building nests on the sea-facing cliffs. "Spring is upon us—the birds proclaim it," noted Bruce.

By the end of August, female Weddell seals had given birth. The Adélies, "the most pugnacious animals," noted Brown, ended winter's silence with raucous squawks. "The only hubbub at all comparable to that of a penguin rookery is the shrill clatter and chatter that rises at a dinner-party," he added. The less-noisy giant petrels "hovered around like black ghouls, to see what carrion they could pick up," Bruce remarked in the *Scotia*'s logbook. On October 21, Bruce filmed Adélie penguin antics near Cape Martin. The short movie was, in his words, "tolerably successful," considering the subject matter.

BRAVING THE UNKNOWN

On November 22, 1903, the pack ice retreated from the bay, freeing the *Scotia* from its grip. Less than one week later, the ship sailed northward.

Six men had volunteered to remain on the island to continue hourly meteorological observations until the *Scotia*'s return.

Bruce reached Buenos Aires on Christmas Eve, and for the next few weeks tried to give Omond House to the British. When officials refused to assume responsibility for Antarctica's first meteorological station because of associated costs, Bruce offered Omond House to the Argentines, who accepted with alacrity. Three Argentines boarded the *Scotia* for the journey back to the South Orkneys to take over the meteorological station on Laurie Island.

When the ship sailed into Scotia Bay on February 14, 1904, the shore party hoisted the Scottish standard to welcome Bruce and the crew back to Omond House. Their stay was brief, for Bruce was anxious to explore the Weddell Sea as far south as possible. One man and the cook volunteered to spend the winter on Laurie Island with the Argentines.

By March 3, the *Scotia* had penetrated the floes as far south as 72° 18' before thick pack ice blocked further progress. The men checked the depth of the sea and discovered unexpectedly shallow water. This reading "sent the skipper up to the crow's nest with a run," noted Pirie. Cheers rang out over the ice when Captain Robertson shouted, "Land ahead!"

The men saw an impressive 100-to-150-foot ice cliff that stretched for many miles from the northeast to southwest. "We found it to be a lofty ice barrier similar to that first discovered by Ross on the other side of the Pole," wrote Pirie. Although birds had been very scarce during the last few days, giant petrels now glided beside the ship and snow petrels fluttered like white butterflies above the mammoth tabular bergs. The crew dredged the ocean floor and examined rocks composed of granite, quartzite, and slate, all characteristic of a very old landmass. Bruce named the discovery Coats Land, for the two brothers who had contributed generously to the expedition.

The *Scotia* skirted the ice shelf for about 150 miles southwest. On March 7 a blizzard engulfed the ship, and the sea turned into icy slush as the temperature dropped. The vessel took refuge in a large cove in the ice shelf. "The sludge drove up and became closer and thicker," wrote Bruce, "and the hope of extricating the ship became less and less." Pack ice pressed against the ship's sides and the *Scotia* "shivered from stem to stern, ground and creaked in a most alarming degree."

Pirie worried that ice would bury the ship, "a fate which has overtaken many a ship ere now."

The icebound scientists passed the time by studying the effects of music on emperor penguins. Piper Kerr, in full Scottish regalia, disembarked onto the ice and tethered an emperor to his leg with a leash. "He played on his pipes," wrote Pirie, "but neither rousing marches, lively reels, nor melancholy laments seemed to have any effect on these lethargic, phlegmatic birds, there was no excitement, no sign of appreciation or disapproval, only sleepy indifference."

On March 12, a thin crack in the ice widened enough for the *Scotia* to plow through, if the ship could reach it. They tried to blast the slabs of ice to smithereens with gunpowder. Then, like Newfoundland sealers beset by Arctic ice, the men jumped simultaneously on portions of the ice. "Nature smiled at our puny efforts," wrote Bruce.

Hours later, Captain Robertson heard a crack and shouted, "She's free!" Masses of slush churned up from beneath the ship as the vessel settled into the water, for the slabs had heaved the *Scotia* four feet above the surface. The men hoisted the Scottish standard and the Union Jack as they calculated their latitude: 74° 01' south.

They sailed northward into the Furious Fifties, the winds shrieking like "the dismal wail of a lost soul," noted the seasick Pirie. The sturdy ship arrived in Kingstown Harbour, Northern Ireland, on July 15, 1904, the end of a thirty-thousand-mile journey. Foghorns blared and the crowd cheered the Scottish heroes. During the endless round of dinners in their honor and lectures to the public, a few men longed to return to Antarctica. "Ay, the ice had cast its indescribable mysterious glamour over our souls," wrote Pirie.

The *Scotia* was sold to pay the expedition's many debts. Beginning in 1912, just months after the *Titanic* sank, the stout little vessel tracked iceberg movements in the North Atlantic Ocean for several years. During World War I, while hauling a load of coal, the *Scotia* caught fire and sank.

William S. Bruce died an impoverished man in 1921. One year before his death, he presented his books and charts to several prestigious organizations, including the Royal Scottish Geographical Society. Bruce was cremated and his ashes were scattered over Antarctic waters, just south of South Georgia.

Catastrophe on an Ice Shelf: Filchner's Expedition

In January 1912, the *Deutschland* skirted a formidable ice shelf deep in the heart of the Weddell Sea. Wilhelm Filchner, the expedition leader, stood on deck to gaze at the alabaster wall that would later be named for him. Behind this barrier, the snow-covered land rose in gentle folds like a rumpled white blanket covering a sleeping body. The ship anchored beside a massive grounded iceberg that was wedged into the ice shelf. This was the site that Filchner selected to spend the winter in preparation for sledging expeditions the following spring. Although he had originally planned to cross the continent from the Weddell Sea to the inner shore of the Ross Sea and then board a second ship, he had only enough money to finance the *Deutschland*.

Work on the station was almost complete when disaster struck on February 17. Unusually high spring tides had floated the iceberg, and the upward movement had fractured the snow-filled seams that bound it to the ice shelf. The entire iceberg with the expedition's winter base had rotated about sixty degrees and was drifting out to sea. Fresh fissures opened near the building while the men frantically dismantled it to salvage as much as possible. Filchner tacked an explanatory note on a plank of wood near the remnants of the station and sailed into the thickening pack ice. The *Deutschland* was soon trapped.

The men built huts on the ice and passed a pleasant winter. By the beginning of October, wide stretches of open water surrounded the ship. The pack ice continued to deteriorate and on November 26, the ship plowed through the last few feet of ice to freedom. Filchner and his men reached South Georgia in December and arrived in Germany in 1913.

High Hopes: Shackleton's Endurance *Expedition*

When news about Wilhelm Filchner's aborted attempt to cross Antarctica hit London, Ernest Shackleton once again felt the lure of the white continent, the same force that had drawn him in 1902 and 1908 to do what no man had done. Filchner's original plan, with its sense of drama, appealed to him and he believed it would capture the public's imagination. This time, Shackleton declared, he would lead the first transantarctic traverse from the Weddell Sea to the Ross Sea, a distance of eighteen hundred miles. "Every step will be an advance in geographical science," he promised his backers. The British, still grieving for the deaths of Robert Scott and his men in 1912, donated money to the charismatic Irishman, a leader who quoted Robert Browning's poetry and spun humorous yarns by the hour.

To succeed with his venture Shackleton, like Filchner, needed two ships, each with enough supplies to last two years. Shackleton purchased the *Polaris*, a 144-foot steamship owned by Belgian explorer Adrian de Gerlache, and renamed it *Endurance*. Then he bought the *Aurora* from Douglas Mawson, recovering from his disastrous expedition to Commonwealth Bay.

Shackleton's plan was bold. With the *Endurance*, he would penetrate the heart of the Weddell Sea and establish a base on Coats Land to winter. Meanwhile, the *Aurora* would sail to the Ross Sea, and a party of men would winter at Hut Point on Ross Island. In early spring, both parties would trek inland from the opposite ends of the continent to rendezvous at the top of the Beardmore Glacier.

When Shackleton announced the expedition's goals in January 1914, more than five thousand would-be adventurers applied for positions. Even the anticipation of war didn't deter the eager young men. "The Boss" selected fifty-six for the adventure of their lives.

On August 4, 1914, just days before departure, Shackleton read the order for general mobilization on the front page of every newspaper. War was no longer a threat but a reality. Although he offered his two ships, supplies, and men to Great Britain's war effort, the admiralty wired a one-word response: "Proceed" and Winston Churchill wrote his thanks. Four days later, the *Endurance* sailed from Plymouth, leaving behind a world soon to be engulfed in violence and horror.

SQUEEZED TO DEATH

On October 26, 1914, the *Endurance* sailed from Buenos Aires toward South Georgia, which now supported several whaling stations. Although mist shrouded the island, the stench was pervasive. "We were able to navigate a course by the aid of our noses," wrote laconic Frank Hurley, Australian photographer for the expedition, "for derelict carcasses, stripped of blubber, drifted about the greasy waters."

As the ship steamed toward Grytviken, the pristine mountains that flanked the buildings belied the hellish operations near the shore. During their brief stay, Shackleton learned from the whalers that ice conditions further south were the worst in recent memory. Undeterred, he continued with his plans. On December 5, 1914, the *Endurance* headed out to sea.

The South Atlantic lived up to its frightening reputation, and most of the men were seasick. Only the sixty-nine sled dogs, lashed to the deck, were unperturbed by the pronounced roll of the ship; for their collective eyes focused on huge slabs of whale meat that hung from the rigging. With each hard lurch, the slabs slapped together and bits of blubber rained down on the dogs, which howled and whined for more tidbits.

Two days later, at about 57° south latitude, the ship steamed through a heavy belt of rotten floes, ominous harbingers of dense pack ice. Skipper Frank Worsley climbed into the crow's nest to direct the *Endurance* through ever-narrowing leads. "The ship became a floating ram," noted Hurley.

Christmas Day 1914 was celebrated with turtle soup, pudding flaming with brandy, mincemeat pies, and rum and ale for many toasts. When curious Adélie penguins waddled over to the ship from a nearby floe, meteorologist Hussey entertained the feathered compatriots with his banjo. The stodgy birds "expressed their feeling with croaks," wrote the unsentimental Hurley.

Five days later the ship steamed across the Antarctic Circle. On January 10, 1915, the high, ice-capped Coats Land appeared in the distance. The *Endurance* zigzagged through the fractured floes toward the coast, but the weather deteriorated. Tiny wind-driven ice crystals rasped against the sails and stung the men's faces as they peered through the haze to catch a glimpse of the land Bruce had discovered in 1904. When the weather cleared, the high seaward face of an ice shelf sparkled like a wall of prisms, refracting light into jewel-like colors. For the next five days, the men skirted Caird Coast, passing the enormous Dawson-Lambton Glacier toward the southwest.

On January 15 wave after wave of northbound crabeater seals rushed past the ship. Most of the men believed the spectacle was a terrible omen for an especially bitter winter. Secretly, they hoped the ship would follow the seals' lead; but Shackleton did not turn north, for the season was still young.

The search continued for a winter base. On January 18 the temperature plunged as they reached 76° 34' south at 31° 30' west. During the next few days, the leads disappeared as new ice seamed the floes together. Thomas Orde-Lees, in charge of supplies, wrote in his diary that the ship was besieged, "frozen, like an almond in the middle of a white chocolate bar." Ever the optimist, Shackleton didn't accept the inevitable until the end of February.

During March, the men converted the ship into warm, comfortable quarters for the winter. The dogs also benefited from the men's creativity. Elaborate "dogloos" with turrets, steeples, and archways promised to shelter the dogs from blizzards and cold winds. The huskies, however, preferred to curl up nose to tail and let the wind bank the snow around them. Only during the most ferocious storms did the dogs condescend to crawl inside their shelters.

"And now the long winter night closes upon us," Shackleton wrote in early May as the sun set for seventy days. He had just finished reading Otto Nordenskjöld's two-volume narrative of the ill-fated Swedish expedition, and now worried about the *Endurance*'s fate. When sleep evaded the leader, he listened to the distant grind of moving pack ice, a constant reminder of their precarious position.

In June the sounds grew louder, "like an enormous train with squeaky axles being shunted with much bumping and clattering," noted Worsley. Far away to the northeast, a horrific storm with hurricane force winds roared for days, exerting a tremendous force on the pack ice. Waves of pressure traveled toward the *Endurance*, buckling the floes ten to twenty feet high. Within days, the endless white field was partitioned into random rooms wherever the skewed walls intersected. On July 22, pressure uplifted tons of ice in a zigzagging line that moved toward the ship. Fortunately, disaster stopped four hundred yards away, much to Shackleton's relief. To Hurley, the colossal upended blocks resembled "very good imitations of Stonehenge in ice."

As spring approached and the days lengthened, the pack ice once more flexed its body in a surge that squeezed the *Endurance*. Recognizing the danger, the men started to dig a canal toward distant open water, but low temperatures doomed the attempt. "We were helpless intruders in a terrifying world," wrote Shackleton, "our lives dependent on the play of elementary forces that made a mock of our efforts."

On October 18, an extreme pressure wave rolled the ship about thirty degrees to port and demolished the dogloos. The animals survived, but even their howls of terror could not drown the sounds of twisting lumber and splitting beams. Six days later Shackleton ordered the men to lower provisions, boats, and personal gear down to the ice as the pressure continued to increase. "It is hard to write what I feel," noted Shackleton. "To a sailor his ship is more than a floating home, and in the *Endurance* I had centered ambitions, hopes and desires." Now the decks splintered and snapped beneath his feet as the pack ice annihilated his ship. The *Endurance*'s prolonged groans of agony, like a mortally wounded creature, prickled the hairs on many necks and scalps.

Several men wrote in their diaries that eight emperor penguins marched toward the ship in the late afternoon on October 26. They formed a tight cluster and lifted their heads skyward, waving their necks back and forth. They launched their double-noted trumpeting calls like a Greek chorus wailing a dirge for the ship. When their song ended, they disappeared behind a curtain of fog. "She is doomed," wrote Shackleton. That night the men huddled on the ice.

The next day the pressure rafted huge slabs of ice that for a moment balanced on end, and then toppled down in an explosion of crushed fragments very close to the ship. Splintering timbers told in vivid sound "the awful calamity that has overtaken the ship that has been our home for over 12 months," wrote Hurley. On the wrecked deck the emergency battery light flashed a weak signal of farewell. Throughout that long night, Shackleton paced on the ice, alone, pausing beside the broken ship. In the early morning hours, a pressure surge drove a thick wedge of ice through the ship's middle, skewering the broken vessel to the pack.

On October 28, each man selected two pounds of essential personal gear to pack on the sledges. Diaries, scientific specimens, and miscellaneous equipment were discarded. Frank Hurley searched his half-submerged cabin for his irreplaceable treasures: four hundred photographic plates that told the expedition's story. He and Shackleton then huddled on the ice and selected 150 to save. Hurley smashed each rejected image so that he would have no reason to return.

Later that same day, Shackleton gathered his men and thanked them for their support. The immediate goal, he said, was to reach Paulet Island and the large food cache left by Nordenskjöld's 1903 rescuers. "His simple words, nobly spoken, touched the heart," wrote Hurley. "It was a sad scene." Piles of abandoned gear resembled funeral pyres, ready for the torch. A short distance away, the stark outline of the twisted ship was a grim reminder that they were castaways in the most inhospitable region on earth.

Rotten ice, thigh-deep snow, and impassible hummocks defeated the Paulet Island attempt. The men trudged back to within one mile of the *Endurance* and established Ocean Camp on November 1, 1915. Three weeks later, Shackleton climbed a small hummock to scan the ice for seals. A slight movement caught his attention, and he peered through binoculars toward the broken ship. "She's going, boys!" he shouted. The men tumbled out of the tents and raced up the hill just as the *Endurance* raised its stern skyward and slipped beneath the ice. "I cannot write about it," he scrawled in his diary.

GOING WITH THE FLOE

In Ocean Camp the twenty-eight men lived under the same terrible conditions as prospectors in a mining camp, only without the hope of discovering gold. Few grumbled or complained, but Charles Green, the cook, was especially unflappable. In the most daunting circumstances imaginable, he calmly stirred his thin soups in crusty black pots over a blubber fire. "The flying soot gave him the appearance of a merry chimney-sweep who had not washed for many months," wrote Hurley.

To supplement their meager meals of salt beef and canned vegetables, the men hitched the dogs to the sledges and scouted for future rations of seal and penguin. As Hurley drove his team over a stretch of thin ice, three killer whales crashed through behind him, blowing noisily. The dogs understood the threat at once. "No need to shout 'Mush!' and swing the lash," wrote Hurley. "The whip of terror had already cracked over their heads, and they flew before it." The sight of a man and dogs scampering over the ice must have surprised or confused the whales, for the mammals hesitated just long enough for Hurley and the terrified dogs to reach thicker ice and safety.

The dogs gnawed on seal bones and penguin carcasses until mid-January 1916, when the scarcity of fresh food doomed them. One by one they were led behind a hummock and shot. On that afternoon the men burrowed deep within their sleeping bags, trying to block the revolver's sharp reports.

The steady but slow northwesterly flow of the ice was undetectable until one night in March, when Orde-Lees woke up feeling nauseous and disoriented. Although he was obviously seasick, none of his tent-mates sensed any movement. Measurements confirmed Orde-Lees accurate motion detection sensitivity: the floe rose and fell about one inch, the result of swells from the open sea.

On March 17 they drifted past Paulet Island, just sixty miles away but unreachable because of slushy ice. Six days later Joinville Island faded from view. In early April, "the white floor began to shatter and to disintegrate beneath our feet," wrote Hurley. Night after night Shackleton woke from hideous nightmares about buckling floes or surging swells. Each dream, he declared, was a warning. As he walked through the camp, he rehearsed procedures for each disaster in his mind.

Their ice floe continued to shrink. Shackleton ordered the three whaleboats packed and readied to launch in an instant. That moment arrived at 1 P.M. on April 10, when a wide lead snaked through the camp. Twenty-eight ragged men piled into the *James Caird*, *Dudley Docker*, and the *Stancomb-Wills* in a desperate attempt to reach a desolate, unexplored Antarctic outpost: Elephant Island, one of the South Shetlands.

To reach this forbidding refuge, the men faced a maze of floes and towering icebergs with grim determination. Raw, bare hands gripped the oars and spray saturated their clothes. Jackets stiffened and crackled with ice, and tissue-thin sheets of ice fell from their clothing to the bottom of the boat. "Each man's breath formed clouds of vapor, showing white against his grubby face," noted Worsley in the *Dudley Docker*.

During the second night the men rested in the boats, holding each other for warmth against the wind-driven sleet. To add to their troubles, "all around us we could hear the killer whales blowing, their short, sharp hisses sounding like sudden escapes of steam," wrote Shackleton in the *James Caird*. Dawn illuminated haggard faces and vacant eyes. Soft snow showers covered the silent sea and the still men with a shroud of white glitter.

On April 12, they reached open water and the enormous swells of the South Atlantic. During the next three days, all the men shared the same miserable nightmare: drenched clothing, frostbitten hands and feet, and unquenched thirst. Painful boils erupted where filthy clothes chafed their skin. Some of the men were seasick from the high waves that rolled beneath their boats. Shackleton, leading in the *James Caird*, stood by the mizzenmast and shouted encouragement to the men in the other two boats. Frank Worsley, in charge of the *Dudley Docker*, could check the accuracy of their course only when the sun peeked through the low clouds, once every few days. At that time, he wrapped one arm around the mast to steady himself and, with his other hand, maneuvered the sextant to determine their position. The *Stancomb-Wills*, with Hubert Hudson and Tom Crean in charge, trailed behind the other two boats.

The temperature dropped rapidly on April 13. Blisters on the palms of hands froze and felt like embedded stones with each pull of the oars. During the night, uncontrollable shivering robbed the men of any possible rest. The sea was calm and they listened to the soft rustle of new ice forming on the surface. After what seemed a lifetime of spasmodic shaking, the men faced the dawn, only to discover a thick sheath of ice had coated the boats during the night. The transition from water to mushy ice was so quick that not even fish escaped. Frank Worsley glanced over the side of his boat. "Lying about in the slush . . . were countless thousands of dead fish, some of which were eight inches long and looked like splashes and bars of silver glistening in the sun."

As the swells increased, the heavily iced vessels floundered. Although Orde-Lees suffered acute seasickness in the *Dudley Docker*, he grabbed a bucket and bailed with superhuman stamina all day to keep the boat from sinking. More ice formed during the long cold night. The exhausted men scraped the sides while Orde-Lees bailed. "They were hideous hours," wrote Hurley, "and the flame of hope all but died in many a heart."

The next day, April 15, 1916, Elephant Island's bleak basalt cliffs infused each man with hope and strength for the last few miles. Later that day they staggered ashore. "Everyone had red-rimmed eyes," Worsley later wrote. "Shackleton was bowed down and aged by the ordeal, accentuated by the responsibility and strain of holding the boats together and keeping his men alive." Several men scooped up pebbles in cupped hands, then laughed with joy as the stones trickled through their fingers. The last time they had touched rocks or walked on firm ground was on December 5, 1914.

THE LONG JOURNEY TO SOUTH GEORGIA

On April 16, after a refreshing eighteen-hour dreamless sleep, second-in-command Frank Wild took a few men in a boat to look for a safer location for their camp. Wild returned with the news that seven miles away was a better haven, later dubbed Cape Wild, complete with penguin colony and elephant seal wallow. The tired men climbed into the boats and rowed to the new site.

Although the island fulfilled their immediate needs for food and rest, Shackleton recognized that the odds for outside rescue were remote. Since 1819, sealers and whalers had avoided Elephant Island because of its strong currents and formidable cliffs. No one would ever search for them there; survival depended upon contacting the outside world. Shackleton's choices were limited; after much discussion, he settled on South Georgia, eight hundred miles to the northeast. To reach one of the year-round whaling stations, he would have to cross the stormiest sea in the world. As he and Worsley evaluated the twenty-two-foot *James Caird* and their chances, the boat "seemed in some mysterious way to have shrunk," Shackleton wrote.

The most urgent priority now was to make the *James Caird* seaworthy. The dour carpenter, Harry McNeish, and assistant Timothy McCarty used sledge runners, crate lids, and the cook's canvas windscreen to build a cover for the boat. The finished product didn't inspire Shackleton's confidence. "I had an uneasy feeling that it bore a strong likeness to stage scenery, which may look like a granite wall and is in fact nothing better than canvas and lath."

Nevertheless, on April 24, Frank Worsley, Thomas Crean, Harry McNeish, Timothy McCarty, John Vincent, and Ernest Shackleton cast off in the seething surf. The men on the beach waved and gave the pathetic little boat and its crew three hearty cheers. No one doubted that "The Boss" would return to rescue them.

But the men sailing the tiny craft wondered if they would survive the first few days of the voyage. Continuous gale-force winds whipped the sea into a frenzy of mammoth swells that curled over the vessel or rushed across the makeshift deck. "We leaped on the swells, danced on them, flew over them, and dived into them," wrote Worsley, undaunted by the sea's fury. "We wagged like a dog's tail, shook like a flag in a gale, and switchbacked over hills and dales. We were sore all over."

Shackleton divided the men into two watches, four hours on and four hours off. Those on duty either worked the tiller or sail, or bailed. Streamers of spray lashed their faces and drenched their wool clothes. Thighs, arms, necks, and faces were chafed raw, yet the men found humor during the terrible hours on deck. Each time Crean took his turn at the tiller, he sang a tuneless song "as monotonous as the chanting of a monk at his prayers," wrote Shackleton, "yet somehow it was cheerful."

The miserable hours on top were no worse than the four hours spent below the deck, in the five-by-seven-foot hold. When they had finished with their shift, the cold, wet men crawled on bloodied knees over ballast rocks to their sour-smelling reindeer sleeping bags. Meals were slurped from tin cups while rivulets of seawater dribbled down their necks from leaks in the canvas cover. Within a few days enough water had accumulated that mini-waves sloshed over their stretched-out legs as they fished out wads of reindeer hair from mugs of hot milk and mush.

On May 1, the sun finally appeared through the clouds. Worsley calculated their longitude and announced that they were just beyond the halfway point to South Georgia, or about four hundred miles. To celebrate, "all hands basked in the sun," noted Shackleton. The men stripped and soon tattered pants and jackets decorated the rigging.

At midnight four days later, Shackleton was operating the tiller when he noticed a line of clear sky on the horizon. He grinned and called to the men that there was good weather ahead. His smile froze when he recognized the pale line for what it was: the foaming crest of a monstrous rogue wave, the largest he had ever seen. "For God's sake, hold on! It's got us!" he shouted. White foam surged as the wave lifted the boat and flung it "like a cork in breaking surf," Shackleton wrote. Half filled with water, the sluggish boat lurched heavily. The men grabbed every available container and "steadily bailed death overboard," wrote Worsley. Slowly, the *James Caird* revived.

On May 6, they were less than a hundred miles from South Georgia and almost out of fresh water. Salt spray stung raw faces and dripped into mouths that burned with dryness. Two days later, tongues were so swollen no one could swallow food. That same morning two blue-eyed shags, resting on thick strands of kelp, drifted by the boat. The birds were a sure sign that land was close. Within a few hours, the rugged glacier-clad mountains of South Georgia's western coast filled the horizon. Worsley's inspired navigation had found the proverbial needle in the haystack. Now he steered toward King Haakon Bay—the least explored part of the coast.

They could not land, because partially submerged rocks in the shallows barricaded the entrance to the bay. As they searched for a landing site, the wind shifted abruptly and the sky darkened. The hurricane struck the next morning with shrieking winds that churned the sea into a white haze of spray. All day and night five men bailed while Worsley worked the tiller to steer the boat away from the rocky reef. That they should die within sight of land and safety filled the men with rage, for the *James Caird* could not hold together much longer. "Just when things looked their worst, they changed for the best," wrote Shackleton. "So thin is the line which divides success from failure." The hurricane began to dissipate that night.

The next morning, May 10, they sighted a calm bay and a wide beach through an opening in the reef. Within minutes of landing, the men listened to one of the sweetest sounds in nature: the gurgling of a stream. They drank the cold water in long draughts, reveling in its purity and taste. "It was a splendid moment," wrote the Boss.

TREKKING ACROSS SOUTH GEORGIA

The next few days Shackleton and his men rested beside a warm fire and dined on albatross from a nearby colony. For the moment, life was good; but their ordeal was not over. They had landed on the island's unpopulated side, opposite the whaling stations. The quickest way to reach civilization was to cross the unexplored island, over unknown mountains and glaciers to Stromness, about twenty miles away. Shackleton selected Worsley and Crean for the journey across a landscape never before seen by humans. McCarthy would stay behind to care for the ailing Vincent and McNeish. Although they were ready to start on May 15, a series of storms delayed the men until May 19. At 1 A.M., Shackleton watched as the bright full moon emerged from the clouds. He gently shook Worsley awake and said, "We'll get under way now, Skipper."

One hour later, Shackleton, Crean, and Worsley hiked toward the mountains with a rope, an adze, and their most precious possession, the ship's logbook wrapped in a tattered shirt. Each man stuffed three days' worth of food into a long woolen sock, knotted it, and slung it around his neck. At the last minute, the men discarded the heavy sleeping bags. They would not rest, the trio vowed, until they reached Stromness.

Moonlight illuminated their footsteps as they plodded up and down the snowfield saddles between the mountains. "The only sounds were the crunch of our feet through the snow, the soft swish of the rope—we were roped now, ready for crevasses," Worsley later wrote. Sometimes the surface dropped eight inches, hissing with the sudden release of trapped air.

By late afternoon, a thick fog from the west inched toward the men just as they reached the top of a precipitous ridge. They peered down into the gloom, but couldn't see if the drop sloped more gently at the bottom or ended on exposed rocks. If they retreated, Shackleton reasoned, they risked losing their way in the fog; yet, to go forward was just as dangerous. With only a moment's hesitation, Shackleton nodded toward the drop and said, "Let's do it."

Like bobsledders without a sled, the men sat on coiled rope and wrapped their arms and legs around the person in front. "I was never more scared in my life than for the first thirty seconds," wrote Worsley. "The speed was terrific." The three men gasped in terror, but then shrieked with delight. The joyride ended abruptly in a soft snowbank after a twelve-hundred-foot descent.

Darkness fell; they marched beneath a splendid moonrise, a radiant light that guided their footsteps once more. At about 5 A.M. they rested beneath an overhanging rock. With their arms wrapped around each other for warmth, Worsley and Crean fell asleep. Shackleton, however, forced himself to stay awake, "for sleep under such conditions merges into death," he wrote. Five minutes later he woke his companions and told them that they had slept for thirty minutes.

One hour later the men gazed toward Stromness Bay. Worsley shook his head and said, "It looks too good to be true, Boss." Within minutes the steam whistle called the whalers from their beds. "Never had any of us heard sweeter music," wrote Shackleton. "It was the first sound created by outside human agency that had come to our ears since we left Stromness Bay in December 1914." Seventeen months had passed since they had sailed away from Grytviken.

Worsley "adjusted his rags" and pulled out four safety pins he had hoarded for one year. He carefully pinned together the worst rips in his pants. The men then walked toward Stromness with only the rancid clothes on their backs, the adze, bits of rope, and the ship's logbook. "That was all the tangible things," wrote Shackleton. "We had reached the naked soul of man."

Two small boys stared at the strange apparitions and then fled without a word. The manager, too, stared with horror at the three strangers standing before him with matted hair, greasy soot-covered faces, and ragged clothes. "Do you know me?" the Boss asked. The manager shook his head. "My name is Shackleton." Only then did the man gasp with recognition.

Later, after a meal and a hot bath, Worsley and a crew steamed to King Haakon Bay for the other three men. Meanwhile, Shackleton planned the first of four attempts to rescue the twenty-two men who waited for him on Elephant Island.

LIFE IN A PENGUIN COLONY

On April 24, 1916, the men on shore had cheered until the *James Caird* disappeared among the icebergs and swells. Frank Wild, Shackleton's second-in-command, now bore full responsibility for the men's welfare. While blizzards and gales raged for the next two weeks, the men constructed an unusual shelter from very limited materials. Reasoning that penguins nest in relatively safe sites, Wild selected the least odoriferous real estate in the neighborhood penguin colony. Next, the men built two parallel stone walls to support the ends of the two overturned boats. Canvas hung from the boats' sides to block snow and the winds that rushed down the nearby glacier. Hurley donated a piece of celluloid from a photograph for a cut-to-size window, and second engineer Kerr crafted a tin chimney for the seal blubber stove from a large biscuit can. The finished hut measured ten by eighteen feet and was about five feet high.

The men quickly recovered strength and optimism "in what was surely the weirdest and most unfavorable convalescent home that can be imagined," commented Hurley. Yet, the strange-looking hut was cozy, especially in the evenings. The flickering fire illuminated their soot-covered faces while shadows danced on the walls. The scene reminded Hurley of "a council of brigands... after an escape in a chimney or a coalmine."

Routine filled each day, which usually began with Wild's cheerful, "Lash up and stow, boys, the Boss may come today." After a penguin or seal steak breakfast, the men hunted or patched their clothes. In the evenings they wrote raucous lyrics about each other's quirks and habits to the familiar tunes that Hussey played on his banjo.

Sleep brought dreams of unlimited bread and butter, suet-filled biscuits, and luscious desserts. The feasts were always interrupted by loud grunts, wheezes, and coughs. Although everyone snored, Orde-Lees was the undisputed champion of prolonged rattles and snorts. "His consistent efforts outrivaled those of a wandering minstrel with a trombone," noted Hurley.

August 30 began as usual, with Wild's wake-up call followed by a meager breakfast. Their food supply was nearly exhausted, for the penguins had not yet returned to the island. Desperate, the men dug up discarded seal bones and boiled them with seaweed for their lunchtime soup. Just before the meal was served, one man climbed to the top of a small hill to scan the horizon. Suddenly, he saw a ship emerge from the mist. "Ship-O!" he shouted, as he ran back to the hut. The other men, in a flash of comedic movement, all tried to dive through the canvas flap at the same time.

Outside, Wild grabbed the pickax and struck a hole in their last can of petroleum. Then he soaked coats, shirts, mitts, and socks with the fuel. Carrying the saturated garments to the top of Penguin Hill, Wild piled them in a heap and started a bonfire to attract attention. The men shouted and waved their arms as a small boat pulled away from the ship. Hearty cheers rang out as the vessel headed toward Elephant Island with a man standing at the bow. It was the Boss.

After reaching South Georgia, Shackleton had made three previous attempts to rescue the men but impenetrable pack ice had blocked his way. Now, as the boat approached the men, Shackleton shouted, "Are you all well?" The years seemed to melt from his face when Frank Wild answered, "We are all well, Boss."

A few hours later all were aboard the *Yelcho*, a Chilean vessel that Shackleton had leased. As the ship sailed away, the men stared for the last time at Elephant Island, their home for 138 days, until the bleak headlands disappeared into the mist.

The men returned to a world gone mad. Most served in the trenches during World War I. Shackleton immediately headed to the Ross Sea to find the *Aurora* party, the other half of his expedition.

• 10 •

ROSS SEA EXPLORATION

On January 10, 1841, James Ross stared with astonishment at land that "rose in lofty peaks, entirely covered with perennial snow." During subsequent days, panoramas of never-before-seen mountains and glacier-filled valleys unfolded before his and his men's eyes as they sailed southward. No further expeditions penetrated the Ross Sea region until 1895, when Henryk Bull searched for whales near Cape Adare and Possession Island. Robert Scott, Ernest Shackleton, and Roald Amundsen began their separate journeys to the South Pole from Ross Island; and between 1899 and 1917 five less well-known expeditions explored this region.

As the 1800s drew to a close, Carsten Borchgrevink and his men were the first to spend the winter on the continent. Nine years later, Professor Edgeworth David and two companions waved farewell and turned away from the Cape Royds hut to begin a grueling 1,260-mile round-trip journey to the South Magnetic Pole. In 1911 another trio, led by Edward Wilson, trudged away from Cape Evans in the dead of winter to collect emperor penguin eggs. The next year, while Scott and his men struggled on their return trip from the South Pole, Victor Campbell's party of six men was forced to endure the Antarctic winter inside an ice cave. In 1915 the other half of Shackleton's *Endurance* expedition arrived at Hut Point to build food depots for the Boss before tragedy doomed their efforts.

In 1899, a lone ship slowly pushed through the Ross Sea pack ice and inched its way toward Cape Adare. The vessel was the *Southern Cross*, and it carried Carsten Borchgrevink and his men toward their home for the next year.

First Winter on the Continent

When Henryk Bull docked the *Antarctic* in Melbourne on March 12, 1895, after his unsuccessful whaling venture, an unknown but ambitious member of the crew rushed off to London. There, thirty-one-year-old Carsten Borchgrevink, a Norwegian who had immigrated to Australia, presented a paper to the Sixth International Geographical Congress. During this time, he also indicated his willingness to winter on the Antarctic mainland at Cape Adare. Dogs, he declared, would perform the work of many men and haul sledges to distant regions. Borchgrevink's proposal for financial support was rejected. "It was up a steep hill I had to roll my Antarctic boulder!" he later wrote.

Three years later, wealthy publisher Sir George Newnes underwrote most of the costs of the expedition. His only request was that Borchgrevink sail beneath the Union Jack, even though few crew members were British citizens. On December 19, 1898, the *Southern Cross* steamed from Hobart, Tasmania, its deck crowded with lashed-down kayaks, sledges, provisions, skis, and eighty howling dogs destined to be the first canines on the mainland.

As the ship maneuvered through thick pack ice, Borchgrevink climbed to the crow's nest, the lofty perch where "you seem lifted above all the pettiness and difficulties of the small world below you," he wrote. Personality conflicts had already surfaced, and whispered grievances had reached Borchgrevink's ears. Lacking leadership skills to implement suggestions or resolve disagreements effectively, he withdrew from his men.

On February 17, 1899, the *Southern Cross* sailed into Robertson Bay. As the men struggled toward shore in small boats, fighting the pull of terrible currents, Borchgrevink stared at the beach where they would "live or perish, under conditions which were as an unopened book to ourselves and to the world." Louis Bernacchi, astronomer and physicist, also studied the wild landscape. "The heavens were as black as death," he wrote. "The land was terrible in its austerity; truly, a place of unsurpassed desolation." It was to be home to ten men and many dogs for the next twelve months.

On February 23, a furious gale descended on Cape Adare as the men continued to unload supplies. Fist-size stones from the black cliffs rained down and pounded the ship. Seven men were trapped on shore, including Finns Ole Must and Per Savio, hired to care for the dogs. The group huddled inside the Finns' small tent, which was erected beneath a low rock overhang. The dogs crawled inside, too, and slept on top of the men. Warmth from the dogs' bodies saved the men from freezing, and they inadvertently became the first humans to overnight on the continent.

Despite the bitter cold and sudden storms during the following weeks, the men erected two prefabricated huts—connected for easy access to instruments and equipment—which they called Camp Ridley. Their living quarters were dark and claustrophobic; the ten bunks that lined two walls were enclosed boxes that "could hold their own with modern coffins," wrote Borchgrevink. At first he despised the bunks, but later did almost all his writing inside his own wooden vault. "To work at the table with nine hungry minds, starved by the monotony of the Antarctic night, glaring at you through nine pairs of eyes at once vacant and intense, was impossible."

On March 2, the *Southern Cross* turned northward, and ten men stood on shore until the ship's sails disappeared beyond the horizon.

ALONE ON THE CONTINENT

Hard work was a good foil for boredom. The men sledged along the shores of Robertson Bay, collecting plenty of birds, seals, and Adélie penguins to prepare for European museums. Storms provided the men with the most unusual specimens. On April 2, 1899, the wind gusted to eighty-two miles per hour. Fierce snow squalls wrapped Cape Adare in dense whorls of snowdrift that sloped around the huts. Pulverized chunks of ice in the bay dashed against the rocks and flew over the huts, several hundred feet from the shore. Six days later, when the storm had blown itself out, zoologist Nicolai Hanson and Borchgrevink scoured the beach for "numberless specimens of medusae, hydroids, star-fish and algae."

On May 15, the rim of the sun slipped below the horizon and darkness rolled over the ice and Camp Ridley. Borchgrevink described the sunless days as "a depressing feeling, like looking at oneself getting old." Tension filled the hut, and most of the men avoided looking into each other's eyes. "We were getting sick of one another's company; we knew each line in each other's faces," Borchgrevink wrote. "Each one knew what the other one had to say."

Neither silent nights nor shrieking winds lifted the men's spirits. Boredom dulled their minds "like a sneaking evil spirit," and conversations were reduced to single sentences. Quarrels erupted among the members "whenever a change was necessary," noted Borchgrevink. Several men hunched over the table and played solitaire with a worn deck of cards. Even the music box was silent, for the men had grown weary of the same tunes played again and again.

Per Savio dug a small cavern in the snow and furnished it with an iron stove and tin chimney. He covered the entrance with sealskins and soon a roaring fire warmed the tiny room. Savio then undressed and relaxed for an hour, luxuriating as the thick steam swirled around his body. For the rest of the winter the men took turns inside Antarctica's first sauna.

One man who could not enjoy the warmth of a steambath was Nicolai Hanson. In early July the zoologist lost all feeling in his legs. Although confined to his bunk, he nevertheless was cheerful and worked on his notes for long hours.

On July 26, Borchgrevink, Hugh Blackwall-Evans, and the two Finns packed the sledges and harnessed twenty-nine dogs for the trek inland from the western coast of Robertson Bay. When the temperature fell to -52° F, the men dug deep into the snow to keep from freezing to death. At the end of each day, the dogs thawed out the sleeping bags. Since canines loved a good roll on anything except ice, the men simply laid the frozen bags on the snow and then unhitched the eager charges. They raced to the bags and rolled to their hearts' content. "Half-an-hour later we could get into our bags," noted Borchgrevink.

While Borchgrevink and other expedition members explored the coast, Hanson's health continued to deteriorate. On October 14, Borchgrevink and his men returned to Camp Ridley just in time for the zoologist to say his final good-byes. On that day, the first Adélie penguin waddled ashore near the huts. Knowing how much Hanson loved the feisty Adélies, one man carried the lone penguin to his dying friend, who smiled and stroked his favorite bird one last time. The men hammered scraps of lumber together for a coffin and hauled it up the steep slope to the top. There, they dynamited a hole in the ground near a large greenish boulder and buried their gentle friend, the first man to be interred on the continent.

On New Year's Day, 1900, Borchgrevink wrote that he felt proud of the men's accomplishments, that they had helped to illuminate "the hidden mysteries of the last *Terra Incognita* on the face of the globe." On January 29, the *Southern Cross* returned for the men and brought letters and other news from home. By February 2, provisions, dogs, sledges, and specimens were lashed and stowed, but the ship did not yet turn northward. Instead, the vessel sailed south and skirted the irregular seaward face of the Ross Ice Shelf. On February 16, Borchgrevink discovered a deep, mile-wide indentation within the ice, very close to the later-named Bay of Whales. Borchgrevink, Savio, and Lt. William Colbeck hitched the dogs to a sledge and trekked southward until they reached 78° 50', the farthest south any human had traveled in the Southern Hemisphere.

When the ship turned northward in late February and sailed past Cape Adare, no one regretted watching the black headland fade into the distance. "A strong longing for sunny shores came over everyone," Borchgrevink wrote, "a longing to see something else but bare rocks and snow, to see other colors, real green grass, and above all, trees."

Louis Bernacchi and Lt. Colbeck sailed back to Antarctica with Robert Scott in 1902, but Borchgrevink never returned. His reception in England was cool, even though the expedition had collected specimens for museums and surveyed many miles of the coast. Norway and the United States rewarded his efforts, but Borchgrevink had to wait until 1930 for recognition in England. Four years later, he died.

Per Savio and Ole Must, so instrumental to the expedition's success, returned home and died in separate fishing accidents. To date, no Antarctic landmarks have been named for the Finns.

Triumph at the South Magnetic Pole

In 1908 at Cape Royds, Ernest Shackleton's written instructions to Professor Edgeworth David were straightforward: "You will try and reach the Magnetic Pole." Their route would take them northward along the sea ice that flanked the mountains of Victoria Land. They would then turn inland, climb to the Polar Plateau, and claim the Magnetic Pole.

On October 5, 1908, David, Douglas Mawson, and Alistair Mackay attached their sledge to the back of the Arrol-Johnston automobile and climbed in with driver Bernard Day. With cheers from the Cape Royds men ringing in their ears, they set off on the first leg of their 1,260-mile round-trip journey. Two miles later, the automobile stalled in deep snow. David, Mawson, and Mackay harnessed themselves to the sledge and disappeared into a blizzard.

Day after grueling day they marched north, hugging the Victoria Land coast as close as possible, stopping only to rest and lay depots for their return trip. Emperor penguins sometimes investigated the strange green tent, uttering sounds like "something between the cackle of a goose and the chortle of a kookaburra," noted David.

Progress was slow because they worked their way inland to collect rock samples as often as possible. On October 23, the three men cut their daily food rations in half. To supplement their meager diet, they killed seals and stockpiled the meat. Still, hunger pangs plagued the trio. Ever careful not to waste a gram of anything edible, they tapped their biscuits against the inside of their mugs so that any loose crumbs fell into the "hoosh." The crumb-salvaging act reminded Mackay of old-time sailors who rapped their biscuits to dislodge weevils.

In early December they faced the formidable Drygalski Ice Tongue, a thirty-mile-long glacial finger that jutted into the Ross Sea. Its contorted surface was webbed with crevasses, and the men faced oblivion with each step. On December 20 David heard "a slight crash." When he turned, Mawson had disappeared. David and Mackay peered into the hole and saw Mawson dangling eight feet below the surface over empty space, held by a rope attached to his sledge harness. While David pawed through their bags for another rope, Mawson examined large ice-crystal formations on the crevasse walls and blithely tossed specimens to the surface for later scrutiny. The two men finally hauled a reluctant Mawson out of the chasm.

The next ten days were filled with the mind-numbing agony of relaying heavy loads up the steep icy slopes and moraines beside Larsen Glacier. Sleep eluded them in the tent as they listened to avalanches, "booming like distant artillery" from the surrounding mountains.

On December 26, cold winds and low temperatures welcomed the men to the Polar Plateau. Each time they exhaled, frozen puffs of vapor formed frosty layers that slowly cemented headgear to beards and mustaches. Yet, they marveled at the beauty of ice with its perfect crystalline structure. In bright sunlight, the dense granular snow that covered the plateau "glittered like a sea of diamonds," wrote David. "The heavy runners of our sledge rustled gently as they crushed the crystals by the thousand. It seemed a sacrilege."

On January 16, 1909, the men reached the South Magnetic Pole at 72° 25' south latitude and 155° 16' east longitude, at an altitude of seventy-three hundred feet. They bared their heads, smiled at the camera lens, cheered three times for the King, and then backtracked twenty-four miles before they slept.

They reached the Drygalski Ice Tongue on February 2 and camped in a conspicuous place where they had unobstructed views of the Ross Sea. The *Nimrod*, scheduled to pick them up in the vicinity, was nowhere in sight. Two days later, as the three men finished lunch, two blasts from the ship's gun reverberated around them. Mawson dove through the door, knocking down Professor David. "Then Mackay made a wild charge, rode me down and trampled over my prostrate body," noted David.

Mawson raced ahead, waving frantically toward the ship—and promptly dropped into a crevasse. Mackay kneeled down and called to him. Although he had fallen twenty feet, Mawson was only stunned. Mackay ran toward the *Nimrod*, which was easing up close to the ice and within hailing distance. He shouted two pieces of pertinent information to the crew: "Mawson has fallen down a crevasse, and we got to the Magnetic Pole!" The crew pulled Mawson to safety, and within an hour all the men were on board, steaming toward Cape Royds.

"And we kept our tempers—even with God."

On June 27, 1911, at Cape Evans on Ross Island, three men finished tying down supplies on two sledges. It was the middle of the Antarctic winter, and darkness draped the land like a black velvet cloth, soft and still. The squeaky crunch of cold snow trailed after each footstep. The tethered dogs howled with interest and strained at the lines, but no one approached to harness them to the sledges.

The men planned to man-haul the loads themselves to Cape Crozier, about seventy miles to the west. The trio strapped on the leather harnesses and then posed for photographer Herbert Ponting's camera. Physician, artist, and naturalist Edward A. Wilson stood in the center with hands on hips and confidently stared into the camera. On his right was the energetic Henry ("Birdie") Bowers, a short man with unequaled physical stamina. The youngest was twenty-four-year-old Apsley Cherry-Garrard, who had paid a thousand pounds to be a member of Scott's *Terra Nova* expedition and now stood on Wilson's left. He had taken off his glasses for the picture and glanced sideways, an uncertain smile on his lips.

On that dark midwinter morning the men started "on the weirdest bird-nesting expedition that ever has been or ever will be," Cherry-Garrard later wrote. The purpose was to observe emperor penguins brooding eggs during the harsh Antarctic winter. Wilson's plan was to collect a few eggs for experts to examine to learn if feathers had evolved from reptilian scales.

The men rounded Cape Armitage the next day and faced the cold air rolling across the Ross Ice Shelf. Temperatures dropped to ⁻56° F that night. "They talk of chattering teeth: but when your body chatters you may call yourself cold," Cherry-Garrard wrote.

The next morning, two men struggled to get the third into the sledge harness. Their outerwear had frozen into strange twisted forms and had to be thawed against their bodies. Once dressed and hitched to the sledge, the men crouched and allowed their clothes to refreeze into man-hauling position. Then they were ready to give three or four coordinated jerks to move the sledge forward. That day they reached a windless area where the ice granules on the surface shared the same characteristics as deep desert sand. They were forced to pull one sledge for about a mile and then return for the second. By the end of the day they had traveled just three miles toward Cape Crozier.

On July 5, the temperature plunged to ⁻77.5° F, the lowest of the trip. Cherry-Garrard was awake for most of the night, enduring a succession of uncontrollable shivering fits so violent that he believed his back would break from the strain. "I will not pretend that it did not convince me that Dante was right when he placed the circles of ice below the circles of fire," he wrote.

Although Cherry-Garrard had poor eyesight, he could not wear his spectacles. Moisture from his breath condensed on the glasses and froze within seconds. Throughout the journey, he stumbled across a torturous landscape riddled with wind-carved sastrugi and felt his way through a maze of pressure ridges. His nearsighted trek in the dark seemed like "a game of blind man's bluff with emperor penguins among the crevasses of Cape Crozier."

On July 15, eighteen days after leaving Cape Evans, they reached Cape Crozier and pitched their tent beside a glacial moraine at the base of Mount Terror. The view was glorious in the bright moonlight, and Cherry-Garrard put on his spectacles and rubbed away the film of ice again and again. "To the east a great field of pressure ridges below, looking in the moonlight as if giants had been plowing with plows which made furrows fifty or sixty feet deep.... Behind us Mount Terror on which we stood, and over all the grey limitless Barrier [Ross Ice Shelf] seemed to cast a spell of cold immensity, vague, ponderous, a breeding-place of wind and drift and darkness."

The trio began to build a small stone hut on a level stretch of the moraine. Since ice had cemented each rock into the ground, many whacks with their axes were required to free a single stone. Gradually, rock by rock, the walls grew taller. To form the roof, they inverted a sledge and stretched a canvas floor cloth over it. On July 19, they struggled the last few miles across crevasses and pressure ridges toward the penguin colony. Exhausted, they were just about ready to turn back when at last they heard the emperors' two-note trumpeting call drift across the ice. They pushed on and saw the tightly huddled birds, "a marvel of the natural world and we were the first and only men who had ever done so." They quickly collected five of the huge, one-pound eggs and headed back to the tiny stone hut. Cherry-Garrard carried two eggs in his mittens, but because he could not see, he tripped twice and smashed them. "Things must improve," wrote Wilson back at their makeshift shelter, but they didn't.

That night, as the men cooked their food over an improvised blubber burner, a morsel of fat popped and hot droplets hit Wilson in the eye. The pain was excruciating, and he moaned softly throughout the long night. Early the next morning, Cherry-Garrard woke to a dead silence outside the hut, an unsettled calm that wasn't natural. "Then there came a sob of wind, and all was still again," he wrote. Minutes later, the wind roared around them and the canvas roof strained against its rock anchors. Yesterday, they had pitched the tent next to the hut and now they wondered if it would survive the gale.

All that day and the next, the men huddled inside their reindeer bags and endured the storm's fury. Suddenly, with a great swoosh a gust shredded the canvas roof. Bowers peeked outside and shouted that the tent was gone. Gravel from the moraine rained down on the men, and fine drift snow collected around their bodies. For forty-eight hours they lay in their bags and shouted encouragement to each other over the din. Bowers occasionally thumped Wilson, "and as he still moved I knew he was alive all right," he wrote in his diary. Their moisture-laden bags stiffened with cold and looked like "squashed coffins, and were probably a great deal harder," noted Cherry-Garrard.

At 6:30 A.M. on July 24, the storm ended. Although Wilson and Bowers tried to act cheerful, "without the tent we were dead men," wrote Cherry-Garrard. "And we kept our tempers, even with God." For the first time in three days, they lit the Primus and cooked porridge laced with reindeer hairs from their bags and feathers from the penguin skins. Still, the hot meal filled their stomachs and restored their hope.

They found the tent, folded up like a green umbrella, between two boulders about one-half mile from the hut. "Our lives had been taken away and given back to us," declared Cherry-Garrard. They carried the tent reverently. Wilson felt that their luck had changed for the better and decided to start back to Cape Evans the next day. That night, and all subsequent nights during the rest of the journey, Bowers tied the tent to himself before he slept. "If the tent went he was going too," noted Cherry-Garrard.

They faced the same awful conditions homeward that they experienced outward, and the weary men dozed in their harnesses. As they neared Cape Evans, they pulled with new resolve toward a warm fire, hot meals, and beds with eiderdown covers. They stumbled through the hut's door on August 1 with three frozen but intact emperor penguin eggs.

Cherry-Garrard carried the eggs back to England and presented them to several distinguished university researchers. The scholars studied the eggs with indifference before concluding that no additional knowledge about bird evolution could be gleaned from the three hard-won prizes. Birdie Bowers and Edward Wilson died in March 1912 while returning from the South Pole.

Winter in an Ice Cave

As a member of Scott's 1910–12 *Terra Nova* expedition, Victor Campbell selected five men to explore Cape Adare. Within days after arriving at Cape Evans, his northern party reboarded the ship. On February 9, 1911, they unloaded a prefabricated hut and supplies near Borchgrevink's dilapidated shelter at Cape Adare. At first the weather favored the men, but within a week hurricane-force winds hurled pebbles from the top of Cape Adare, pelting the hut "until it shook and quivered like a thing alive," wrote geologist Raymond Priestley. The hut withstood the terrible onslaught of pebbles and wind throughout the otherwise uneventful winter. Short sledge journeys began at the end of July and continued until the *Terra Nova*, en route from New Zealand to Cape Evans, picked up the group on January 3, 1912.

So successful was their first year together that Campbell and the others decided to explore Terra Nova Bay, about midway between Cape Adare and Cape Evans. The men landed on the coast on January 8 and arranged with the skipper to pick them up at the same location about February 18. With only a tent, the clothes on their backs, and enough food for ten weeks, they surveyed the area near Mount Melbourne. The men endured snow blindness and close calls with dangerous crevasses as they crossed a string of glaciers.

By February 18 they had returned to the place where the ship was scheduled to pick them up. However, unknown to the men, the *Terra Nova* was unable to push through the thick offshore pack ice. Three days later, high winds damaged the men's tent. Still hoping that the ship would return, they postponed killing the last few molting Adélie penguins until it was too late and they too fled before the cold breath of winter. Each man now wondered how any of them would survive the winter without shelter, warm clothing, food, or fuel.

Only a large snowdrift near an outcrop of rocks offered the stranded men any promise of protection. After much discussion, they began to hollow out an underground room in the drift. Blocks of packed snow reinforced the walls; layers of pebbles and dried seaweed insulated the icy floor. On March 17 the men moved into the nine-by-twelve-foot shelter on Inexpressible Island, the name they had selected for their desolate location. Inside the ice cave the men arranged their reindeer bags according to rank: officers on one side and sailors on the other. Fortunately, no one minded Campbell's insistence on following navy protocol, and the men worked well together under very difficult circumstances.

On March 21, the men killed a Weddell seal and found thirty-six edible fish in its stomach. "It was a red-letter day for us, and the discovery threw a glow of cheerfulness over the party," wrote Priestley. That night the cave almost felt cozy. The four tiny blubber lamps softly illuminated six soot-covered faces with halos of yellow light. As the small stove rattled and clouds of steam drifted toward the roof, one of the men broke into a rousing song. "One by one we go through our favorites," noted Priestley. The spontaneous sing-along lasted well into the night.

Another favorite way to fill the empty hours was to listen to Murray Levick read a chapter or two from Charles Dickens's *David Copperfield.* His expressive voice breathed life into the characters, and the men looked forward to evenings for the next several months.

The long Antarctic night settled over Inexpressible Island, and brief periods of depression gripped everyone at one time or another during the months that followed. Filthy clothes, the omnipresent cold, lack of exercise, and a diet high in fat contributed to abrupt mood swings. A despondent man would draw his sleeping bag over his head and mentally slip away from the others. "I am content to lie and dream about past times, and not worry my head about the future," wrote Priestley in his diary. Days later, alert and talkative once more, he spent hours memorizing intricate patterns of black soot and gleaming ice crystals on the walls.

As the months passed, their grimy, oil-splattered clothes rotted to such a degree that the men risked serious frostbite each time they crawled from their ice palace. Pants and jackets were so encrusted with blubber that the garments became stiff, despite many knife scrapings or vigorous rubs with penguin skins to absorb the grease.

On September 30, 1912, the men emerged from their icy hovel for the last time, determined to reach Cape Evans or die in the attempt. The sledge was packed with seal meat and the tattered tent. Slowly, day by day, the men trudged southward.

Four weeks later, they found a food cache near Granite Harbor, left by a mapping team one year earlier. "Hurrah! Hurrah! Hurrah! Good news and plenty of biscuits," Priestley wrote in his diary. More food was found at Butter Point: cases of sugar, bacon, hams, tea, chocolate, and jams. The men ate their fill for the next two days and slept with full stomachs for the first time in seven months. Long banquet tables laden with favorite foods haunted their dreams no more.

The men arrived at Cape Evans on November 7, after a grueling thirty-nine-day, 225-mile journey. There, they learned the fate of Scott and his team from the men who had volunteered to remain another year to search for the long overdue South Pole party. On January 18, 1913, the *Terra Nova* returned to carry the men home, as arranged the previous year.

The winter of 1912 had marked each man. "None of us would care to repeat the experience," Priestley later wrote, "but in my own case . . . the 'Call of the South' remains a force to be reckoned with." Raymond Priestley returned to Antarctica two more times before he died in 1974. Throughout their lives, the six men of Campbell's party remained friends. "And well it might be," Priestley wrote, "for if ever men knew each other inside and out, it was the six of us who had dwelt together for seven months in a hole in the snow."

The Forgotten Heroes: Shackleton's Ross Sea Party

In the late summer and early fall of 1914, on the eve of the Great War, two ships sailed from London. The *Endurance* carried Ernest Shackleton and his twenty-eight men to their rendezvous with the Weddell Sea pack ice. Another ship transported the second part of the expedition to Sydney, Australia, where the men boarded the *Aurora* for its journey to Ross Island. Under the command of Mackintosh, the *Aurora* pushed through the ice to Hut Point, the site where Scott's storage hut from the 1901–04 *Discovery* expedition stood abandoned.

On January 21, 1915, twelve men landed with dogs and supplies. The first order of business was to make the hut habitable for the coming winter. Next, the men planned to sledge food and fuel to 80° south during the summer and build a food depot for Shackleton's trek across the continent from the Weddell Sea. More depots as far south as the Beardmore Glacier would be established during the 1915–16 season.

Mackintosh and two other men left Hut Point on January 28, 1915, with twelve hundred pounds of supplies. The dogs were exuberant after their long confinement on the ship, but soon deep snow discouraged even the hardiest canine. One by one, they sat down on their haunches and refused to pull. The men had no choice but to relay the heavy loads, a back-breaking struggle for the few miles gained. "Surface too dreadful for words," Mackintosh wrote in his diary on January 31. With each step the men sank to their knees; the dogs, panting and wild-eyed, jerked the sledge forward mere inches for each effort. After twelve hours of mind-numbing drudgery, men and dogs had covered only two miles.

During the next few weeks, the men slowly advanced toward 80° south and reached it on February 21. The journey back to the hut was even more horrific. Two dogs died during a blizzard that confined the men inside their tents for several days. "The poor dogs are feeling hungry," wrote Mackintosh. Only a minimal amount of dog food had been packed on the overloaded sledge, and now the dogs gnawed on leather harnesses and straps for nourishment.

The men, too, were short of food and reduced their own rations to one meal per day. When the weather cleared, they tried to make up for lost time but each step was like wading through deep sand. They made little headway. On March 2, the dogs collapsed and curled up beneath a shroud of snow to wait for death. The men had no choice but to haul the sledge themselves or die beside the dogs.

Blizzards, frostbitten toes, and fear of starvation plagued the discouraged men. On March 24, there was nothing left in their food bag but mealy biscuit crumbs. Fortunately, later that day they saw the flag from a food cache they had built earlier. "We simply ate till we were full, mug after mug," Mackintosh wrote in his diary. Only then did they tend to frostbitten faces, hands, and toes. The next day, the party reached Hut Point and safety.

Meanwhile, at nearby Cape Evans, the *Aurora* was tethered to the shore with two anchors and seven steel hawsers. Four men lived in Scott's spacious hut, built in 1910, while other crew members stayed on the ship. By the end of March the *Aurora* was iced in as planned and its boilers were dismantled.

Uneventful weeks passed until May 7. On that day, ferocious winds roared around Cape Evans. At 11 P.M. the *Aurora* lost its moorings; four hours later, the pack ice and the ship had disappeared, blown out to sea by the tremendous force of the wind. When the Cape Evans men discovered that the ship had vanished, they believed it had sunk with all hands and mourned the loss of their friends.

The six men at Hut Point had by this time marched to Cape Evans and joined the others. The winter passed peacefully and days were filled with small tasks, such as mending clothes, hunting seals, and recording weather observations. Only six dogs had survived the strenuous earlier sledging treks, and these were now well fed and fussed over.

By October the men had begun the serious work of depot laying in anticipation of Shackleton's overland journey. Throughout the following months the men sledged supplies almost continuously. They ignored the pain of frostbite, snow blindness, and strained muscles in bodies pushed to the limits of human endurance. "Everybody is up to his eyes in work," wrote Mackintosh.

By January 12, 1916, six men still had enough reserve energy to continue to 81°, where they built a depot. Six days later they reached 82°. On January 19, Spencer-Smith, who was the first ordained clergyman on the continent, complained of swollen legs. Extreme exhaustion, bleeding gums, and other symptoms of scurvy also marked Mackintosh. The next day Spencer-Smith told the others that he could not go further. The men left him with a tent and provisions and then pushed ahead to Mount Hope at the base of the Beardmore Glacier.

On January 26 they reached the glacier, established the depot, and then climbed a ridge. "Imagine thousands of tons of ice churned up to a depth of about 300 feet," wrote Joyce, filled with awe. The men then returned for Spencer-Smith, who was too sick with scurvy to walk. They bundled him up in his sleeping bag and hauled him on the sledge. Mackintosh also grew weaker with each mile.

By February 18, they had reached 80°. Six days later, a blizzard forced them to remain in their tents with short rations for the next five days. When the storm finally ended, they were so weak with hunger that digging out the sledge took two hours. After one hour of sledging, Mackintosh could walk no further. The three strongest men forged ahead after pitching a tent for Spencer-Smith and Mackintosh. One man volunteered to stay behind and look after the two desperately sick men.

The other three men and four remaining dogs struggled in the cold face of a blizzard to reach the depot. Three days later they spotted the black flag. The starving dogs strained in the harness and pulled with their last scrap of strength toward the flag that meant a meal. At the depot men and dogs ate their fill. "If ever dogs saved the lives of any one, they have saved ours," Joyce wrote in his diary.

Another blizzard confined the men to the tent, but on February 27 they harnessed the dogs. The huskies lost all heart when they realized they were going away from food and home. Two days later, the men spotted the invalids' camp. With two sick men on the sledge, Joyce and the other three men struggled northward while Spencer-Smith grew sicker each day. "He is still cheerful," wrote Joyce, "but he has hardly moved for weeks and he has to have everything done for him."

On March 8, Mackintosh volunteered to stay behind in a tent with enough provisions for three weeks. The next day Spencer-Smith died without uttering a harsh word or complaint during his forty-day illness. The men buried the preacher in a cairn of ice marked with a cross of bamboo poles.

A few days later, the men reached Hut Point, rested for three days, packed the sledge with seal meat to combat Mackintosh's raging scurvy, and started on the thirty-mile journey back for the sick man. On March 16, Mackintosh crawled outside his tent to welcome the men. "The Skipper shook us by the hand with great emotion, thanking us for saving his life," wrote Joyce. Two days later they arrived back at Hut Point.

Thin sea ice prevented the men from returning to Cape Evans, but they found enough penguins and seals to supplement the tins of vegetables stacked inside the hut. Fuel was a problem until they rigged a blubber stove and burned seal fat for warmth. They wore the only clothes they had, and threadbare jackets soon stiffened with soot and grease.

Mackintosh improved to the point that he insisted on trekking to Cape Evans to see if a message had been left by the *Aurora*, for he refused to believe that the ship had sunk. On May 8, he and Haywood started toward Cape Evans across thin sea ice. Two hours after their departure, a terrible blizzard struck with high winds. The two men were never seen again.

Unknown to the stranded men, the *Aurora* had drifted out to sea in the clutch of pack ice during the blizzard on May 7, 1915. Day by day during the following March, the crew watched a lead widen in the pack ice. When the ship broke free and reached the open sea, the men raised their cups of grog as the *Aurora*'s whistle pierced the stillness with three farewell blasts to the pack ice. With little coal for its engines, the battered, rudderless ship headed northward. At Bluff, New Zealand, a tugboat towed the crippled ship the last few miles to port on April 3, 1916.

After his legendary ordeal in the Weddell Sea sector, Ernest Shackleton arrived in New Zealand the following December. By that time the *Aurora* had been repaired, and after discussions with the *Aurora*'s captain, Shackleton immediately sailed to Ross Island to bring his men home. On January 10, 1917, the ship pulled alongside the pack ice near Cape Royds and worked its way to Cape Evans. One week later, Shackleton and the Ross Sea Party survivors were headed back to Wellington, New Zealand.

Although none of Shackleton's men returned to the Ross Sea region, others were lured by its sense of mystery and beauty. In 1928, Admiral Richard E. Byrd arrived at the Bay of Whales and established the first of five Little Americas. Six years later, weary of the hoopla surrounding his second expedition, Byrd decided to stay alone at Bolling Advance Weather Station on the Ross Ice Shelf. The experience nearly killed him.

In 1935, adventurers Lincoln Ellsworth and Herbert Hollick-Kenyon flew the *Polar Star* from the tip of the Antarctic Peninsula to Little America on the Ross Ice Shelf. The year 1947 marks the beginning of American military involvement with Operation Highjump and Operation Windmill. In the middle of the Cold War, a group of scientists ignored politics and planned the International Geophysical Year (IGY). From July 1, 1957, through December 31, 1958, twelve countries forgot their ideological differences and established research bases on the mainland and surrounding islands. Information, building materials, and personnel were shared; a can-do spirit permeated each station. While the world watched and waited, scientists and support personnel proved that they could, indeed, work together without the fuss and furor of diplomatic brigades.

At least *men* had the opportunity to prove that they could get along. Since scientists and the military controlled who was allowed on the continent, they barred women from research bases because of the added expense to build separate housing. In 1969, a female biologist was allowed on the continent to study penguins. That same year, women scientists from Ohio State University arrived at McMurdo Station for the summer season and later were flown to the South Pole.

Since the mid-1970s, scientists, support personnel, and a few tourist ships have converged each year on the Ross Sea region to experience the quiet poignancy of the four historic huts. The names of many geographic features, such as Inexpressible Island, Ross Ice Shelf, Cape Crozier, Beardmore Glacier, and Cape Evans are forever linked in history, for they represent the stories of human triumph and tragedy, of survival and self-sacrifice, of fulfillment and loss.

• 11 •

EAST ANTARCTICA EXPLORATION

On January 27, 1820, Russian Thaddeus Bellingshausen peered through his field glasses at "a solid stretch of ice" and distant snow-covered hills. Without realizing it, he was looking at the continent near Princess Martha Coast. Eleven years later, sealer John Biscoe guided the *Tula* and the *Lively* through broken floes. To his astonishment, he saw the tops of mountains poking through the ice. In February 1840, Charles Wilkes explored sixteen hundred miles of East Antarctic coast before reaching a great barrier of ice. Meanwhile, the irrepressible Dumont d'Urville celebrated his expedition's discovery of Adélie Land with a glass of wine.

For the next sixty-two years, East Antarctica remained an unexplored mystery until Germany's Erich von Drygalski led an expedition to a blank area on the Antarctic map. Ten years later Australia's Douglas Mawson, one of three men who trekked to the South Magnetic Pole in 1909, returned to the continent to face the ultimate horror.

Voyage to the Unknown: The Drygalski Expedition

Erich Dagobert von Drygalski was thirty-six years old when the *Gauss* sailed from Kiel bound for Antarctica. Born in Germany in 1865, he had studied mathematics, physics, and geology before receiving a doctorate in geophysics in 1887. During the early 1890s Drygalski had traveled twice to Greenland to study the physics of ice formation.

In 1895, he attended the German Geographical Conference, where the topic of the moment in learned circles was Antarctica. Secretly, Drygalski wanted to lead an expedition, but first published his Greenland results to show that he was intellectually well equipped for research on the ice continent. Drygalski was the polar committee's unanimous choice.

The *Gauss*, specially built for Antarctic conditions, was named for the great German mathematician and magnetism theorist Johann Karl Friedrich Gauss, whose South Magnetic Pole calculation had launched three expeditions in the late 1830s. Now German scientists, too, would explore the most formidable region on Earth.

On August 11, 1901, thirty-three expedition members sailed away from all that was familiar. What they would find was anyone's guess, for Drygalski had elected to explore the continent's unknown eastern side. After spending a month on Îles Kerguélen, Drygalski and his men loaded forty previously shipped sledge dogs and set sail. The *Gauss* then stopped at Heard Island before heading toward a section of the Antarctic coast sandwiched between Knox Land and Kemp Land, sighted during the 1830s but unexplored.

By the time the expedition had reached Antarctic waters, the season was well advanced. As the men maneuvered the ship into the pack ice, the icebergs appeared fresher and less eroded the further south the *Gauss* sailed. On February 21, 1902, an undulating ice-capped white expanse appeared in the distance. Later that evening, a blizzard engulfed the ship and by the next morning, the *Gauss* was trapped. Only one other ship, the *Belgica*, had drifted with the pack ice and survived.

The next day Drygalski and his men watched with alarm as massive bergs bore down on the imprisoned ship, plowing through the floes and leaving a wake of churned-up bits of ice. "They seemed to be coming toward us as if driven by magic, without any movement of the air," he wrote. By 11 P.M. the icebergs surrounded the ship. Now there was no retreat, even if new leads opened in the pack ice. "Nature had drawn massive bolts across the only outlet remaining to us," Drygalski wrote with resignation.

In mid-March, three men and eighteen dogs sledged southward about fifty miles and discovered a solitary volcanic cone rising from the surrounding blanket of ice to about a thousand feet. They returned not only with news of the extinct volcano but also a mummified Weddell seal that sensitive canine noses had detected. Intrigued, Drygalski ordered the expedition's massive balloon to be filled with hydrogen. He gingerly climbed into the small basket and ascended fifteen hundred feet. For two hours he photographed the surrounding ice and the lone volcanic cone, now named Gaussberg, the only black speck on the monotonous white plain. Drygalski was enthralled. "The view was so extensive that it was like looking into infinity." While he worked, he reported his impressions to the men on the ground via a crude telephone, the first in Antarctica.

By the first of April all the Adélie penguins had disappeared. Two weeks later, northward-bound skuas from Gaussberg flew over the ship to escape winter's cold breath. Emperor penguins popped out of narrow leads and marched toward the south, a sight that puzzled Drygalski. "Groups of more than 200 individuals were seen again, slowly waddling across the ice in philosophical fashion," he noted in his journal.

Even though winter was fast approaching, Drygalski and five others sledged to Gaussberg. The dogs, eager to investigate anything dark and solid looking, needed no shouts of encouragement or cracks of the whip to drive them toward the now visible volcano. For the next week the men surveyed Gaussberg. As they investigated the higher ledges, they found many snow petrels deep inside crevices, sitting on nests that were lined with sand and feathers. On May 2, the birds suddenly scurried from their hiding places and took wing, darting across the sky like windblown confetti. Drygalski wondered if the birds had sensed bad weather. "This was not a good sign," he concluded.

Within hours blustery winds buffeted the men. By the next day, they reluctantly packed away their surveying instruments. Although great clouds of drift snow swirled around Gaussberg, the men still climbed its lower flanks to collect rocks. On May 8, snow petrels once again returned with the dying winds. Two days later, the group headed back to the ship. On May 15 strange mirages and many "ship masts" danced in the distance. A few hours later, Halle Gate, two mighty grounded icebergs standing together like portal sentinels, slowly came into focus. Within minutes the *Gauss*'s masts appeared to the men like welcoming friends.

PROBLEM-SOLVING WITH RUBBISH

As the nights lengthened, more emperor penguins marched across the ice to investigate the *Gauss* and the men, who whistled tunes for their appreciative audience. Even the expedition's most dour members perked up when the penguins arrived. Although several scientists tried to capture the emperors' voices on an Edison phonograph, the trumpeting calls were too soft to record. Only the very perturbed squawks of Adélie penguins came through loud and clear.

During the winter, activities filled most evenings of the week. On Sunday nights the men drank drafts of grog and shared stories. When the temperature suddenly plunged to ten below zero, dozens of bottles of hearty dark ale burst and were duly mourned. Wednesday night lectures were well attended and applauded, no matter what the subject. Saturday evenings were reserved for games, singing, and camaraderie. Clubs formed whenever two men shared a common passion for cigars, chess, or chorale music. An unusual band also entertained regularly, its members playing tunes of the day on a harmonica, flute, and triangle. Two hefty pot lids functioned as rousing cymbals.

The men also had access to a variety of books. Drygalski pored over reports of the *Belgica*, the first ship to drift with the pack ice for an Antarctic winter. He put some of the lessons learned by the *Belgica*'s crew to practical use, such as the observation that dark objects sink into the ice faster than lighter-colored ones. Drygalski, too, had noticed that stove soot deposited on the ice soon pockmarked the surface. He ordered the men to collect ash and other dark-colored rubbish to spread around the *Gauss*. In theory, he explained to the men, paths of slush would form and weaken the ice's grip on the vessel. "This regulation engendered nothing but mirth and incredulity," noted Drygalski. Throughout the winter, the garbage-strewn path grew and stretched beyond the ship, slowly turning into a slurry of ice and rubbish by early spring.

Soon, signs of spring were unmistakable. Flocks of Antarctic petrels circled the *Gauss* and then continued southward toward Gaussberg. Drygalski and seven companions also headed toward the lone outcrop in mid-September 1902. The forty-five-mile trek took ten days because of persistent blizzards. Clear, sunny weather welcomed the men when they finally reached Gaussberg. For the next two weeks, they folded their heavy furs on the ice and basked in the sun's warmth. The ancient lava flows, interspersed with glassy black bands of obsidian, intrigued the men, and they collected samples from different levels.

Although they had more than enough supplies to push further south, Drygalski decided against the effort. Nothing more would be gained from enduring blizzards, cold, and hunger to trudge across "featureless wastes," he declared. The men started back on October 9 but were tentbound until October 14. Late in that afternoon, the huskies pulled with renewed vigor when they caught the distant scent of penguins mustering near the ship.

In early December the men staked out a two-thousand-foot-long path to the edge of the pack ice and heaped ash, rotted peas, rancid fish, and penguin blood along the trail. Day after long day they watched for signs that the ice would release the ship. Desperate schemes were discussed in detail. Should they trek to open water and toss bottles into the sea with the ship's coordinates neatly printed on paper inside? Or would it be better to launch the bottles in the balloon basket to drift over the sea? Drygalski knew that only the power of nature could free the ship, but a little dynamite and sawing with custom-forged tools would speed the process.

On February 8, 1903, the ice creaked and groaned "like a thing alive." Later that afternoon, Drygalski had retreated to his cabin to drink a cup of hot cocoa in peace when two sharp jolts rocked the ship. Stunned silence was quickly followed by shouts of joy and excitement.

The ice had split along the sludge paths, and seawater now sloshed over the edges of the deep gullies. For quick intense heat, the stokers tossed penguin skins into the boilers. Within an hour the *Gauss* sailed down Garbage Lane to the open sea and freedom.

When the expedition reached Cape Town, South Africa, the Kaiser denied Drygalski a second year in Antarctica because so little new territory had been discovered during the first. The men were ordered home. During the next thirty years, Drygalski worked on the expedition's scientific results. Twenty volumes were finally published in 1931.

No one sailed again to East Antarctica until 1912, when Douglas Mawson and the crew of the *Aurora* followed leads through the pack ice toward "the home of the blizzard."

An Incredible Will to Live: Mawson's Expedition

Douglas Mawson understood the perils of Antarctic exploration, for he had been a member of Shackleton's 1907–1909 expedition. On January 16, 1909, he planted the Union Jack at the South Magnetic Pole with Professor Edgeworth David and Alistair Mackay. When the *Aurora* sailed from Hobart, on December 2, 1911, the twenty-nine-year-old Mawson considered himself an old Antarctic hand.

A geologist by training, Mawson had the survival instincts of an explorer and the organizational skills of a leader, traits that would later help him endure the unimaginable. Although Robert Scott offered Mawson a place on the 1910–13 *Terra Nova* expedition, the Australian turned him down. Mawson wanted to lead his own trek to Adélie Land, an area that had not been explored since Dumont d'Urville and Charles Wilkes had sailed among the massive icebergs in 1840.

On January 8, 1912, the *Aurora* sailed into Commonwealth Bay. How very different this monotonous snow-covered plateau must have seemed to Mawson when he remembered Victoria Land's rugged mountains flanked by convoluted glaciers. Here, only an irregular wall of ice, marred by blue-gray crevasses, stretched as far as the eye could see. Just as the ship was about to leave the ten-mile-wide bay, the men spotted a rocky promontory jutting out from the shore. Relieved to see gray and brown boulders and an ice-free beach, Mawson selected Cape Denison for his base. That same night, chill winds steadily increased to 70 MPH; two days later, the men still waited for calmer conditions. Whenever they ventured outside the tents, the wind whirled dense clouds of drift snow around their bodies. But overhead, blue sky and bright sunlight belied the blizzard's fury at ground level.

Although Mawson had misgivings about the location, he did not change his plans. Later, he would discover that this site was home to some of the most violent winds on earth. Unlike the Antarctic Peninsula or the Ross Sea region, no coastal mountains broke the force of the katabatic winds that rolled down the nine-thousand-foot Polar Plateau. Mawson and his crew quickly learned to tie everything down and to wear their crampons.

For the first few weeks, backs and legs ached with unrelenting pain, for the men were forced to walk in a crouched position whenever they ventured outside the tents. Bending at the waist and leaning into the wind at impossible angles, they seemed to defy gravity. With each step, they swung their ice axes into the frozen ground to momentarily anchor their bodies. If there was a sudden lull, they fell forward; then powerful gusts once more tumbled them like so many bowling pins and rolled them toward the sea. Inch by inch the unlucky men had to claw their way back to safety with frostbitten fingers. Gradually, they adapted to the persistent 60-MPH winds and communicated with hand gestures and pantomime.

While the *Aurora* sailed to the Shackleton Ice Shelf to drop off eight men who would set up a second base and endure winter fifteen hundred miles away, Mawson and his group built the main hut and smaller structures to store supplies and scientific equipment. When the construction projects were finished, the men worked on the radio masts, a job that lasted until September because the wind toppled each effort.

As winter descended at Cape Denison, the wind gained strength and ice granules rasped against the hut like shotgun pellets. Gray wooden planks were planed satin smooth, and the denser knots and whorls stood out in stark relief. Raised nail heads, once hammered flush with the wood, looked like decorative iron studs outlining the hut's seams. Even footsteps in the snow were eroded into boot-shaped pedestals when the wind carried away the looser snow around the compacted prints.

At night the wind lulled them to sleep. Sudden calm, however, startled the men into wakefulness, and they fretted away the hours huddled before the stove until the wind once more shrieked its song. One night, gusts to 200 MPH lambasted the walls. "Everyone was tuned up to a nervous pitch as the Hut creaked and shuddered under successive blows," noted Mawson, "and the feasibility of the meat cellar as a last haven of refuge was discussed."

Restless with inactivity during the long winter, Mawson and two companions faced gale-force winds and headed south. After sledging for about five and a half miles, the men dug out and stocked an ice shelter, dubbing it Aladdin's Cave. During September, teams ferried food and other supplies to the depot. No sledging was done throughout October, for the weather "fluctuated between a gale and a hurricane." Mawson used the time to plan five spring expeditions.

In early November 1912 three teams headed east, another struck out to the west, and the last group trudged south toward the Magnetic Pole. Mawson, Lieutenant B. E. S. Ninnis, and Swiss mountaineer Xavier Mertz headed east with eighteen dogs and three heavily laden sledges. For Mawson, it was to be the worst trek of his life.

A TERRIBLE CHAPTER IN MAWSON'S LIFE

The three men headed for Oates Land. To get there, they would have to cross two perilous glaciers that poured down from the Polar Plateau like "a solid ocean rising and falling in billows 250 feet high," wrote Mawson. By the end of November, the first glacier, later named for Mertz, had been crossed, but hair-raising escapes and close calls were now the rule, not the exception. On December 6, 70-MPH winds drifted mounds of snow on all sides of their tent. Three days later, they dug their way out and searched for the dogs. Only black noses were visible, looking like randomly cast pebbles on the snow.

Unstable snow bridges spanned treacherous crevasses. Still, they trudged across the terrible surface, pushing the sledges from behind while the dogs strained at their traces in front. Dull, booming sounds from the glacier's icy bowels raised canine hackles and unnerved the men. Mawson speculated that caverns had formed beneath the crust, and as the glacier crept toward the sea, each movement echoed deep inside the caverns. At other times, the disturbances of men and heavy sledges moving across the unstable surface caused sections of the crust to sink and release trapped air that sounded like a prolonged sigh.

On December 13, one sledge was discarded. Most of the supplies, food, and the tent were then packed on Ninnis's sledge. The six fittest dogs were selected and harnessed to his sledge to pull the load over the treacherous glacial surface, webbed with snow bridges.

The next day was a warm 21° F with clear, perfect Antarctic weather. "Everything was for once in harmony," Mawson wrote. The men continued across hazardous ice. Mertz skied ahead, singing student songs from his university days, as he prodded the ice with his pole for telltale weaknesses. Suddenly, he held up one ski pole, a signal that yet another hidden crevasse had to be crossed. Mawson mushed the dogs over the snow bridge without any trouble but stopped abruptly when Mertz, who had been watching the two men, shouted in horror and pointed. Mawson whirled around. "Nothing met the eye but my own sledge tracks running back in the distance," he wrote. Ninnis, the sledge, and the dogs had disappeared without a sound.

Mawson and Mertz rushed back and crawled on their stomachs toward the gaping hole where Ninnis had hurtled down into the cold black abyss. The two survivors peered into the chasm and saw a ledge about 150 feet below the surface. A whimpering dog tried to stand but collapsed, its back broken from the fall. For the next three hours, the men shouted into the depths, but only silence answered their calls as cold drafts of air billowed out from the mouth of the crevasse.

Mawson and Mertz mourned their friend beside the jagged hole and tried not to think about their own chances of survival. Ninnis's sledge had carried the food, tent, spare clothing, and all of the dogs' biscuits. If they drastically reduced their meals, the men could stretch the remaining food for ten or eleven days, if they were careful. But they were 315 miles from safety, and not a single depot lay between them and Cape Denison except Aladdin's Cave, just five and a half miles from the base.

They ate just six ounces of food per day. Those meager meals did little to silence the gnawing pains in their stomachs. The dogs, too, were slowly starving. On December 15, Mawson killed the weakest husky to feed the other five and themselves. During the next three days, Mawson and Mertz covered many miles, but only three dogs now dragged the sledge. These were given worn-out fur mitts and spare rawhide strips to chew. "Outside the hungry huskies moaned unceasingly until we could bear to hear them no longer. The tent was struck and we set off once more."

On Christmas Day only one dog still limped beside the sledge. Mawson and Mertz tried to be cheerful, for they had reached the halfway point in the long homeward journey, but they were very weak and sick. Although the dogs' livers had a peculiar fishy taste, neither man could afford to be picky. With each bite, toxic quantities of vitamin A poisoned their starving bodies.

Mertz was not his usual optimistic self on December 30. He stopped eating dog meat, but on New Year's Day 1913, he doubled over with severe abdominal pain. He refused any portion of the last butchered dog. By January 5, Mertz was too weak to sit up and just lay in his sleeping bag. "Both our chances are going now," Mawson recorded in his diary.

The next day Mertz became delirious and mumbled incoherently until midnight. When the sick man grew quiet, Mawson fastened his companion's sleeping bag and crawled into his own. Several hours later, he felt no movement from his friend. "I stretched out an arm and found that he was stiff." Shocked and despondent, he lay for hours wondering how he could reach Cape Denison, still at least a hundred miles away. "I seemed to stand alone on the wide shores of the world."

The next morning Mawson rallied and did what had to be done. He dragged Mertz's body, still in the sleeping bag, outside and piled blocks of snow around it. He then cut the remaining sledge in half with his small pocketknife and packed only what he could drag. He made a sail from Mertz's jacket.

On January 9 Mawson cooked the rest of the last dog, but could not travel because of gale-force winds. When he removed his boots, the sight of his black toes sickened him. Two days later, the gale abated. He again examined his feet, only to find that the thick skin on the soles had separated from each foot. He smeared lanolin on the raw skin and then tied the soles back on with strips of cloth. Outside, the weather was calm and warm. Mawson undressed "in the glorious heat of the sun." He felt cleansed and at peace for the first time in many, many days.

On January 17, he climbed a steep slope covered with fresh snow. He had almost reached the top when the ground opened like a trapdoor beneath his feet with a loud *whoosh*. Seconds later Mawson found himself dangling at the end of his rope, fourteen feet below the lip of the crevasse. As the sledge inched closer to the mouth, he whispered, "So this is the end." But suddenly, the sledge stopped its fateful advance with him hanging and slowly turning above the dark chasm. "It would be but the work of a moment to slip from the harness," Mawson reasoned, "then all the pain and toil would be over." But he could not let go. Inch by inch he hauled himself toward the light, away from the black mouth that had almost swallowed him. He dragged his body over the lip of the crevasse and lay on the ground, his body shaking violently. When the terrible spasms had stopped, he made a rope ladder and fastened it to the sledge and his harness to protect him should this ever happen again. As he crossed the Mertz Glacier, the rope ladder saved him from sudden death many times.

On January 25, Mawson did not feel well enough to continue the struggle. Outside, a blizzard raged and drifts pressed against the tent. Soon his space was no bigger than a coffin and he imagined what death would be like. Once more he willed himself to survive. When the storm ended, he crawled out and slipped the harness on, shouting at himself to lift one leg and the other until moments blurred into hours and hours into days.

His ordeal ended on January 29 when he discovered a cairn filled with food that a search party had stocked. After he had eaten his fill, he strapped nail-studded wooden planks to his feet and crossed a broad field of very old blue glacial ice. On February 1 he found Aladdin's Cave and devoured oranges and a pineapple that had been left. Inside the cozy ice shelter, Mawson rested and regained his strength while a blizzard raged outside.

One week later, he emerged and pulled the sledge the last five miles to Cape Denison. Soon he recognized the rocks around the hut and watched as members of the search party raced to help him home. "The long journey was at an end," he wrote, "a terrible chapter of my life was finished."

He had just missed the *Aurora*, which had sailed away the previous day. Mawson immediately radioed the sad news of the death of his two companions to the ship that was on its way to pick up Frank Wild's party fifteen hundred miles to the west. Although the vessel tried to return, heavy seas and high winds prevented it from reaching Cape Denison. Mawson and the search party were forced to spend another winter. They had plenty of food and maintained radio contact with Australia via Macquarie Island, which served as a relay station.

The *Aurora* returned on December 12, 1913. As the ship sailed away, Mawson watched Cape Denison disappear behind a billowy curtain of blowing snow, a last farewell from the home of the blizzard.

Later Explorers

Mawson returned to East Antarctica in 1929 to help Australia's territorial claim. For two years, he and his crew surveyed more than twenty-five hundred miles of the coast from ship and by small plane. At the same time Lars Christensen and Captain Riiser-Larsen surveyed areas between 20° west and 45° east longitude. During the early 1930s, at least two hundred whaling ships plied Antarctic waters in search of blue, fin, and sei whales. A few vessels were dedicated to mapping East Antarctica and carried aircraft for that task. Vast tracts of Dronning Maud Land were thus reconnoitered from the air.

In 1938, Hermann Goering approved the *Schwabenland* expedition for Germany. Nine years later, the United States launched Operation Highjump with 4,700 men (just twenty-four were scientists), thirteen ships, and twenty-three planes. Helicopters were used for the first time in Antarctica for aerial photography and survey work. When the ships headed home, more than 60 percent of the coastline, including large sections of East Antarctica, had been mapped.

Many countries participated during the 1957–1958 International Geophysical Year (IGY) in East Antarctica, including the Soviet Union, Australia, the United States, and France. The countries' respective governments financed the construction of the bases and furnished the supplies; the military controlled the transportation of scientists and other logistic considerations. This odd pairing of science and the military in Antarctica continues today.

• 12 •

EXPLORING SELECTED SUBANTARCTIC ISLANDS

Even the hardiest mariners dreaded that swath of ocean lying between 50° and 60° south latitude, naming it the "Furious Fifties." Cyclones, born from the clash of frigid Antarctic air masses with tropical systems, swirl into great climatic depressions with winds heaving ocean swells sixty feet or higher. Gales battered ships that ventured into those forbidding seas and drove them toward uncharted islands.

The history of the subantarctic islands is a prelude to the discovery and exploration of Antarctica. Reports of vast fur seal colonies spurred intrepid captains to weather the storms and seek their fortunes on these secret and forlorn islands. Within a few short years, fur seals were slaughtered to the brink of extinction, and the world market for skins collapsed.

Entries from Captain Robert Fildes's 1821 journal, Thaddeus Bellingshausen's 1831 published narrative, and an 1844 memoir by Thomas W. Smith paint vivid word pictures of the sealers' lives and the appalling conditions the men endured. When the population of seals was decimated, the islands were abandoned. For almost sixty years, they were home only to the wild pigs and rabbits that were released to breed on the islands.

Carl A. Larsen established the first whaling station in 1904 on South Georgia Island. Enterprising shipping firms that had invested in whaling ships to satisfy the demand for baleen and oil now built floating factory ships and stations in sheltered harbors of the subantarctic islands. After World War II, profits dwindled and the stations were abandoned one by one.

During the International Geophysical Year, scientists converged on the continent and the subantarctic islands and remained there. Researchers and their bases became a scientific and political presence for their respective countries. Political disputes simmered until one April morning in 1982, when war came to South Georgia and the Falkland Islands.

Today rusty boilers, digesters, and try-pots illustrate the carnage of the past. Elephant seals wallow in the wreckage, a territory of twisted rods, skewed oil drums, grooved slipways, and silent buildings. Fur seals haul out on beaches where pale whalebones mark the dark sand like gravestones in a cemetery. On all but the most forbidding subantarctic islands, these and other creatures are a reminder of human excesses of the past and symbols of Antarctica's future.

South Shetland Islands

Soon after crossing the Antarctic Convergence, the boundary where cold southern waters clash with warm tropical seas, the first iceberg emerges from the fog like a ghostly guide to a world beyond imagination. The South Shetland Islands are wild and raw, with snowcapped mountains, black basalt cliffs, and glaciers scored with cobalt blue fissures that seem to pulsate from deep within. The shadow of a giant petrel, gliding over the dark volcanic sand, gives life and movement to a deserted beach. The enraged shriek of a skua warns an intruder of nests and territorial violations. On these desolate shores human voices seldom break the silence, but piles of bleached whalebones and solitary graves of long-ago sealers hint at the islands' past.

In 1598 Isaac Le Maire, an Amsterdam merchant, financed five Dutch ships that sailed for the East Indies. Pilot Dirck Gerritsz guided the *Flying Heart*, also known as *Glad Tidings*, through the Strait of Magellan in September 1599. For days, Gerritsz and his men fought a violent storm in the Pacific that pushed the *Flying Heart* to about 64° south latitude. When the sky cleared, Gerritsz gazed at "a mountainous land, like Norway, entirely white with snow," wrote Le Maire in 1622. Although several Antarctic historians interpret Gerritsz's discovery as the Antarctic continent, others argue that the description fits the South Shetland Islands.

Sealers from other countries may have known about the Shetlands, but Englishman William Smith was the first to report his find to his superiors. Like Gerritsz, he was blown southward by a fierce storm. Between violent squalls of snow with gale-force winds, Smith saw a frightful landscape on February 19, 1819. Although his superiors scoffed at the notion of new islands, he persevered and returned the following October. "The Land was barren and covered with Snow; Seals in abundance," he wrote after landing on King George Island.

The following year, Shetland Island beaches were inundated with sealers wielding cudgels and knives for the indiscriminate slaughter of males, females, and pups. Joining the frenzy was Thomas W. Smith, a British sailor with a sharp eye for detail. He wrote a vivid autobiographical account of sealing, published in 1844 and titled *A Narrative of the Life, Travels and Sufferings of Thomas W. Smith, Comprising an Account of his Early Life, Adoption by the Gypsies, his Travels during Eighteen Voyages to Various Parts of the World, during which he was Shipwrecked Five Times, thrice on a Desolate Island near the South Pole, once on the coasts of England and once on the coast of Africa . . . Written by Himself.*

Thomas Smith signed on with the *Hetty* in 1820 and headed to the newly discovered South Shetland Islands. After searching a few islands and finding nothing but fresh carcasses on the shores, the captain finally located a beach alive with fur seals. "But on our landing, we were met by the crews of three vessels, who forbid us from taking any, claiming the beach as theirs, as they had first taken possession of it," Smith wrote. *Hetty*'s crew reluctantly worked on another beach.

While Smith and several men slept under an overturned boat, sealers from a competitive ship rowed to a site farther down the beach and killed eight thousand fur seals. Smith's group tried to stop the intruders from skinning their bounty, but the result was "a Bloody engagement, in which several men were severely Injured." The skins were a total loss to both parties because copious quantities of blood had ruined the fur. "They were left to rot upon the Beach," wrote Smith.

Other ships and crews in the vicinity also had problems. On December 7, 1820, the *Clothier* was wrecked in a harbor at Robert Island, a popular anchorage for sealing ships. Less than two months later, Captain Alexander Clark and the crew organized an auction to sell anything salvageable from the wrecked *Clothier* to other ships in the area. Antarctica's version of a yard sale was a success, and the shipwrecked men stuffed a few extra coins into their pockets. They returned home on the American vessel *O'Cain*.

Captain Robert Fildes and the *Cora* met misfortune on Desolation Island in January 1821, when the ship sank after a violent storm. Fildes and his stranded crew pitched a tent on the beach and built bunks from salvaged wooden casks. The ship's feisty cat also survived and, like a typical feline, soon claimed a cask for herself. Two days later, two penguins popped out of the surf, waddled up to the cat "and took their station alongside of her in the cask," wrote Fildes, "they neither minding the people in the tent or the Cat, nor the Cat them, poor shipwrecked puss used to sit purring with their company." Each day the penguins left to feed at sea for several hours. When they returned, they waddled into the men's makeshift shelter and joined the cat in the cask. This odd trio enjoyed one another's company until Fildes, his men, and the cat were rescued and taken to Liverpool. The penguins then had the cask to themselves.

While Fildes and his men waited for better times on Desolation Island, British and American sealers almost came to blows at Cape Shirreff on Livingston Island over the right to possess all the fur seals on shore. When the Americans rowed close to the beach, the British blocked the boats and seized about eighty sealskins. In retaliation, Captain Burdick and his Yankee crew planned to "muster all our men from our Several Camps and as one body to go on to said beach at Shirreff's Cape and to take Seal by fair means if we Could but at all Events to take them." Several men were injured during the ensuing free-for-all.

Disgusted with the crowded and volatile conditions at Cape Shirreff, American John Davis recorded his plans in the *Cecilia*'s logbook. "Concluded to make the best of our way for our People that is stationed on the South Beach and then to go on a cruise to find new Lands as the Seal is done here." Davis and his crew set out on January 30, found new land, and became the first men to step ashore on the Antarctic Peninsula.

Russian explorer Thaddeus von Bellingshausen sailed to the Shetlands on the final leg of his Antarctic circumnavigation. On February 6, 1821, near Deception Island, he discussed sealing with American Nathaniel Palmer. When the young American told the Russian commander that over eighty thousand fur seals had been killed so far that season, the news saddened Bellingshausen. "There could be no doubt that round the South Shetland Islands just as at South Georgia and Macquarie Islands the number of these sea animals will rapidly decrease." He was right, for within two years South Shetland Island beaches were deserted.

James Weddell mourned the fur seals, too, as he stared at the barren beaches in 1823 and remembered how the gentle seals "would lie still while their neighbors were killed and skinned." He estimated that during the 1821–23 seasons, over 320,000 seals were slaughtered. "The system of extermination was practiced at Shetland Islands," he wrote.

Conditions on the South Shetland Islands were harsh for the sealers, men that had few options and even fewer prospects. They lived under overturned boats, ate whatever they could kill, shivered by fires fueled with penguin skins, and sometimes starved to death on the dangerous shores. One man scrawled a message on a cave wall that read, "J. Macey 1820–21 & 22 but never more."

In 1830 American geologist Dr. James Eights arrived at the South Shetlands with a sealing fleet. During his forays ashore, the massive rounded boulders that littered the beaches intrigued him. Since most of the islands were composed of dark volcanic basalt, the pinkish granite boulders were clearly out of place, "for we saw no rocks of this nature *in situ* on these islands," he later wrote.

At the end of February, Eights and several others "towed clear of several Ice bergs. Landed on one & got a specimen of rocks which are not found at Shetlands" as stated in the *Penguin*'s logbook. Eights concluded in his 1833 published report that the boulders had been "brought unquestionably by the icebergs from their parent hills on some far more southern land." Although Eights was the first to speculate that ice had transported the boulders to distant locations, his published report was ignored. Another naturalist, Charles Darwin, reached the same conclusions in 1839, six years after Eights, and named the out-of-place boulders *erratics*.

Sealers returned to the South Shetlands when news spread that the fur seals had repopulated the islands. In 1877, Captain James W. Buddington sailed the *Florence* to Rugged Island and ordered his crew to land and collect fur seal skins. The men killed every seal they found and stacked the pelts on the beach, but the ship never returned. Desperate, they rowed to Potter Cove with the hopes that a late-season ship would rescue them. When pack ice formed on the bay, the men lost heart and died one by one, until just Mr. King, the mate, remained. He managed to survive until a ship picked him up the following spring.

When Otto Nordenskjöld, leader of the Swedish expedition, gazed at King George Island in 1902, he wrote that "a loneliness and a wildness reigned here such as could, perhaps, be found nowhere else on earth. . . . I had never expected to find so much ice and snow." Although Nordenskjöld had read most of the available books about the Antarctic and the South Shetlands, the unimaginable quantities of ice still overwhelmed him.

As Ernest Shackleton and his men rowed ashore on Elephant Island in the South Shetlands on April 15, 1916, they had not seen anything except ice and snow for the past sixteen months. They leaped out of the boats onto the dark volcanic sands, grateful to be alive and on solid land once more.

When Shackleton and his five companions sailed to South Georgia for help, the remaining twenty-two men camped under an overturned boat in the middle of a chinstrap penguin colony, to be close to their food source. One morning, however, the birds suddenly gathered en masse near the water to seek their collective fortunes elsewhere. Several days later, hundreds of plump gentoos popped out of the surf and claimed the colony for a community molt. Grateful once more for the continued food bank, the men watched over the penguins with care.

Eight years earlier, Jean-Baptiste Charcot had anchored the *Pourquois Pas?* at Deception Island "in the midst of a veritable flotilla of boats, all at work as though in some busy Norwegian port." The water was coated with slime, and large pieces of whale drifted by the ship. The smell was unbearable. To pass the time, Charcot accepted an invitation to dine with the head of the Norwegian whaling station. Inside the man's house an unexpected guest perched beside the dinner table. "A parrot, which ought to be feeling very much out of it in the Antarctic, is talking solemnly," Charcot noted.

Although World War II had little effect on the South Shetlands, Great Britain and Argentina established an ongoing rivalry that stemmed from their overlapping territorial claims. The British built Base B on Deception Island in 1943 during Operation Tabarin, a military plan to guard against German occupation in the Antarctic. In 1947 Argentina established a base about four miles away from the British station. Relations were said to be friendly, even during football games.

On December 4, 1967, Deception Island erupted with fire and brimstone, the first volcanic flare-up in more than one hundred years. Flaming bombs, ejected from smaller craters around the lagoon, damaged the Chilean and British bases. Personnel were forced to evacuate their respective bases after grabbing a few essentials: research notes and clothes. Although the island settled down for the next year, eruptions in 1969 and 1970 convinced the British that there were better—and more stable—sites for a permanent base.

South Georgia Island

Long before South Georgia glistens on the horizon, the wind blows raw and cold. Wandering albatross crisscross the ship's wake with elaborate figure-eight patterns. Homeward bound king penguins crowd the broad beaches while blue-eyed shags drift by on thick tangles of kelp. When the gray mist evaporates, the sun illuminates wild glacier-clad mountains and hills covered with tussock grass.

The Allardyce Range forms the curved spine down the middle of the hundred-mile-long island, and several other parallel mountain chains rise to the northeast. The valleys between these ranges are filled with permanent ice fields. Glaciers creep down to sea level and fill the heads of the narrow fjords along the serrated coasts. Permanent snow clings to mountain slopes, starting at about 650 feet on south-facing peaks.

In spite of all the glaciers and snowfields, parts of South Georgia have been ice free for thousands of years. Hills that rise from boulder-strewn beaches are covered with tussock grass, mosses, ferns, and sedges. Over the years, these plants have slowly decomposed and formed deep peat bogs. The savage southwest side of the island is lined with steep cliffs and massive talus slopes, pounded by crashing surf.

These same treacherous waters may have discouraged the earliest seafarers from exploring this tiny island in the southern Atlantic. In April 1675, a terrific storm blew London merchant Antonine de la Roche's ship east of Cape Horn. When the sky briefly cleared, he saw "some Snow Mountains near the Coast, with much bad Weather." He found a bay and anchored near a cape. Two weeks later the weather improved, and Roche looked to the southeast toward "another High Land covered with snow." A relatively narrow strait separated the two snow-covered mountainous lands, according to a translation of his records.

No such waterway exists on South Georgia, and it is for this reason that a few polar historians doubt that Roche discovered the island. Exactly what he did see is a mystery. A relatively short strait divides East Island from West Island in the Falklands, but it's difficult to imagine this island group being described as "High Land covered with snow." However, the passage between Elephant Island and Clarence Island in the South Shetlands fits the description, for both landscapes are snow-covered and mountainous. Unless his original records are found, the question whether Roche discovered South Georgia, the South Shetlands, or even the Antarctic Peninsula will continue to be debated.

On June 26, 1756, severe storms battered the *Leon* as it rounded Cape Horn. The ship was blown eastward, past the Falkland Islands. Three days later, Commander Gregorio Jerez "beheld a Continent of land full of sharp and craggy mountains of frightful aspect, and of such extraordinary height that scarcely could we discern the summits," wrote one of the travelers.

When Captain James Cook sailed eastward from Cape Horn, he carried the *Leon's* account, originally published by Alexander Dalrymple. Cook sighted the island on January 14, 1775; the scenery did not impress him. "Wild rocks rasied their lofty summits till they were lost in the Clouds and the Vallies laid buried in everlasting Snow," he later wrote. "Not a tree or shrub was to be seen, no not even big enough to make a tooth-pick." Cook landed at Possession Bay, and claimed South Georgia for Great Britain.

The Forsters, father and son naturalists sailing with Cook, also expressed disdain at the British Empire's newest acquisition. The twosome concluded that perhaps the island could serve a purpose. "If a Capt, some Officers & Crew were convicted of some heinous crimes, they ought to be sent by way of punishment to these inhospitable cursed Regions," wrote J. R. Forster, the father. "The very thought to live here a year fills the whole Soul with horror & despair." The naturalists, crew, and Cook still recognized the island's one bountiful attraction: numerous fur seals.

The 1777 publication of Cook's narrative of his second voyage caused a sensation around London's better banquet tables. Society reveled in tales of scandalous behavior among the South Sea islanders. Owners of whaling firms, too, read Cook's book—but for a different reason. The meticulous explorer had carefully noted the locations of vast colonies of fur seals throughout his journey, including those on South Georgia. Beginning in the late 1780s and continuing until the mid-1820s, sealers plundered the island, butchering fur seals for skins and stripping the blubber from elephant seals to be rendered into oil.

Sealer Thomas W. Smith signed on a South Georgia-bound ship as an apprentice in 1815, four years before the South Shetland Islands were discovered. As soon as the *Norfolk* anchored in Royal Bay in December, the crews rowed to shore to search the high tussock grass for elephant seals. "Poor innocent animals! I could not but pity them," wrote Smith, "seeing the large tears rolling down from their eyes; they were slaughtered without mercy."

The men dragged their boats to higher ground and then turned the vessels over for makeshift shelters. Elephant seal hides and penguin skins were burned for warmth and for cooking meager meals. One night, the boatswain plucked white feathers from a sheathbill and presented each man with two. He told his companions to place both under their bodies so that they could "on some future day say we had slept on feathers while engaged in sealing on the coast of South Georgia."

When they sailed for England in December 1816, the ship was loaded with thirty-five hundred barrels of elephant seal oil. By that time Smith's clothes were in tatters. "I suffered much with the severity of the cold, which affected my feet to a great degree," he wrote in his autobiography in 1844. "I had neither shoes nor stockings, having previously worn them out."

Having few options, Smith signed on again and returned to Cooper's Bay with the *Norfolk* and a small open boat. One evening a terrible gale battered the smaller craft, tethered with a few feeble cables to beach boulders. The men on board stared in horror as three enormous rogue waves rolled toward them. When the first wave hit, the captain shouted, "There it is, boys! She's gone!" The shallop dragged its anchor toward rocks that looked like dragon teeth, waiting to splinter the vessel. One man was washed overboard and drowned; another was tossed among the rocks. The men who had jumped and reached the shore now tied ropes around their waists. They crawled out on the rocks and stretched their bodies, straining to reach the floundering man. Suddenly, a huge wave broke over the boulders. The man was last seen "waving his hand as he was carried away by the back [ebb] sea," Smith wrote.

Sealing relics are scattered around the island and hidden in the tall tussock grass. Derelict try-pots and broken barrels, warped planks from wrecked boats, caves with blackened hearths, bleached human bones, and a skull with a bullet hole testify to the harsh living conditions, conflicts, and violence endured by sealers.

In December 1819 Captain Thaddeus von Bellingshausen arrived at Undine Harbour, where he was welcomed by sealers who had lived on the island for the past four months. Bellingshausen gave the sealers grog, sugar, and butter in exchange for information about the South Georgian wildlife and plants.

Scottish explorer James Weddell was no stranger to either South Georgia or sealing. After his voyage deep into the heart of the Weddell

Sea in February 1823, Weddell arrived on South Georgia in March. Without adequate clothes, the crew had suffered from icy winds and seas, in spite of extra servings of rum and the cut-up blankets they used for rough coats. On shore, the men feasted on young albatross and bitter greens.

During the next sixty years, ships continued to stop at South Georgia, primarily for elephant seal oil. Fur seal populations had been devastated, but hidden coves still sheltered small groups that reproduced until they were once more profitable to hunt. In 1882, during the first International Polar Year, commercial interests were interrupted when a German expedition arrived and set up the first land-based scientific station at Royal Bay. The eleven-man team stayed for one year. Remains of the hut and other artifacts can still be seen today.

For years, sealers had talked about the many whales that return year after year to South Georgian waters. In 1894 Norwegian Carl A. Larsen visited the island in the *Jason* and harpooned the first right whale in Royal Bay. He returned in 1902 as commander of the *Antarctic* with the ill-fated Nordenskjöld expedition. Two years later, Larsen established the most successful whaling station on any subantarctic island. He named the small community Grytviken, which means "cauldron cove" in Norwegian, a reference to all the try-pots that dotted the shore. Within a few years, six more whaling stations were operating at full capacity. Between 1904 and 1966, a staggering 175,250 whales were slaughtered, flensed, and rendered into oil and other products.

On October 4, 1911, Dr. Wilhelm Filchner and his German expedition arrived at King Edward Cove. Just six months prior, an outbreak of beriberi at Husvik Company had caused strikes and mass resignations; but at Grytviken, three hundred men worked together without any serious problems. On one of the rare days when the sun broke through the "wreaths of cloud," the Germans were rewarded with magnificent views of mountains that seemed "to hoist their heads defiantly into the sky," noted Filchner. The expedition left on December 11, 1911, to keep their disastrous date with the Weddell Sea pack ice.

Ernest Shackleton's *Endurance* expedition also stopped at South Georgia before it sailed into history. Full of hope and high spirits, Shackleton and his men arrived on November 5, 1914. The foul odor of the whaling factories was unmistakable as stripped carcasses drifted in the water and bumped against the sides of the ship. Shackleton spent one month at Grytviken.

In 1921 Shackleton sailed once more for Antarctica on the *Quest*. The man who had never tasted victory but had achieved glory in defeat was now forty-seven years old. After a rough passage, the *Quest* arrived at Grytviken on January 4, 1922. All that day Shackleton stood on the bridge, gazing through his binoculars toward the familiar coast where he and two companions had crossed the mountains with almost superhuman strength. As the shore passed before his eyes, gusts of wind carried the stale odor of dead whale.

Then, just after 2 A.M., terrible, vise-like chest pains gripped Shackleton and surged up his neck and across his shoulders. Gasping, he blew the small whistle beside his bed for Dr. Macklin. When the physician rushed into the room, Shackleton asked for pain pills, but it was too late. He died of a massive heart attack in the helpless arms of his friend.

When Lady Shackleton received word that her husband had died, she requested that his body be buried on South Georgia. It was here that he and his two companions "had reached the naked soul of man" during their arduous crossing of the island in May 1916. The gray stone is a slab of rough-cut granite that suits the man who never forgot his men or his humanity. Inscribed on the back of the granite marker is a line from Robert Browning's *The Statue and the Bust*: "A man should strive to the uttermost for his life's set prize."

In 1908, British officials consolidated various subantarctic and Antarctic territorial claims into one document, and the just-staffed Falkland Islands Dependencies assumed administrative duties over the new lands. South Georgia magistrates continued issuing leases to the seven whaling stations that supplied the world's demand for high-grade oil and baleen until the last one closed in the 1960s.

Grytviken whaling station is typical. Located within Cumberland Bay, it is not only a ghost town but also a fitting memorial to the whales that were slaughtered to within a breath of extinction. Here are the rusty boilers, bone and blubber cookers, and grinders that now creak and groan with every strong breeze. The heavy chain cables that once clattered against the flensing platform are silent. The fertilizer, or guano, factory still smells of ground bone, sweat, and dried blood. All of these buildings and equipment served one purpose: the systematic slaughter of humpbacks, blues, and fins until the mid-1960s. Today, the absence of whales breaching and blowing in South Georgia waters tells the sad tale better than words and adds to the poignancy of this lost place.

The iron-plated slipway, rubbed raw by the thousands of whales hoisted tail-first up the ramp, was called the Gates to Hell. On the flensing platform, blubber was peeled and rolled like a curl from the skin of an apple. The men wore long spikes on the soles of their boots to keep from slipping in the blood, guts, and muck. After the blubber was removed, the bright red carcass was dumped into the water to make room for the next whale.

Sheds filled with pigs, sheep, and chickens provided fresh meat and eggs for the men. The hospital had surprisingly modern orthopedic equipment and a physician to repair the inevitable broken or crushed bones. Organized slop chests grew into a general store, its shelves stacked with tobacco, socks, stationery, buttons, and needles. Bakers, laundrymen, physicians, electricians, engineers, stokers, telegraph operators, and administrators worked side by side without too many complaints or arguments.

Most of the two hundred graves scattered over South Georgia date from the 1800s. Injuries from fights or cuts from flensing knives sometimes did not heal properly and festered with bacteria. Within days infection overwhelmed weakened bodies, and the men died gruesome deaths.

Not far from Grytviken is Maiviken, or May Cove, named by members of the ill-fated Nordenskjöld expedition in honor of May Day. They found a large green boat buried in the tussock grass, and six iron boilers used by sealers to render elephant seal blubber into oil. On the scenic south side of the bay, they found several graves "near a rippling brook, which fell, a fleecy-white force, down the side of the mountain," wrote Gunnar Andersson. He found the first fossil on South Georgia, a mollusk embedded in a block of stone. The men also explored a nearby cave. Inside were two hearths where sealers had warmed themselves, away from the inclement weather.

On the western side of an islet near the head of the bay is a graveyard for wrecked ships, according to Frank Worsley, Shackleton's navigator. Here were the "lower masts, topmasts, a great mainyard, ships' timbers, bones of brave ships, and the bones of brave men," he later wrote. Broken by the relentless sea, the carved figureheads, teak cabin doors, and skylights were cast ashore at that spot by "some strange freak of eddies." Worsley was convinced that parts of the wreckage dated back to the late 1500s, when Francis Drake battled a Cape Horn storm and one of his ships disappeared among the monstrous waves. Shackleton and Worsley promised each other to return and search for treasure among the rubble. Neither man ever got the chance, however, for Ernest Shackleton died on the *Quest* in 1922 and Frank Worsley, who accompanied the Boss, was too grief-stricken to return without his friend.

Although the island had escaped unscathed from two world wars and seemed destined to avoid conflicts by its isolation, all that changed on March 18, 1982. On that day an Argentine naval ship, the *Bahia Buen Suceso*, arrived with "salvage workers" to collect scrap metal at an abandoned whaling station at Leith Harbor. Supplies were unloaded and the refuge hut was soon occupied with troops instead of pseudo-workers. On April 2, after Argentine forces had captured the Falkland Islands and imprisoned the governor, the *Bahia Paraiso* sailed into Cumberland Bay.

On April 4, two hundred Argentine troops landed at King Edward Point. Twenty-two Royal Marines defended the station for two hours while civilians took refuge inside the church at Grytviken. British personnel surrendered the island and were shipped to Argentina. Parliament declared war, and within days six Royal Navy ships, including one nuclear submarine, sped toward the Falklands and South Georgia. On April 25 the Argentines surrendered the base, followed by the garrison at Leith Harbor the next day.

During the rest of the brief war, action centered around the Falkland Islands. The Royal Navy used South Georgia as a resupply base until the end of hostilities on June 14, 1982.

Macquarie Island

On most days, a gray mist hovers over Macquarie Island and hides the green hills and valleys. Remote yet accessible, this island is a wildlife sanctuary. At least twenty-five bird species, including albatross, penguins, and small petrels, build nests on hilltops and rocky ledges or burrow beneath the tussock grass.

Macquarie is geologically distinct from the other subantarctic islands because it was formed from an uplifted oceanic ridge. The island is relatively young, a mere eleven million years old, and its bedrock is composed of basalt flows and other ocean-floor sediments. When seen from a distance, the island rises stark and steep to an undulating plateau, dissected by deep gullies and streamlets. None of the hills are glaciated.

Macquarie's history, like the other subantarctic islands, is a microcosm of wildlife exploitation. In 1810, Sydney sealer Captain Frederick Hasselborough discovered the island and named it for the governor of New South Wales. News of the island's fur seal bonanza spread quickly, and within ten years no seals returned to Macquarie beaches. The next victims were the elephant seals, hunted for blubber that was then rendered into high-quality oil. When these ponderous seals became scarce, the crews turned to penguins as a new source of oil and eliminated the king penguin colony at Lusitania Bay.

The first Antarctic explorer to visit Macquarie Island was Thaddeus von Bellingshausen, who admired its lush green hills and dark cliffs when he arrived in November 1820. Sealers invited the men into their sod huts, desperate hovels built into the hillside near stored casks of oil, the rewards of their work. Inside, the huts were so dark with smoke from the constantly smoldering fires that the visitors had to be led around by the hand. Low light penetrated windows covered with stretched seal bladders. In gratitude for the sealers' hospitality, Bellingshausen ordered that "grog should be offered to them, as for several months they had not had this their favorite drink."

One evening, violent shocks rocked Bellingshausen and his men near Nugget Point. At first the leader thought a whale had struck the *Vostok*, but the *Mirnyi*'s alarmed captain also reported the jolts. Since it was unlikely that a luckless whale had hit both ships, Bellingshausen concluded that the shocks were earthquakes.

Bellingshausen did not see a single fur seal during his stay, but there were plenty of king, rockhopper, gentoo, and royal penguins. Wild dogs, abandoned by sealing crews, also roamed the island. Before sailing away, Bellingshausen pickled several barrels of "wild cabbage" to make into *shtshi*, a tasty Russian cabbage soup.

In early January 1840, only one ship arrived at Macquarie Island, the American expedition's first rendezvous spot for its five vessels. While leader Charles Wilkes and the other three ships continued towards East Antarctica, Captain Hudson steered the *Peacock* toward Macquarie, accompanied by several royal albatross that glided beside the lone vessel. The pounding surf and long strands of kelp convinced Hudson that this was not the ideal landing spot; after more than a dozen attempts, midshipman Henry Eld and another man struggled ashore to find the rugged hills covered with penguins. "Such a din of squeaking, squalling, and gabbling I never before heard, or dreamed could be made by any of the feathered tribe. It was impossible to hear one's self speak," wrote Eld in his report to Wilkes. The rockhopper penguins "stand erect in rows, which gives them the appearance of Lilliputian soldiers." The men bagged a few penguins and, after a cold dunking in the breaking waves, they returned to the *Peacock*.

Many years passed before another explorer set foot on the island. In the late 1800s, the sealer *Eagle* was wrecked in a gale on Macquarie Island's western coast. Nine men and one woman survived two years on a diet of penguins, skuas, and cormorants. They lined a cave with tussock grass for comfort and waited for a ship to rescue them. On the day when a vessel did spot them, the woman died under suspicious circumstances and was buried on the island.

Gangs of sealers continued to work on the island, rendering penguins into oil. In 1891 New Zealander Joseph Hatch was granted a lease to collect oil on Macquarie. For twenty-five years he ran a successful business, until the public learned how he obtained his high-quality oil—by clubbing at least 150,000 penguins per year. Royal penguins were herded into mazelike corrals that ended near the great boiling vats. Workers dealt the coup de grace to each feathered head and tossed the body into the closest vat. So efficient was this operation that by the early 1900s, over two thousand penguins could be processed at one time. When Hatch reported that the royal population had *increased* since the day he first fired his digesters, the public branded the man a vile liar. But Hatch was correct. The royals had increased by 10 percent because he had not slaughtered mated pairs; instead, he killed a percentage of the "fats," or juveniles, who had swum to shore to molt.

From 1895 to 1900, taxidermist Joseph R. Burton sailed to Macquarie several times to collect and prepare wildlife specimens for museum and private collections. He lived in a small hut where he carefully preserved penguin skins, including an albino, and other birds. In 1900 he closed the door for the last time and left many exquisite specimens inside the dwelling.

One year later, the *Discovery* sailed to the island en route to Antarctica. Robert Scott, Ernest Shackleton, and Edward Wilson spent an enjoyable late afternoon and evening trundling through the tussock grass, surprising the elephant seals that were snoozing in the muck. The men wandered into an abandoned hut and examined moldy crumbs of bread and cheese, utensils and, to Wilson's delight, a fine collection of bird skins. Each was carefully wrapped in brown paper, numbered, and dated. One skin was that of a pure white penguin with a gold crest. Wilson wondered who could have left such treasures, and was convinced that the man must have died. "We looked in all the bunks to see if they contained a corpse, but we couldn't find one," he wrote. Wilson resisted the impulse to carry the skins back to the ship, perhaps hoping that the owner would return one day to claim his work, but Burton never had the opportunity to return for his specimens.

The next Antarctic explorer to land on the island was Australia's irrepressible Douglas Mawson in December 1911. The expedition sailed to Caroline Cove at the southwest tip of the island. As they passed through a narrow channel, a beautiful inlet opened up before the men. Green tussock grass matted the hills, and long streamers of kelp gently undulated with the swells. Glistening penguins porpoised towards the shore, reminding Mawson of "shoals of fish chased by sharks."

Then came "the incessant din" from the penguin colonies on the steep hillsides. They followed the noise along a tumbling stream until they reached the top of a hill. There, spreading out before the men, were fields of birds and eggs. When the penguins marched down the slope and jumped into the foaming breakers, squawking and splashing, Mawson thought they looked like "a crowd of miniature surf-bathers" on a holiday.

Mawson sailed into North-East Bay near a flat-topped slope later named Wireless Hill. A shipwreck with timbers askew seemed to warn the men to stay away, and onshore were two huts but no sign of life. Suddenly, one man dashed out of a hovel, stood on a boulder, and signaled a welcoming message with flags. Mawson later learned that the wreck was the sealer *Clyde*, and that all hands had escaped to shore. Within minutes the rest of the crew crept out of the huts to greet Mawson and his men.

While Mawson prepared to sail to Cape Denison in East Antarctica, five men from the expedition unloaded supplies on the island that was to be their voluntary home until 1913. The group's goal was to build a tower for wireless communication with Hobart, Tasmania, and relay messages from Mawson's Antarctic base. The tower was completed several days later with the help of the stranded sealers, and Mawson arranged for them to return to Hobart on an auxiliary vessel, the *Toroa*.

The newly constructed hut was first named George V Villa, but the more descriptive title, "The Shack," prevailed. The wildlife amused the men, especially the king penguins. Sooty albatross soared above the eastern side of Wireless Hill, and their haunting singular cries emphasized the men's isolation. No fur seals were seen during the entire stay, and many more years would pass before the species recovered from near extinction. However, one introduced species thrived and created havoc—rats scuttled across the roof of the Shack and slid down the space between the outside and inside walls.

For the next two years, the men collected data on botany, zoology, geology, and meteorology while enduring the island's cold, wet climate. The first message they received from Mawson at Cape Denison and transmitted to Hobart was the news that two expedition members, Xavier Mertz and B. E. S. Ninnis, had died during a horrific sledging trek.

In 1933 Macquarie was declared a wildlife sanctuary, and in 1998 the island was granted World Heritage status for its wildlife, geology, and aesthetic values.

Auckland, Enderby, and Campbell Islands

Auckland, Enderby, and Campbell Islands share similar histories of discovery followed by brutal wildlife exploitation. The saga continues with failed financial ventures, abandonment, and finally rediscovery by scientists and tourists. Relics of bygone days—the tiny hovels that sealers called home for many months, the rusty try-pots and the wooden bones of wrecked ships—give visual evidence of the harsh era. Broad sweeps of tussock grass and other formerly pristine vegetation were trampled by cattle and cropped to the tender roots by sheep. Sealers set fire to the tussock grass to flush out wild pigs for fresh meat and elephant seals for oil. Damage to the landscape from these practices still mars the sweeping vistas.

On calm sunny days, when the beautifully clear notes of the bellbird sound like a call for celebration, it's easy to understand American explorer Captain Benjamin Morrell's unbridled enthusiasm. After his 1830 visit, he declared that Auckland Island was the perfect choice for a small settlement, with its hills covered with "lofty trees" whose leaves made "an excellent substitute for spruce in making that pleasant and wholesome beverage, spruce-beer." In reality, gnarled and stunted rata forests, peat bogs, and impenetrable scrub belied his descriptions. For those who were cast ashore against their will, the islands' true nature became all too apparent when bone-chilling gales, pelting rain, and intense hail squalls battered sealers and castaways for days without a break. Men who had sailed around Cape Horn and Cape of Good Hope swore that they had never experienced such violent storms as those that batter these islands.

AUCKLAND ISLAND

In August 1806 Captain Bristow sighted Auckland Island, but high swells and thunderous breakers beating against the basalt cliffs made landing impossible. The next year he returned and released many pigs, which dashed for shelter in the rata forest and for food in the tussock grass. The fur seals that crowded the shore didn't sense the danger until it was too late. Captain Bristow returned with thousands of silky pelts.

During the next ten to fifteen years, many sealing crews lived on the island in mud and tussock grass hovels, grimy with grease splatters from fried seal steaks. Untold thousands of fur seals were slaughtered and stripped of skins, the carcasses left to feed giant petrels, skuas, and albatross. The fur seal bonanza collapsed when the number of animals dwindled until none returned to the island. Even today, the population is small.

From the late 1820s through the early 1830s whalers came for the calving season, when right whales gathered in the sheltered bays. At least thirty years before Norwegian Svend Foyn perfected the harpoon gun, a French whaler in the 1830s became obsessed with designing a gun that would shoot a harpoon. When test after test failed, he killed himself in a fit of despair and was buried on Auckland Island.

On March 7, 1840, American Captain Ringgold steered the *Porpoise* into a harbor known then as Sarah's Bosom. The men made several excursions on land, but the rata forest and peat bogs made hiking tiresome and dangerous. Yet, the island delighted the men. Small birds sang with such clarity that the sounds reminded the men of distant bells tolling across the valleys. "Hawks, too, were numerous," wrote Assistant Surgeon Holmes, "and might be seen on almost all the dead trees, in pairs." The men found two huts and several gardens brimming with turnips, cabbages, and potatoes. Nearby was the grave of the failed French inventor. On March 10, after just three days on the island, the Americans sailed away.

The very next day, Dumont d'Urville anchored his two ships in the same harbor. During the entire week, storms assaulted the island and the men were forced to huddle inside a leaky hut. Whenever the rain ceased for a few minutes, clouds of sand flies bit the men and burrowed into their wool clothes. At night, they listened to the rats chew on the ropes of hammocks the tired men had slung inside the hut. During the week, a whaling captain showed d'Urville the smudged letter Captain Ringgold had attached to a brightly painted plank stuck into the ground. D'Urville then nailed his own plaque with a record of his Antarctic achievements.

When the British expedition arrived in November 1840, leader James Ross found Ringgold's and d'Urville's plaques. He also saw many wild pigs, pointing out in his narrative that they had "increased in numbers in a surprising manner." Botanist Joseph Hooker was especially delighted with Auckland vegetation and managed to collect two hundred representatives, at least two-thirds of the total number of plant species. "Perhaps no place in the course of our projected voyage in the southern ocean promised more novelty to the botanist than Auckland Islands," he wrote.

Not too long after Ross left, a group of Chatham Island Maoris paddled their canoes into the same harbor and settled on the shore in 1842. When Ross arrived in England, he discussed the whaling prospects around the Auckland Islands in glowing terms with Charles Enderby. By 1850 Enderby and three hundred British colonists settled at Sarah's Bosom, now named Port Ross, and surprised the Maoris, who were still there. Within a few years, the Enderby settlement was abandoned; the disillusioned farmers and merchants opted for Sydney, Australia, and a fresh start. The Maoris, too, left a few years later for another island group, east of New Zealand.

Sealers and whalers continued to anchor at Port Ross and other deeply indented harbors and waterways. At least seven shipwrecks added to the island's grim reputation as a geographical siren to the unwary.

One ship that continues to inspire landlubbers' romance with the island is the *General Grant* and its lost gold. On May 14, 1866, the American vessel drifted towards the western cliffs while the captain waited anxiously for a wind to fill the sails and carry the ship out to sea. During the night strong currents swept the vessel inside a huge cavern within 1,510-foot walls. Suddenly, a surging wave rolled like thunder into the cave and lifted the ship until its foremast struck the roof hard enough to drive the wooden mast through the hull. Before the crew could repair the damage, a terrible storm whipped up the swells that seethed inside the cavern. The *General Grant* sank, carrying 70 to 80 kilograms (about 150 to 175 pounds) of gold and sixty-eight people to the bottom of the cave.

Fourteen men and one woman—a stewardess—survived the wreck. They rowed two boats to Disappointment Island, rested, and then continued to Port Ross where they built a rough shelter. Albatross bones were whittled into needles and flax was used for thread. They tied rabbit skins on their feet and wrapped up in sealskins for warmth and protection against the pounding rain. After several months, the survivors moved to Enderby Island and tamed the wild goats and pigs. They were finally rescued eighteen months after the *General Grant* sank. To this day, the gold has never been found.

New Zealand authorities addressed the island's sinister reputation by establishing several depots filled with blankets, clothing, tools, weapons, and food for shipwrecked survivors. More pigs, goats, and rabbits were cast on shore as potential food sources.

After two years on Antarctica's Ross Island, Wilson was once more delighted with the color green and the feel of warm sunshine on his back. On shore, the beaches were littered with wreckage from ships, wooden planks, and other rubbish that marked the high tide zone along a few beaches. As the *Discovery* passed Enderby Island, Wilson watched cattle grazing near the boathouse on Sandy Beach while Hooker sea lions rested on the white sand.

ENDERBY ISLAND

Enderby Island is a tiny treasure that lies just northeast of Auckland Island. Enderby's green grass, warm sandy beaches, and picturesque rata forests seduce most people who explore the island. The six-foot-high stunted trees, gnarled and twisted from the persistent winds, support a dense red-flowering canopy that shelters rare yellow-eyed penguins, Hooker sea lions, and red-crowned parakeets. The crisp notes of bellbirds, the protracted rumblings of adult sea lions, and the catlike calls of their pups blend into a pleasing harmony. Fresh, spring-green mosses and grass with pockets of tiny wildflowers cover the low island. The undercut banks of steep-sided gullies shelter yellow-eyed penguin nests.

In March 1904, Edward Wilson, Robert Scott, and eight others landed at Sandy Beach on a warm sunny day. The idyllic scenery and Hooker's sea lions enthralled Wilson. "Why the island has never been colonized is hard to understand," he later wrote in his diary. There were plenty of grazing cattle "all left to fend for themselves and they seem to do very well."

In the mid-1800s, one hundred head of cattle were shipped as an experiment with ranching on Enderby. Although the owners soon left because of terrible weather and hardship, the black and white cattle thrived on the local vegetation. Over the years the animals developed long bodies and short muscular legs that suited the harsh environment. In an attempt to return the island to its former pristine state, New Zealand ecological authorities destroyed the herd in 1992 but saved one cow named Lady. In 1998 New Zealand scientists at the Ruakura Research Center successfully cloned Lady to perpetuate the Enderby herd's hardy genes.

CAMPBELL ISLAND

Campbell Island is untamed and timeless, a wild Scottish Brigadoon that awakens when shafts of sunlight penetrate the heavy gray mist. The eleven-by-nine-mile island wears green well, a vibrant color that intensifies on gloomy days. The island enjoyed splendid isolation until January 4, 1810, when Captain Frederick Hasselborough sailed the *Perseverance* into a long, dogleg-shaped harbor, later named for his ship. Seven sealers disembarked with enough supplies to last several months while they slaughtered thousands of fur seals. Hasselborough also discovered Macquarie Island, far richer in fur seals, that same year. In 1828, the *Perseverance* was wrecked in its namesake harbor.

In 1839 John Biscoe visited Campbell Island and endured a week of summer gales and pelting rain. That same year, John Balleny sailed into Perseverance Harbor and rescued three men and one woman, abandoned by their ship, in exchange for a mere seventy fur seal skins, all that they had collected over four years. Unfortunately, the woman drowned under suspicious circumstances the same day she was rescued.

In December 1840 explorer James Ross anchored in the harbor and found the crosses of a few sealers and the lone woman. Perhaps that grim discovery darkened Ross's mood when he described the island as having a "desolate appearance," but the wild landscape worked its magic on botanist Joseph Hooker, who roamed over the island and collected sixty-six plants.

Beginning in 1909, shore-based whalers butchered southern right whales and continued the slaughter until too few returned. Remnants of the huts and sheds of an early farming attempt at Tucker Cove are still visible today. Started in 1895, the farm was abandoned in the early 1930s and four thousand sheep were left to fend for themselves. Surprisingly, the ruminants flourished. Concern for the native plants triggered conservation efforts, and the island was declared a nature reserve in the 1950s. The sheep were destroyed.

One of the few surviving introduced plants is a Sitka spruce at Camp Cove, dubbed "the loneliest tree on earth." It was planted in the early 1900s in a sheltered site. Although it is green and looks healthy, the spruce has grown just twenty feet in almost a century, a paltry figure compared to the 150 to 200 feet the tree achieves in southern Alaska.

Another supposedly introduced plant was Scottish heather. According to legend a lone woman, dressed in a Royal Stuart tartan, lived in a sod hovel on Campbell for a number of years. Next to the door was a shrub of heather that she had planted when she first arrived on the island. Whenever whalers spotted the cloaked woman walking near shore, her Glengarry bonnet always had a fresh sprig of heather stuck in its band. Although she hid in the rata forest if someone came to her hut, the men concluded that she was the granddaughter of Bonnie Prince Charles, the result of her grandmother's affair. Fearing a threat to the throne, enemies had kidnapped the young girl and gave her a choice: exile or death. She was abandoned on Campbell Island. Legend has it that the lone heather shrub still blooms somewhere on Campbell Island, and the wind carries its scent each summer.

• PART III •

EXPLORING ANTARCTICA'S GEOGRAPHY AND WILDLIFE

• 13 •

THE HUB OF GONDWANA

MAPS OF THE KNOWN WORLD INTRIGUED SIR FRANCIS BACON, English philosopher and intellectual. In 1620 he noted that structural similarities among the continents defied the odds of mere chance. "The New and Old World are both broad and expanded toward the north, and narrow and pointed toward the south," he wrote in *Novum Organum*. Certain continents seemed to belong together, like "Africa and the Peruvian continent, which reaches to the Straits of Magellan, both of which possess a similar isthmus and similar capes, a circumstance not to be attributed to mere accident."

Just forty-six years later French monk François Placet studied a globe of the world and concluded that the continents were once joined, but had broken apart during the Great Flood. He titled his work *The Breaking Up of the Greater and Lesser Worlds: Or, It Is Shown That Before the Deluge, America Was Not Separated from the Other Parts of the World.* With the discovery and subsequent mapping of Australia and New Zealand from the late 1600s through the 1700s, a new problem puzzled the monk's followers. The "fit" of the continents in the Southern Hemisphere was not perfect. A pivotal chunk, like the hub of a wheel, was missing.

In January 1820, the hub was discovered. Thaddeus von Bellingshausen was the first to view the Antarctic continent; three days later Edward Bransfield sighted the Antarctic Peninsula. For theologians, the jigsaw-puzzle fit of the continents now solved a metaphysical problem or two. Scientists, however, scoffed at the notion of one large landmass in the Southern Hemisphere. They argued that the supposed "fit" of the continents was nothing more than a coincidence. If one landmass had existed eons ago, they reasoned, then similar rocks and plants should be found on the remnant continents, linking these parts to the whole. It would not be long before evidence was found to refute their comfortable steady-state view of the continents.

The Fossil and the Theory

In 1829 American Dr. James Eights sailed on the *Seraph* with two other ships to explore the South Shetland Islands and to search for new sealing grounds. There, Eights discovered a treasure of another kind—a fine piece of fossilized wood, the first indication that these islands had not always been "cold and cheerless."

Botanist Sir Joseph Hooker, who traveled with Sir James Ross's 1840–44 voyage, was puzzled by the similarities among the plants on Îles Kerguélen with those found on the Falkland Islands, Tierra del Fuego, Tasmania, and New Zealand. "I am becoming slowly more convinced of the Southern Flora being a fragmentary one—all that remains of a great Southern continent," he wrote in an 1851 letter to Charles Darwin.

Fossil plants interested Austrian geologist Eduard Suess in 1885—especially a deciduous species with fernlike leaves. The fossil was *Glossopteris*. How could this distinctive plant, he wondered, have thrived in India, Africa, and South America at the same time, millions of years ago? Even the sandstone and limestone rocks that contained the fossils shared remarkably similar characteristics. For Suess, the only reasonable explanation was that the continents had been joined at the time

Glossopteris flourished. Then, in the dim geologic past, the massive continent broke up and the pieces drifted apart. He named the supercontinent Gondwanaland, now referred to as Gondwana, after an ancient tribe that had lived in central India. Suess was promptly dismissed as a quack because he had not suggested any mechanism powerful enough to move continents.

Still, the idea of a great southern continent, a concept that echoed the works of the ancient Greeks, intrigued more open-minded scientists. In 1898, Henryk Arctowski, meteorologist with the *Belgica* expedition, concluded that the Antarctic Peninsula "is connected with Patagonia by a submarine ridge...and that the chain of the Andes reappears in Graham Land [northern section of the peninsula]."

In 1901 a young scientist had just finished reading a paper on the geology of Victoria Land to the Royal Geographical Society. During the discussion afterward, a member commented on the similarities between Victoria Land rocks and those of eastern Australia's Great Dividing Mountains. He then suggested that Antarctica must have linked South America with Australia. How else, the member mused, could scientists explain the same allied marsupial species on both continents—unless the continents had been joined?

In February 1912 Robert Scott and his men stood near the banded perpendicular walls of the Beacon Sandstone, near the top of the Beardmore Glacier. There, the men gathered thirty-five pounds of rock specimens—including fossils. The doomed party placed such value on their collection that they hauled the extra weight on their sledge to their final camp on the Ross Ice Shelf. Two years later a botanist examined the fossil plants: the leaves and stems were *Glossopteris*, a legacy of Gondwana.

Although the concept of continents torn asunder was not seriously discussed in learned circles, annoying pro-Gondwana questions continued to surface. Summarizing the views of several maverick geologists and botanists, Alfred Wegener solved these and other puzzles in his 1915 book, *The Origin of Continents and Oceans*. Nothing short of outrage greeted its publication. In his book Wegener proposed that a single supercontinent, Pangaea, existed until two hundred million years ago, when it fractured into at least two substantial continents, Laurasia and Gondwana. Over time, Gondwana broke apart, and oceans filled the ever-widening gaps as these landmasses drifted to their present positions. He justified his revolutionary concept by pointing to the worldwide distribution of *Glossopteris*, ancient glacial erosion features and thick silt deposits in today's tropical areas, and the superb fit of the North and South Atlantic continental coastlines. He also referred to Ernest Shackleton's discovery of coal and coniferous fossils during the explorer's attempt on the South Pole in 1908.

Unlike his predecessor Eduard Suess, Wegener suggested a mechanism that caused the continents to break apart and drift toward the equator: the earth's centrifugal force. Tidal drag and the gravitational effects of the sun and moon also influenced this drift, he reasoned. Scientific communities on both sides of the Atlantic tromped on Wegener's theory in lecture halls and scholarly journals; within a few years, his professional reputation lay in ruins.

During World War II, the mechanism that had fractured Pangaea and Gondwana was inadvertently discovered when American geologist Harry Hess mapped the seafloor near Iwo Jima with sonar. He then charted sections of the Mid-Atlantic Ridge, a meandering mountain range that flanks a deep rift valley. Hess and other geologists suspected that the rift allowed molten material from the earth's mantle to rise and push older crusts away laterally on both sides of the Mid-Atlantic Ridge. If the hypothesis was correct, the seafloor should be thinnest near the rift and thickest at the edge of the continents, Hess reasoned. This proved to be true, and scientists later determined that the oldest seafloor material was two hundred million years old—the time of the supposed breakup of Pangaea.

Another geologist, Robert Dietz, popularized the term "seafloor spreading." This concept not only provided the mechanics for continental drift but was also the foundation of a new discipline that revolutionized the earth sciences—plate tectonics. According to the new theory the earth's crust, or *lithosphere*, is a mosaic of twelve plates that slide over the upper mantle, or *asthenosphere*. Throughout the eons, movement of these plates—their speed, direction, and force—determined the geological face of the earth.

Although evidence for plate tectonics was irrefutable by the mid-1960s, many earth scientists resisted the integration of continental drift into the theory. Giving up the notion of stable continents and permanent ocean basins was especially difficult. A few geologists suggested that strong winds and waves carried *Glossopteris* seeds to today's continents instead of the plants propagating across the single vast Gondwana landscape.

That attitude changed in 1967 when a geologist from Ohio State University, Dr. Peter Barrett of New Zealand, made a profound discovery near Graphite Peak in the Transantarctic Mountains. There, half buried in an ancient river deposit, lay the fossil jaw fragments of a four-legged lizardlike amphibian. Like *Glossopteris*, this animal, called a labyrinthodont, was another Gondwanian legacy. Since the amphibian was a shallow-water specialist and incapable of swimming long distances, its presence in Antarctica was viewed as irrefutable proof of the existence of Gondwana. Within a few years, hundreds of amphibian and reptilian fossils were recovered—and earth science textbooks soon included hefty chapters on plate tectonics and continental drift. Antarctica was now the accepted hub of fractured Gondwana.

In 1982 another riddle was solved when a fossil of a rat-size marsupial was found on Seymour Island, near the tip of the Antarctic Peninsula. This discovery proved that marsupials had traveled from South America to Australia via Antarctica, as speculated in 1901.

Today, Antarctica's role as the centerpiece for Gondwana reconstruction models is unquestioned. Although geologists are still tweaking certain geological time sequences, the continent's physical history, as recorded in layers of rocks, can now be read.

WHEN DINOSAURS ROAMED AND *GLOSSOPTERIS* GREW

More than four billion years ago a massive shield of sediments formed over Eastern Antarctica, sharing structural characteristics with Australia, Africa, and peninsular India in the Southern Hemisphere. Throughout the ensuing eons, tectonic plates of the earth's crust, some embedded with continents, drifted in a choreography of collisions, rifts, breakups, and reconstructions. Freshwater and marine sediments were deposited in thick layers and then uplifted into mountains and high plateaus, only to be leveled again by erosion. Based on the configuration of the continental plates, seas were born and later vanished, leaving invertebrate fossil records in the marine sediments.

Beginning about four hundred million years ago, fine and coarse-grained river sand, gravel, mud, peat, and swamp flora filled broad basins for hundreds of miles. Over time the sediments formed layers of pink and yellow sandstones, interspersed with dark shales and thick seams of low-grade coal. This sequence of sediments, famous for its layer-cake appearance and its abundance of fossils in Antarctica, is called the Beacon Supergroup.

Approximately three hundred million years ago, an ice cap formed deep in the interior of Africa and surged across Antarctica, peninsular India, Australia, New Zealand, and South America. Forests of *Glossopteris* shrubs and trees died as the climate chilled and the ice sheet reached its farthest extent. A warming trend followed and the ice retreated, leaving a distinctive trail of pebbles, boulders, and silt in its wake.

Toward the end of the Paleozoic era, about 250 million years ago, just one supercontinent existed. Pangaea stretched from pole to pole and was surrounded by one ocean, Panthalassa. Across the Southern Hemisphere landmass, four-foot-long amphibians resembling oversize newts, roamed near ponds and streams, eating dragonfly-like insects. Small shallow-water reptiles with periscopic eyes mounted on long stalks on the tops of their heads surveyed lands where conifers and cycads flourished. About two hundred million years ago, long rifts split Pangaea into two continents, Laurasia and Gondwana.

A mere twenty million years later Africa separated from Antarctica, the heart of Gondwana. Volcanoes erupted near the rift zones and basalt covered portions of the Beacon Supergroup. India broke away and moved toward Eurasia, a journey that ended in the formation of the Himalayan Mountains. About 140 million years ago South America and Africa rifted apart, and the South Atlantic Ocean was born. New Zealand separated from Gondwana while dinosaurs roamed over the Antarctica–Australia–South America landmass. Southern beech forests and flowering plants flourished, and marsupials scurried in the underbrush. It was an idyllic period in Antarctic geologic history.

The end came swiftly sixty-five million years ago at the close of the Cretaceous period. Within a few million years about 70 percent of the earth's species vanished, including the dinosaurs. During the next twenty million years, the clash of tectonic plates uplifted the Beacon Supergroup, creating the Transantarctic Mountains. Volcanoes erupted along the mountains' seventeen-hundred-mile perimeter, a paradox of fire and ice that continues today.

Australia rifted from East Antarctica about thirty-five million years ago, carrying marsupials that evolved into separate and distinct species. Moisture from the intervening ocean condensed as snow over Antarctica. When South America drifted northward, the Drake Passage opened, allowing the southern ocean current to flow unimpeded around the now isolated continent that rested over the South Pole.

As temperatures fell, far more ice accumulated than was removed by winds and bergs. Interior ice fields merged into a deep mantle that buried most of the Transantarctic Mountains and, as it surged toward the coast, obliterated the nesting sites of six-foot penguins, blue-eyed shags, and skuas. Beyond the shore, the ice eroded the continental shelf, gouging and crumpling it into an abrupt precipice. For more than a hundred miles the ice sheet surged out to sea, carrying boulders and other rock debris plucked from the floors of the now ice-filled valleys.

During interglacial periods, temperatures increased and ice ablation outpaced accumulation. The ice retreated, leaving behind piles of giant erratic boulders, broad U-shaped glacial valleys, and the steepest continental shelf of any landmass.

THE NOT-SO-ANCIENT ARCHIPELAGO

West Antarctica, which includes Marie Byrd Land and the Antarctic Peninsula, has a much younger but more complex geologic history than East Antarctica. Unlike continental East Antarctica, which rides on a single tectonic plate, West Antarctica is an archipelago composed of at least three plates that rotate independently. The oldest rocks date back about six hundred million years and are similar in structure and mineral content to South America's basement rocks.

Until the end of the Cretaceous period, sediments filled several marine troughs that stretched from South America to the Peninsula. Plants and marine life have been preserved as spectacular fossils within the layers of sandstone and shale, especially near the tip and eastern side of the Antarctic Peninsula. During the Cretaceous, volcanoes erupted along the western edge of South America—from Panama to Tierra del Fuego, in the South Shetland Islands, and from the Antarctic Peninsula to Marie Byrd Land. Pressure from several tectonic plates compressed the rock layers like a folding fan. A series of upheavals formed the Andes and the Antarctic Andean Range, the spectacular spine of the Peninsula.

The fossil boundary between the abrupt end of the Cretaceous with its mass extinction and the start of the Tertiary is especially distinct on Seymour Island in the Weddell Sea region. A thin band of fine-grained sediments riddled with the very rare element iridium separates the two geologic periods. Since iridium is associated with meteorite impacts, some scientists believe an asteroid slammed into Earth at the end of the Cretaceous and extinguished the dinosaurs and many other species. Immediately above this unusual band lies a layer of fossilized fish bones, a massive marine graveyard.

Later, during the Tertiary period, ten-foot-tall carnivorous birds chased marsupials and mammals across mud flats, leaving behind deep footprints that became part of Seymour Island's geologic history. Interspersed among the layers of sandstone and shale are fossils of ferns and flowers, plants that thrived until lava flowed over the land and the Drake Passage opened.

As the climate cooled, the beech forests thinned. The trees' growth became stunted, the leaves were smaller, and the shallow root systems shriveled in the cold, dry soil. The trees first disappeared from the interior and then from the coasts. The fibrous structure and texture of trunks, branches, and leaves have been exquisitely preserved as silica replaced the organic matter, molecule by molecule. Even the rings on petrified stumps show temperature fluctuations during the life of the trees.

The East and West Antarctic ice sheets formed at about the same time and merged in Marie Byrd Land. Beneath this seamless white boundary lies an ice-filled trough. The West Antarctic ice sheet is the cement that binds the archipelago to East Antarctica.

White Beacon on a Blue Planet

Astronauts in space describe Antarctica as radiating light "like a great white lantern" across the bottom of Earth, a bright beacon surrounded by blue seas. It is a world of ice that reflects over 80 percent of sunlight back to space. Flat spirals of low-pressure systems swirl around the almost circular continent like a well-choreographed dance, with movement from west to east. Two deep gulfs, lying opposite each other, bite deep into the continent. Each is covered with a floating lid of ice: the Ronne-Filchner Ice Shelf in the Weddell Sea and the Ross Ice Shelf in the Ross Sea.

The thousand-mile-long Antarctic Peninsula juts out into the Drake Passage like a crooked finger, beckoning the curious six hundred miles southward. The Transantarctic Mountains bisect the continent from the tip of Victoria Land to the Theron Mountains in Coats Land at the edge of the Weddell Sea, a distance of about seventeen hundred miles. Most of the peaks east of the Queen Maud Range are buried beneath the ice sheet. Only a few isolated mountaintops, called "nunataks," poke through the white blanket. Although approximately thirteen to fourteen thousand miles of coastline circle Antarctica, pebbly scraps of shore account for less than eight hundred miles. Instead, grounded ice cliffs, glaciers, and ice shelves outline the continent.

Ice defines Antarctica, for it covers more than 98 percent of the continent. The East Antarctic ice sheet is a broad white dome that caps the polar plateau at nine to twelve thousand feet above sea level and merges with the West Antarctic ice sheet in Marie Byrd Land. Together, they blanket about 5.5 million square miles—nearly one and a half times the size of the United States or about half the area of

Africa. The topography of the ice sheet hints at the underlying post-Gondwana landscape. Broad undulations and basins suggest mountain ranges higher than the Alps, separated by deep U-shaped valleys.

Although estimates vary, most scientists believe the ice sheet formed eleven to fifteen million years ago. Since then it has gained and lost most of its mass at least fifteen times, cycling through each Ice Age. Today, the ice sheet's thickness averages about one and a half miles, but reaches nearly three miles in Wilkes Land. Its unimaginable weight has depressed the underlying bedrock; if the ice vanished, the land would rise about three thousand feet in places. East Antarctica would be a kidney-shaped continent about the size of Australia, and West Antarctica would show its true morphology—a mountainous archipelago extending from the Antarctic Peninsula through Edward VII Peninsula.

Like all geographic features, erosion carves and shapes the surface of the ice sheet. Drift ice, tiny windblown grains of ice that have the gritty consistency of baking soda, sculpts the ice into wafflelike formations called sastrugi. Where gales blow incessantly from one direction, the sastrugi looks like a broad field of plowed snow, the rows polished to a high sheen and aligned in the same direction. If the wind changes directions frequently, the sastrugi looks like a stormy sea of frozen whitecaps.

Ice streams surge within the slower-moving ice sheet like swifter ocean currents within the sea. Favorable bedrock conditions, such as broad valleys or troughs, form channels for these ice streams. The areas where they converge show signs of extreme ice-flow pressure and stress—great buckled ridges of ice and gaping crevasses. Ice streams that spill into the sea, called ice tongues, protrude for more than forty miles beyond the shore, spawning huge newsworthy bergs along deep fracture zones.

Gravity pulls the ice sheet downward from an average elevation of about nine thousand feet on the plateau toward sea level. It creeps like a viscous plastic, less than a few feet per year at the South Pole to hundreds of feet per year near its edges. When the ice squeezes through the Transantarctic Mountains, it begins a steeper and faster descent in the form of heavily-crevassed glaciers. These rivers of ice nourish the ice shelves or calve directly into the sea as icebergs. When Douglas Mawson sighted Adélie Land, with its massive ice cliffs along the coast, he felt elated—and overwhelmed. "Here was an ice age in all earnestness," he wrote in 1915, "a picture of Northern Europe during the Great Ice Age."

• 14 •

EXPLORING ANTARCTICA'S WILDLIFE

A SHIFTING BOUNDARY OF FOG MEANDERS AROUND ANTARCTICA between 50° and 63° south latitude. This gray curtain marks the Antarctic Convergence, or Polar Front, a biological barrier formed by the clash of cold, oxygenated, nutrient-rich water from the south with warm unproductive seas from the north. Rapidly falling air and sea temperatures signal the presence of this boundary. Even more convincing, albatross and other seabirds suddenly glide through the gray gloom and hover above the charcoal-colored swells to feast on bits of the bounty: krill. These small, shrimplike crustaceans found in Antarctic oceans form the largest biomass on Earth at one billion tons.

The turbulent southern seas supply the nutritional broth for phytoplankton, tiny microscopic plants that are food for krill. The abundance of nitrates and phosphates, the extended hours of daylight, and the retreat of sun-blocking sea ice provide the essentials for the plants' growth. By early spring the sea south of the Antarctic Convergence is checkered with fresh phytoplankton pastures, extending for miles in all directions.

This abundant food source attracts krill, the foundation of the Antarctic food chain. Without these tiny creatures to provide meals for higher-ranking food chain members, the entire Antarctic ecosystem would collapse. This chapter begins with this smallest but most significant link, followed by descriptions of fish, penguins, other birds, seals, and whales.

Krill

During the austral spring, swarms of two-and-a-half-inch, shrimplike krill (*Euphausia superba*) blanket up to sixty square miles of sea to depths of hundreds of feet and tint the water a deep maroon. Female krill lay several thousand eggs at one time, which sink about one-half mile before hatching. Slowly, the larvae ascend toward the sunlight and surface, feeding on drifting phytoplankton.

In winter, juvenile krill graze on the thick mats of algae on the underside of pack ice. By spring, if penguins or crabeater seals haven't plucked the crustaceans from their algal food fest, the young krill mature "in countless myriads," as whaler Henryk Bull wrote in January 1895, near Cape Adare. "When a floe is broken, they scatter for shelter in millions, reminding one of the animation in a disturbed ant-heap."

The relentless slaughter and subsequent near extinction of the great whales, once major hunters of crustaceans, caused the krill population to explode. In turn, penguins and crabeater seals have especially benefited from the krill bonanza.

Humans have also responded to the krill surplus, but with much less success. Fisheries from several countries believed that the almost limitless protein-packed morsels were the solution to end world hunger. During the 1970s, krill harvesting began in earnest. However, the crustaceans had a few defenses that foiled human predators. First, krill shells contain high levels of fluoride in doses that are poisonous to humans. Shelling the miniature shrimp was labor-intensive for the fisheries; profitability shrank as payroll costs surged. Second, krill turns into a black muck as potent enzymes dissolve its protein; the meat must be processed almost immediately. Because of these difficulties and the tempestuous nature of the seas surrounding Antarctica, krill harvesting has now declined to less than one-quarter million tons per year purchased for farms.

Krill anchors the entire Antarctic ecosystem. Any change in its environment—a decrease in sea ice formation or the potential damaging effects of UV radiation on phytoplankton, for example—may destroy the food web. W. Burn Murdoch, an artist who sailed on the Dundee whaling expedition in 1892, was among the first to recognize the crustacean's important link in the food chain. "We found stones and red shrimps inside the penguins; and penguins, red shrimps, and stones inside the seals."

Fish Among the Floes

At least 220 fish species are found south of the Antarctic Convergence; they include Antarctic cod, dragon, and ice fish. In order to survive in the frigid waters, several species have evolved a curious mechanism to keep from freezing to death. The fish manufacture protein-based antifreeze molecules that stop ice crystals from forming in their tissues. This ice inhibitor allows the fish to thrive in subfreezing temperatures.

One unusual group of fish has no oxygen-carrying red blood cells, or hemoglobin. Transparent blood gives the ice fish an ephemeral, ghostlike quality as it hovers above the seafloor or actively pursues prey. Its large heart responds to the chase by pumping at least ten times the volume of blood per beat as other fish because the absence of hemoglobin makes the ice fish's blood less viscous. Large fins and a body without scales enhance the transfer of oxygen from the water to its cells. Minute amounts are then carried in the blood plasma to vital organs, enabling the ice fish to chase its prey with bursts of energy.

All Antarctic fish mature slowly and generally do not reproduce until they reach five to seven years of age. Unlike many Arctic species that spawn millions of eggs, most Antarctic fish lay just a few thousand larger, yolk-rich eggs. They also deploy one unusual tactic to improve the odds of their offspring's survival: the adults guard and defend their fry long after the eggs have hatched.

Penguins

Antonio Pigafetta, who sailed with Magellan in 1520, wrote one of the earliest descriptions of penguins. "Strange black geese" were spotted on two small islands, and the men stopped to investigate. "They are unable to fly," noted Pigafetta, "and they live on fish, and are so fat that it is necessary to peel them." Magellanic penguins nesting on islands in the Strait of Magellan remained a food staple for later Antarctic expeditions.

Since rotundity was an admired characteristic of a potential food source, Pigafetta's "strange black geese" were soon called *penguigos* by Spanish explorers, modified from the Latin word for fat, *pinguis*. Welsh sailors named the birds *pen gwyns*, which means "white heads." Although penguins have black heads, their size and markings may have reminded the homesick Welshmen of the now-extinct great northern auks.

As travel to Tierra del Fuego and the Falkland Islands increased, descriptions of penguins changed with the times. One captain who visited the Falklands in the late 1760s wrote that "on shore [penguins] walk quite erect with a waddling motion, like a rickety child." In 1825 James Weddell agreed with Sir John Narborough's whimsical description of penguins as "little children standing up in white aprons." Edinburgh artist W. Burn Murdoch watched penguins in 1892. "They scurry over snow in an upright position, like little fat men in black coats and white silk waistcoats," he wrote. One Scots sailor, noted Murdoch, referred to penguins as "thae blasted funny wee beggars."

Although seventeen or eighteen penguin species (ornithologists disagree over the exact number) live in the Southern Hemisphere, only the Adélies and emperors are circumpolar and breed on the continent or offshore ice. Gentoo penguins return to established colonies on subantarctic islands and the Antarctic Peninsula, south to about 65° latitude. Chinstraps, macaronis, rockhoppers, kings, magellanics, and royals nest on the subantarctic islands. The remaining penguin species are found in rookeries in South Africa, Australia, and New Zealand.

Fossil records show that penguins evolved about sixty million years ago from birds that flew. During the ensuing eons, penguin species exchanged broad wings for bony flippers. Their once-hollow bones evolved into solid ones for added ballast as they dove for krill, squid, and fish. Today, the muscles that control the flippers are the largest and strongest in a penguin's body. With powerful flipper thrusts, penguins fly through the water at up to 15 MPH, using tails and feet as modified rudders.

For long distance sea travel, penguins "porpoise" to destinations by leaping from the water, gulping air, and then diving back into the sea in rapid succession. On land penguins walk with a pronounced waddle, as though invisible ropes shackle their ankles. Surface lumps or cracks confound the birds. More times than not, they slip, lurch, or stumble over the least offending irregularity, or direct a myopic glare toward the potential trip-up. For more rapid transit over the ice, penguins flop on their bellies and toboggan across the surface. With flippers spinning like whirligigs and feet kicking up puffs of snow, penguins can easily outdistance humans.

To keep warm as they scoot across the ice or huddle against the bitter wind, penguins sport a double layer of small tightly packed feathers, about seventy per square inch. The outside layer has stiff curved feathers that overlap in rows like tiles on a roof. At the bottom of each shaft is a woolly bit of down, which forms the second barricade against the cold by trapping a warm layer of air near the skin. Penguins waterproof their plumage by applying a coat of oil from a gland near their tails. Meticulous grooming keeps their feathers in tiptop order.

Another defense against the cold is the one- to two-inch layer of blubber that insulates core organs. Because of fat's effectiveness, overheating is a problem for penguins when the temperature soars above 30° F. To rid themselves of excess heat, penguins fluff their feathers, eat snow, and pant. The featherless undersides of flippers flush pink as their blood vessels expand to dissipate heat. Bare feet also help to regulate internal temperatures.

Penguins' two-tone feathered tuxes are effective garb for evading predators, especially leopard seals. When penguins swim among the floes and bergs, white chests make them difficult to see from below. The deeper-swimming leopard seals only see a blur of white as they peer upward toward the diffused light. If the seals peer down into the water from an ice floe, the penguins' black backs blend with the dark sea.

High traffic areas near rookeries are favorite haunts for predators. The brush-tailed penguins—Adélies, chinstraps, and gentoos—travel almost daily to and from the sea to keep their chicks' krill cravings satisfied. When penguins spot a leopard seal prowling near a preferred route back to the colony, the panicky birds race toward the shore and hurl themselves out of the water like catapulted bowling pins to the tops of rocky ledges—and safety.

Penguin nesting materials are mostly flat, round, cracker-sized pebbles, bits of feathers, and ancestral bones. Small stones with just the right heft and shape are sometimes a rare commodity in the mega-rookeries. To overcome this deficiency, penguins resort to thievery. While a neighbor is distracted or sleeping, the culprit sidles up and snatches a choice pebble with its bill. If caught with the goods, the crook sometimes pretends to doze or scrutinizes a bit of dried moss until a resounding flipper thwack convinces the bird to move on. In the meantime, the stone-swiper's own nest has been pilfered by its neighbor.

Other species have solved the pebble problem by eliminating the need for traditional nests. The large gray wrinkled feet of emperor and king penguins serve as ready-made nests for eggs, and later, hatchlings. Magellanic penguins prefer prairie dog–like burrows on the moors of the Falkland Islands or the arid plains of the Patagonian coast.

For those penguins that nest in rookeries with over one million birds, the frequent journeys to and from the sea to feed are arduous. Many short-tempered, brooding birds do not hesitate to cuff or peck passing penguins. Once the penguin has successfully lurched through a gauntlet of flailing flippers to its nest, the bird greets its mate with a complex series of bows and head-swaying. Both penguins then stretch their necks skyward and emit one of the least melodious sounds in the bird kingdom. If the mood is right, the colony soon reverberates with donkeylike brays or kazoolike croaks, depending on the particular penguin species. Minutes later the "ecstatic display," as first described by Edward Wilson, ends and the colony is silent.

BRUSH-TAILED PENGUINS

Adélie (*Pygoscelis adeliae*)
Height: 28 inches
Weight: 9–12 pounds
Distinguishing mark: A white ring around each eye.
Nest: Snow-free ridges and knolls on small islands and the continent.

Navigating by the sun, males arrive at the rookeries as early as mid-October, sometimes traveling fifty or more miles over sea ice. The male claims the previous year's shallow nest of stolen pebbles, and soon sets about refurbishing it by spinning in a circle and kicking debris backward. Within a week or two, the females arrive to claim last year's mate—and nest—or find a new one.

Sometimes fierce late-season blizzards strike the rookery. After one such storm in 1907 at Cape Royds, Ross Island, biologist James Murray walked out of the hut to check on the Adélies. "I could see no penguins; they had entirely disappeared. Then, suddenly, I was surrounded by them; they had sprung up out of the snow."

Two eggs are laid from early to mid-November. The female feeds at sea while the male incubates the eggs for the next week or two. After the female returns, the pair alternate nesting duties during the remaining incubation period. Hatchlings appear by late December, and the parents continue to take turns feeding the chicks in a frenzy of comings and goings. Three weeks later the dark gray chicks huddle together as a group, or crèche, while both parents hunt for krill. A few adults remain behind to guard their charges from the ever-opportunistic skuas.

By late February the adults abandon the chicks and hitch rides on northbound icebergs to molt. Hunger convinces the young penguins that are still on shore to take their first sea plunge. Once beyond the reach of the leopard seals that invariably patrol offshore, the juveniles follow the floes northward to the rich fields of krill.

GENTOO (*Pygoscelis papua*)
Height: 30 inches
Weight: 12–13 pounds
Distinguishing marks: A white band across the head that looks like a set of headphones; orange bill, and red feet.
Nests: Shallow pebble nests on broad rocky beaches and flat terrain; on the subantarctic islands, the birds nest in the tussock grass.

The gentoos are gentle, curious, and photogenic penguins. They cannot resist gripping the toe of a shiny black boot or giving a tug on the hem of a bright red parka. Cameras especially attract them, and they will peer and preen in front of the lens if the visitor sits quietly. Gentoos nest in small groups that make the birds easy to observe.

Although gentoos are not the most populous penguins, they are the most widespread. They breed on the subantarctic islands and on the Antarctic Peninsula to about 65° south. At the northern colonies, the gentoos lay two eggs as early as July; for those returning to the Antarctic Peninsula, egg-laying duty begins about the first week in November. If rookery real estate is in short supply, gentoos will nest on the fringes of Adélie colonies.

Courting gentoos perform the usual bowing, head swaying, and flipper-bobbling to attract a mate. Successful pairs greet each other with calls that sound like a two-note donkey bray. Sometimes the ecstatic display spreads to other pairs, from cluster to cluster until the whole colony is united by tone-on-tone hee-haws.

Incubation and brooding responsibilities are shared. The gray and white chicks move into crèches when they are about four weeks old. Both parents continue feeding their one or two chicks several weeks longer than do Adélies or chinstraps, continuing even after the chicks have fledged. Gentoos sometimes dive as deep as 350 feet in search of fish, krill, and squid to satisfy their offspring's enormous appetite.

Chinstrap (*Pygoscelis antarctica*)
Height: 28–30 inches
Weight: 8–10 pounds
Distinguishing mark: Head markings that look like World War I helmets with black straps.
Nests: High ridges and hills; bones and molted feathers garnish the pebble nests.

Chinstraps are "far more cocky, pugnacious, and swashbuckling" than gentoos, noted ornithologist Robert Cushman Murphy in 1913 on South Georgia Island. These penguins are also accomplished at pebble swiping. Oblique stares directed toward the desired object are followed by a quick grab with its beak. If successful, the penguin nonchalantly saunters back to its own nest and adds the new treasure to its stash of stones. Even unmated chinstraps are avid rock collectors, perhaps honing their skills for future nests. Called "hooligans," these solitary penguins can cause a ruckus in the rookery with their belligerent behavior.

Usually the last penguin species to arrive at multispecies sites, chinstraps claim the highest ridges and rolling hills on subantarctic islands. The birds hoist themselves up steep slopes by using beaks as mini-pickaxes and clawed toes to gouge grooves into the rocky ledges.

Primarily krill eaters, chinstraps have enjoyed more feast than famine since the late 1930s because of the slaughter of the krill-eating whales. Even with this explosive increase in population, chinstraps are not the most populous penguin species. That honor belongs to the macaronis.

CRESTED PENGUINS

Macaroni (*Eudyptes chrysolophus*)
Height: 25–27 inches
Weight: 8–11 pounds
Distinguishing mark: A droopy yellow-orange feather fringe starting between the eyes and sweeping back along the sides of the head.

Nests: Open hillsides and steep rocky ledges on the subantarctic islands; nests are made with pebbles and a few ragged feathers.

These birds reminded British sailors of a group of eighteenth-century fashion-obsessed London dandies, who dyed long locks of their hair yellow and wore exaggerated Italian clothes. This is a penguin species with a short fuse and a preference for pecking at the least disturbance. With its blood-red eyes, slicked-back yellow fringe, and thick pink bill, Macaronis have been compared to old folklore hags, especially when the birds slouch on their nests.

By early spring, subantarctic island ridges are loaded with feisty macaronis. Although two eggs are laid at the beginning of the austral summer, only one is incubated. The first egg is either lost to the ubiquitous skuas or shoved out of the nest when the larger second egg arrives. The surviving hatchling is fed krill by both parents for about two months. By early March, the young birds take to the sea to find their own food.

Rockhopper (*Eudyptes chrysocome*)
Height: 22–24 inches
Weight: 5–7 pounds
Distinguishing mark: Yellow plumes of feathers above its eyes that look like gaudy, overgrown brows.
Nests: Shallow pebble nests on steep rocky ledges and volcanic slopes on the subantarctic islands.

Although smaller than macaronis, rockhoppers are just as aggressive and easily irritated. They respond to petty disturbances with squawks and shrieks until the offender either becomes deaf or moves away. Rockhoppers demonstrate the appropriateness of their name with ebullient vertical jumps through the tussock grass and up precipitous ledges. This ability enables them to spring from the wild, thundering surf to high boulders on coasts inaccessible to other penguin species.

Like macaronis, rockhoppers have bright red eyes and a deep pink, sturdy bill. Differences between the sexes are minor; the males have more robust bills and a slightly larger build. Both share egg incubation, chick brooding, and crèche duties; and both take turns foraging for krill and small fish.

Royal (*Eudyptes schlegeli*)
Height: 27–29 inches
Weight: 11–12 pounds
Distinguishing mark: Similar to a macaroni but with more white on its face and throat.
Nests: Broad sloping beaches on Macquarie Island, the only place they breed; shallow bowl-shaped nests are made with pebbles, bits of tussock grass, and other vegetation.

Royal penguins are huskier than macaronis and share their counterparts' temperament. "Extraordinarily quarrelsome and jealous they are too," wrote Edward Wilson in 1901, "for they nest so close together that when a bird wants to walk about, it gets pecked on each side by the birds squatting on their nests. And as it walks, it keeps its flippers going hard, and hits everybody on its way on both sides." When the flippers of returnees and nest-sitters smack against each other, the loud claps sound like random bursts of applause.

THE LARGE PENGUINS

King (*Aptenodytes patagonicus*)
Height: 36–40 inches
Weight: 26–35 pounds
Distinguishing marks: Orange comma-shaped ear patches and bill.
Nests: Uses its large wrinkled feet to brood a single egg on sub-antarctic islands.

Kings are second in size only to emperors and, like them, aren't the least bit bashful toward people. Black heads, framed with glowing orange ear patches, sway back and forth as the birds scrutinize rubber boots and appraise brightly colored jackets. If visitors pass muster several kings trundle toward the colony, leading the way to a tight cluster of 250 penguins on a Falkland Island moor, or a metropolis of 100,000 on one of South Georgia's broad glacial moraines. At the outskirts of the colony the escorts launch a series of prolonged doubled-noted, trumpetlike trills skyward to announce their return.

Thousands of kings shuffle from one spot to the next, pausing to ponder shiny pebbles or bits of twiggy peat. Some lift a roll of abdominal feathers like a ponderous white curtain, exposing an egg or a small gray bundle balanced on their feet. The parcel moves, and a week-old chick peeks out from its blanket of feathers. King penguins, like emperors, use their feet as nests and incubate one egg for about fifty-five days. Once hatched, the chick is relatively safe from predatory birds, such as the skuas and giant petrels that hover nearby.

Chick-rearing responsibilities last for more than a year. Although meals are regular throughout the spring and summer, the chicks must depend on fat reserves—and sporadic meals from parents—to survive their first winter. During the following spring and summer, the chicks "resemble college boys in raccoon coats," noted Robert Cushman Murphy in 1913. At the end of summer they molt their brown woolly down and take to the sea. Some of the pairs mate again at this time; others wait until the following spring. Because of this difference in breeding cycles, eggs, tiny hatchlings, and older crèche-guarded chicks can be seen at the same time during the summer season.

In the past king penguins have been exploited by sealers, who rendered the birds' fat into oil when elephant seals were in short supply. Also, the magnificent sheen of blue-gray feathers that drape from the kings' shoulders like an expensive cloak attracted merchants who stitched the skins into small rugs, women's muffs, and captains' slippers. Fortunately, feather fashions went out of style. Conservation measures have been successful and the number of breeding pairs has steadily increased.

EMPEROR (*Aptenodytes forsteri*)
Height: 45–48 inches
Weight: up to 90 pounds
Distinguishing marks: Size and lemon-colored ear patches.
Nests: Uses feet as a nest; forms colonies on the fast ice.

If an official mascot were chosen for Antarctica, it would probably be the emperor penguin. With its ramrod posture and courtly bows, the emperor seems to exude dignity in the face of the insurmountable odds against survival in the harshest climate on Earth. Nothing was known about its nesting habits until the *Discovery* expedition in 1902, when Edward Wilson spotted a colony at Cape Crozier on January 31. The following October, a party investigated the rookery and carried back two live chicks for Wilson. The size and age of the chicks convinced the naturalist that the penguins must breed and incubate the eggs during winter, behavior eccentric "to a degree rarely met with even in ornithology," Wilson noted in his diary. Nine years later, he and two others undertook one of the great epic journeys of Antarctic exploration: the quest for emperor penguin eggs at Cape Crozier in the depths of winter.

The emperors' odyssey begins in March, when the fast ice expands from the shores and merges with new sea ice. As they streak through the swells toward the coast, the penguins look like brilliant flashes of sunlight in the slate-colored sea. When the channels between the ice floes close, the emperors spring out of the water and march single file, heads swinging from side to side. Sometimes the somber procession must cover seventy or more miles to reach a particular location where the fast ice is thickest and closest to land. Once they have arrived, emperor pairs perform courtship duets—an orchestration of elaborate bows and swaying heads, punctuated with almost musical double-noted trills.

In late May, the female produces a one-pound egg and carefully transfers it to the male's feet with a series of gentle chest bumps, ceremonial bows, and reassuring calls. The male now cradles the egg and keeps it warm until his mate returns. Free of incubation responsibilities, the females march toward the sea, a much longer trek than the inbound trip. As winter has deepened, more sea ice has formed, adding miles to their journey before they can feed on fish and squid.

Male emperors banish all traces of territorial bravado and hunker together in dense oval-shaped huddles for warmth. With their blue-black backs to the incessant gales, the males face the center and rest their beaks on the hunched shoulders of the penguins in front. They shuffle sideways in a maze of rows toward the center, the warmest part of the huddle. Throughout the sixty-five to seventy-day vigil, each male is careful to keep the egg cushioned with a fold of abdominal skin and feathers. If the egg hatches before his mate's return, the male sustains the chick with an oily secretion from his crop.

By early August the females have replenished their stores of fat and head back to the colony. As soon as the huddled males spot the females trudging across the sea ice, they call continuously. Each voice has a unique nuance and cadence that guides the females to the right male. Deep bows reaffirm the bonds established during courtship. Although he is reluctant at first to part with the naked hatchling, the female persists and loads the tiny chick onto her feet. The starving male is now free to trek to the sea for a month-long feeding frenzy.

During spring and summer, the chicks grow rapidly. Dressed in silver-gray down suits with white masks and black caps, they look like wind-up toys when they bobble their long flippers and whistle for food. Two-month-old chicks huddle in crèches for warmth until they hear a parent's call to dine.

In December the chicks begin to molt their down. A few weeks later, the ice cracks and drifts away from the continent. Only when the floe rots beneath their feet do the juveniles plunge into the sea. They now must learn to dive for squid, fish, and krill to ease their gnawing hunger. By the time they are adults, they will be able to dive to depths greater than fifteen hundred feet for eighteen minutes or longer.

Emperors show a fearless curiosity toward any unfamiliar object or animal. In 1908 at Cape Royds, biologist James Murray watched as a small group of emperors came ashore and inspected the men's new hut, small boat, and piles of supplies with animated interest. Then the penguins discovered the dogs, "and all other interests were swallowed up," wrote Murray. "Crowds came every day for a long time, and from the manner in which they went straight to the kennels, one was tempted to believe that the fame of the dogs had been noised abroad."

TWO SUBANTARCTIC PENGUINS

MAGELLANIC *(Spheniscus magellanicus)*
Height: 25 inches
Weight: 8–9 pounds
Distinguishing mark: White bands that ring the face extend across the chest and run down on the sides of the body; pink skin above the eyes.
Nests: Network of burrows on Falkland moors, arid Patagonian scrublands, and among tussock grass roots.

Magellanic penguins are shy and skittish toward anything new or novel. When frightened, they scurry back to their burrows by tobogganing across peaty moors, sandy beaches, or decayed tussock grass fronds. Paddling frantically with flipper tips and pushing with their feet, they disappear in a puff of dust, sand, or vegetation down a hole like miniature Alices.

In September magellanic pairs return to claim last year's burrows and build fresh nests with a few pebbles, chalky bones, and bits of tattered penguin skins. Eggs are laid in October and incubated for about forty days. Both parents feed the hatchlings a swill of small fish and squid. In January the chicks emerge from the burrows after they acquire their seaworthy feathers.

YELLOW-EYED PENGUIN *(Megadyptes antipodes)*
Height: 28–30 inches
Weight: 11–13 pounds
Distinguishing mark: Its straw-colored eyes.
Nests: Beneath the low canopy of the rata forests or in gullies with heavy vegetation, only on Campbell and Enderby Islands.

These distinctive and nonmigratory penguins do not form colonies. Instead, they lead secretive lives among the gnarled trunks of the rata forest with red-crowned parakeets for company. Their pale amber eyes give these solitary penguins an ephemeral quality, as they inhabit New Zealand's fog-shrouded subantarctic islands.

Sometimes, yellow-eyed penguins must waddle through a predatory gauntlet on Enderby Island as they journey to and from the sea. Rare Hooker sea lions that romp and rest on the buff-colored sands will occasionally grab an unwary penguin.

Albatross

At the Antarctic Convergence, albatross suddenly appear and waltz back and forth across a ship's foamy wake, sometimes close enough for observers to imagine the *swoosh* of air through stiff feathers. Without any detectable wing adjustments, the birds glide among decaying icebergs to the choreography of updrafts and blustery winds. Sometimes they paddle the air with fleshy pink webbed feet, as though swimming on air currents. Albatross need the winds of the "Roaring Forties" and "Furious Fifties" to keep their hefty bodies aloft. On rare calm days, the birds must fold their great wings and settle on the sea, bobbing like fluffy white buoys until the wind freshens enough for successful flight.

Aerial grace doesn't translate to ground maneuvers, however. As they glide toward the land, the birds spread pink toes to slow their descent toward rough hillside terrain. Most landings result in head-over-webbed-feet tumbles through the brush. Unharmed, the birds pick off debris from chests and toddle toward nests. Take-offs are just as inelegant. The birds careen down a steep hill or lurch toward the edge of a cliff, flapping their great wings with each awkward step. If the day's updrafts are strong enough, the albatross launch themselves into the air, soaring above island homes on the back of the wind.

For most albatross species, the edge of the pack ice defines the limit of their southern range. Ice damps the ocean swells; without updrafts, albatross cannot keep aloft. Only the smallest and lightest albatross, the light-mantled sooty, can follow the leads in pack ice southward to search for food.

Although squid, fish, and krill are their staple foods, albatross will never turn down a meal of carrion or galley garbage from the ships they follow. Whaling vessels were especially favored. In 1912 ornithologist

Robert Cushman Murphy boarded a Norwegian whaler near South Georgia to observe the hunt. Soon the sea was covered with whales' entrails, attracting "millions and millions of petrels and albatrosses," he wrote. So bloated were the birds that none could launch itself into the air—even as the ship bore down on them. "Several albatrosses were actually bumped out of the way by the bow of our steamer, but the experience only made them whirl around and look indignant instead of trying to get into the air."

THE GREAT ALBATROSS

WANDERING (*Diomedea exulans*) AND ROYAL (*Diomedea epomophora*)
Wingspan: Up to 12 feet
Distinguishing marks: Both species: White heads and bodies, with black edges on their wings; features turn white with age. Wandering albatross: Black band across the tail; juveniles are brown with white speckles on faces; females have light brown patches on backs and necks.

Weighing twenty to twenty-five pounds, these albatross are the largest bird species to circumnavigate Antarctica. Gliding and soaring north of approximately 60° latitude, wanderers and royals are the master aviators of the stormy southern seas. They hover above the swells, searching for squid, fish, and carrion. When they spot a suitable entrée, they land on the water and seize their prey.

Giant albatross breed biennially with lifelong mates and return to the same island hilltop and, most often, the same nests. The males arrive first to repair the cone-shaped mounds with fresh tussock grass and mud. Within a week or two the females return. Courtship is a slow waltz around the thronelike mound. With their great wings held high and tail feathers spread like an opened fan, they keep time with bill claps, gurgles, and synchronized head-swaying. "Sometimes this will continue for two hours," wrote sealer James Weddell in 1825, "and to a person inclined to be amused, the whole transaction would appear not unlike one of our own formal courtships in pantomime."

A single egg is laid in late October or early November and incubated for about eighty days by both adults in five- to ten-day shifts. Once the chick emerges, it is fed daily for several weeks and then less regularly for the next nine months. Like king penguins, wandering and royal chicks must face winter gales alone, surviving on fat and sporadic meals from their parents. At the arrival of spring the chicks are almost adult size, but still a bundle of tattered downy fluff and gangling wings. When strong updrafts from the nearby cliffs or steep hills ruffle their feathers, the birds stand on top of the mounds and flap their wings to strengthen chest muscles. One spring day when conditions are right they will soar into the arms of the wind and not return to the island for eight to ten years. For the adults, a breeding cycle that lasted about 380 days is now complete.

THE MOLLYMAUKS

BLACK-BROWED (*Diomedea melanophris*), GREY-HEADED (*Diomedea chrysostoma*), AND SHY (*Diomedea cauta*)

Wingspan: Varies from 7 to 8 feet

Distinguishing marks: Black-browed: A black smudge above each eye gives the bird a quizzical expression. Grey-headed: Head a soft pastel gray; lemon yellow streaks and orange tips mark the bill. Shy: Upper wings are a battleship gray rather than black.

Nests: Cone-shaped mounds of mud with bits of tussock grass, moss, and plant stems; a single egg is laid in the shallow depression on top.

Seafarers dubbed these smaller but similar-appearing albatross "mollymauks," a Dutch word meaning silly or foolish gulls. This unflattering nickname seems to fit the birds' bumbling behavior on land rather than the glorious aerial displays at sea.

Black-browed and grey-headed albatross have colonized the subantarctic islands north of 60° south latitude. Shy albatross nest only on the islands south of New Zealand and Tasmania. None of these species has any qualms about sharing prime territory with rockhopper and macaroni penguins. Since albatross chicks must endure long periods alone while the adults forage for food, feisty penguins waddling among the thronelike albatross nests probably provide relief from boredom.

By early April the chicks are nearly as large as the adult birds. Whenever brisk updrafts from nearby cliffs or steep hills gust through the rookery, the albatross chicks stand on their mounds and furiously flap wings to exercise chest muscles. In the early morning light these young albatross look like avian Orpheuses, calling forth the wind with wings instead of the sun with songs.

SOOTY (*Phoebetria fusca*) AND LIGHT-MANTLED SOOTY (*Phoebetria palpebrata*)
Wingspan: 6 to 7 feet
Plumage: Both species: Rich chocolate brown heads and wings; a comma-shaped mark accentuates each eye. Light-mantled sooty: Cream-colored back and chest.
Distinguishing marks: Sootys have a yellow stripe on the lower bill and light-mantles have a blue one.
Nests: Widely scattered shallow nests on inaccessible cliff ledges and in tussock grass on steep hills; colonies are small.

Unsolicited oohs and ahs emanate from even the staunchest nonbirders whenever these albatross soar into view. With their rich colors and graceful aerial displays, sootys and light-mantled sootys are a joy to watch. Light-mantled sootys range far to the south, following the leads in the pack ice to search for krill, fish, and carrion.

Both species are solitary breeders. The males arrive first and choose a lonely site on a cliff ledge or among the mounds of tussock grass. When a female circles overhead, the male stands near his nest and emits a haunting two-note cry, a sorrowful and eerie call that ranks as one of the most forlorn sounds on the subantarctic islands. If the female lands, courtship begins immediately as they spread their tail feathers like unfolding fans. Bill-clacking and head-waving follow as the twosome circle and bow. Then, the pair soar on updrafts and perform a synchronous aerial ballet in silence, high above the nest. To see such perfection against the craggy cliffs of Campbell Island, or above the green tussock grass on Bird Island near South Georgia, is a special moment for birders and nonbirders alike.

Petrels, Fulmars, Shearwaters, and Prions

The family *Procellariidae* is a diverse group of tubenose birds that vary in size from dainty storm petrels to husky southern giants. With their distinctive narrow tubes mounted on top of upper bills, these seabirds have a fine sense of smell that leads them to carrion. It is not unusual for the larger petrels to be first on the scene to feed on a fresh carcass. Smaller petrels, such as cape pigeons, gather en masse to wait their turn or scavenge for tidbits. During the early twentieth century, whaling stations were popular sites for the petrels, and a variety of birds nested in the vicinity to feed on waste blubber and entrails.

The smaller petrels, prions, and shearwaters burrow into volcanic scree hillsides, windswept moors with scrub vegetation, or the tussock grass slopes on the subantarctic islands. Other species nest on cliff ledges or penetrate cracks or cavities in the rock faces on the continent. Most of the smaller petrels forage during the day and return to their burrows under cover of darkness to escape predatory skuas and giant petrels, which are less active at night.

GIANT PETRELS

Southern Giant (*Macronectes giganteus*) and Northern Giant (*Macronectes halli*) Petrels
Wingspan: About 6.5 feet
Distinguishing marks: Southern giant: buff to pale yellow bills with olive green tips; plumage varies from almost pure black to pure white. Northern giant: straw-colored bills with reddish-brown tips.

Nests: Southern giants construct mounds out of mud, shells, pebbles, and moss on exposed and elevated land near penguin and seal colonies on the Antarctic Peninsula and subantarctic islands. Northern giants nest only on islands north of the Antarctic Convergence.

Although both species can be confused with mollymauks, the giant petrels flap their wings stiffly and lack the supreme gliding grace of albatross. Known to the explorers as Nellys, stinkers, or gluttons, giant petrels have few qualities that elicit admiration. These birds are Antarctica's equivalent of vultures; where digestive systems are concerned, no other Antarctic bird is its equal. One distinct advantage they have over other birds is the ability to forage on land and sea. They patrol penguin rookeries and compete with skuas for meals of unprotected or abandoned chicks.

During the days of sealing in the early 1800s the birds gathered on the beaches by the hundreds, waiting to pick blubber from sealskins. "Their fondness for blubber often induces them to eat so much that they are unable to fly," noted explorer James Weddell in 1823. Like weighted brown corks, giant petrels had to bob on the swells until digestion lightened their mass.

Although giant petrels have a limited arsenal of defensive moves, one technique is particularly effective: vomiting oily bile toward any perceived threat with bulls-eye accuracy.

SMALL PETRELS

Cape Petrel (*Daption capense*)
Wingspan: 2.5 to 3 feet
Distinguishing marks: Checkered sooty-brown and white splotches across the back and upper wings.
Nests: Crevices and exposed cliff ledges on the Antarctic Peninsula and subantarctic islands.

Cape petrels, or cape pigeons, are also called *pintados*, a Spanish word for "painted." A very sociable species, hundreds flock together to accompany ships along the Antarctic Peninsula. The birds feed on fish, krill, and squid by dipping close to the surface and seizing prey. Cape petrels' flying technique is similar to that of pigeons, with labored flapping between short glides.

Antarctic Petrel (*Thalassoica antarctica*)
Wingspan: Up to 3.5 feet
Distinguishing marks: Bold coffee-colored and white plumage.
Nests: Saucer-shaped nests on monoliths and seaside cliffs.

Although most Antarctic petrels breed in the mountains of the Ross Sea sector, the largest colonies are found in East Antarctica at Svarthamaren Mountain in the Mühlig-Hofmann range, 125 miles from the Queen Astrid Coast. About one-quarter million Antarctic petrel pairs fill the ledges of two natural rock amphitheaters. Another massive colony is found at Scullin Monolith. There, the birds flutter near the dark-colored rock like windblown brown and white confetti.

One egg is laid in November, and both adults alternately incubate the egg for about forty-five days. By March the chick has fledged and the colonies are deserted. During the austral winter, adults and juveniles forage for fish, squid, and krill just north of the edge of the pack ice.

Snow Petrel (*Pagodroma nivea*)
Wingspan: Up to 2.5 feet
Distinguishing marks: Brilliant white plumage and a black beak; it flies erratically, like a bat.
Nests: Circumpolar established colonies on the continent and islands south of 55° south latitude.

Snow petrels are the only birds in the world that roost on the tops of massive tabular icebergs. Thousands flit and flutter like white butterflies and blend into the scenery, then suddenly whirl high above the bergs, radiant before a storm-gray sky. These pigeon-size birds seem almost too fragile to survive the brutal blizzards and gales that are synonymous with Antarctica, but they thrive in regions where the larger petrels cannot penetrate because of pack ice's damping effect on updrafts.

In early November they nest in large loose colonies on sea cliffs and squeeze into every nook and cranny on exposed peaks and monoliths. One egg is laid in late November; by late March the chick has fledged—if it has escaped the severe weather.

White-headed Petrel (*Pterodroma lessonii*)
Wingspan: About 3.5 feet
Distinguishing marks: A white head, deep gray-brown plumage on its upper body and under its wings, and a white chest and belly; darker feathers form an "M" across its wings and back.
Nests: Pairs burrow on the subantarctic islands.

White-headed petrels breed on Îles Kerguélen, Macquarie, and other islands near the Antarctic Convergence. They return each year and form loose colonies, burrowing among tussock grass roots and scrub plants near the shore. On most islands suitable burrows are in short supply. The petrels must compete with rabbits, introduced by French and English explorers, for well-drained underground sites. Quiet evenings are soon filled with the staccato calls and clicks that emanate from beneath the tussock canopy.

One egg is laid in late October or early November and incubated for about fifty-five to sixty days. If neither feral cats nor rats make a meal of the egg or hatchling, the chick is fed for the next ninety to one hundred days before it fledges. The adults forage for fish, crustaceans, and squid during the night and return to their burrows under cover of darkness to escape from predators, such as giant petrels and skuas.

Wilson's Storm Petrel (*Oceanites oceanicus*)
Wingspan: About 16 inches
Distinguishing marks: Soot-colored with a white band on its tail.
Nests: Volcanic slopes, cracks, and crevices in rocky headlands as far as 76° south on the continent; also favor sealing and whaling debris, such as small wrecked boats, wooden barrows, and long-abandoned huts on the subantarctic islands.

About the size of a sparrow and weighing just thirty to forty grams, Wilson's storm petrels are the tiniest birds to inhabit Antarctic waters. They are surface feeders and flutter just inches above the waves. On long twiggy legs they pitter-patter on the crests of waves, the yellow webbing between their toes flashing against the slate-colored sea. Although krill is their primary food, offal from fishing ships attract frenzied flocks of storm petrels, feeding on the tiny morsels that larger birds have left behind.

Wilson's storm petrels are probably the most populous seabirds in the world. Many millions search for swarms of krill south of the Antarctic Convergence to the edge of the pack ice. In early spring the birds return to their favorite burrow, crevice, or abandoned man-made item to prepare nests. Potential pairs flit just above the ground like evening moths. Eventually, the male scrapes out a nest and lines it with a few penguin feathers, lichens, tiny pebbles, and shell fragments. When the nest is ready, he calls for the female to squeeze between the rocks or debris to join him. One egg is incubated for about forty-five days. For the next two months both parents share feeding responsibilities until the chick fledges in early April.

Antarctic Fulmar (*Fulmarus glacialoides*)
Wingspan: Up to 4 feet
Distinguishing marks: Pale silver-gray feathers on its upper body and white plumage beneath its wings; the trailing edge of the wing is black.
Nests: Large colonies on exposed cliffs on the Antarctic Peninsula; abandoned whaling stations on subantarctic islands.

At first glance an Antarctic fulmar resembles a gull, with its thick neck, pinkish bill, and husky body. The bird is a strong flyer, but must flap its wings a dozen or more times before gliding for a short time over land or sea.

Courtship is boisterous with energetic aerial displays. Noise levels remain high on cliff ledges, where pairs establish nests that are little more than a scrape on the bare rocks. One egg is laid in November and incubated for about forty-five to fifty days. Three months later, the chick fledges and disperses northward to the subtropics before returning to the high southern latitudes to breed.

Sooty Shearwater (*Puffinus griseus*)
Wingspan: Up to 3.5 feet
Distinguishing marks: Soot-colored body with flashes of buff under its long narrow wings; a twisty, wheeling flight with rapid wing strokes to gain altitude.
Nests: Burrows in scrub-covered hillsides on subantarctic islands.

Sooty shearwaters are very sociable birds and gather in large noisy flocks to feed or rest. Although these birds ignore large ships, they seek out and follow fishing boats for galley waste and fresh offal.

A single egg is laid in mid-November and incubated for about fifty-five days. Every morning during the next few months, sooty shearwaters leave their burrows in one enormous flock to feed at sea. This early morning flight is like a swift-moving shadow across the land and sea, an unforgettable spectacle against the rosy gray dawn. By late April the chicks have fledged and the vast colonies are usually deserted.

Great Shearwater (*Puffinus gravis*)
Wingspan: About 3.5 feet
Distinguishing marks: A pronounced hooked bill, dark brown cap, white cheek patches and collar, and a white band across its tail.
Nests: Shallow burrows among tussock grass roots and scrub vegetation.

Like other shearwaters, this species makes twisty wheeling movements in the air. When a bird spots squid or small fish near the surface, it folds its wings at a tight angle and plunges into the sea. A quick grab with its hooked beak secures the meal.

In September the birds return to colonies on the Falklands and other island groups where mated pairs number in the millions. Eggs are laid in late October and incubated for about fifty-five days. The fledglings depart the colony by the end of April.

Prion (Pachyptila sp)

Wingspan: About 2 feet
Distinguishing mark: A distinctive dark "M" across back and wings.
Nests: Crevices at the base of cliffs or shallow burrows on volcanic slopes.

These light blue-gray birds with white chests are also known as whalebirds and scoopers for the way they feed. Like baleen whales, prions have a comb-like fringe on the inside of their bills to filter plankton. If large flocks were spotted, sailors knew from experience that plankton-eating whales were nearby. The birds use their feet to patter rapidly across the water, wings outstretched and submerged bills opened, scooping up tiny bits of food.

Prions have another unusual trait. During the most ferocious storms, prions settle on the surface but do not attempt to dive through the high waves like other birds. Instead, they lift straight up and flutter momentarily as each successive monster rolls beneath them.

Like other Antarctic birds, prions lay one egg in very early spring. Incubation lasts for about fifty days, and caring for the chick requires another fifty to fifty-five days until it fledges. Although skuas and giant petrels prey on the chicks, feral cats and rats do far more damage to prion populations.

Other Birds

Snowy Sheathbill (*Chionis alba*)
Wingspan: About 2.5 feet
Distinguishing marks: White except for black feet and legs; pink and white wattles; bill has green and yellow tints.
Nests: Close to penguin and blue-eyed shag colonies; steps and windowsills of abandoned huts and buildings; among piles of discarded equipment from research bases. Materials for shallow

nests include penguin feathers, bits of dried carrion, tiny bones, limpet shells, and man-made rubbish.

Sheathbills are Antarctica's contribution to evolutionary oddities and look as though they were designed from discarded genetic bits and pieces from other species. They are the only Antarctic birds without webbed feet and must feed almost entirely on shore. Turkey-like wattles hang from their faces, and their characteristic head-bobbing and pecking at the ground for bits of food reminds the causal observer of pigeons in a park.

Sheathbills exhibit the cunning of skuas as they collect blue-eyed shag or penguin eggs from nests. These crafty birds also threaten adult penguins with squawks and aggressive wing flaps while they feed their chicks. When the flustered parent misses the chick's open beak, the regurgitated krill is deposited into the sheathbill's mouth. Elephant seal wallows are good choices for additional culinary tidbits, such as rich milk droplets intended for a nursing pup.

Since sheathbills depend on penguins for food, they breed at the same time as the larger birds. After a brief courtship, sheathbill pairs incubate two to four eggs for a month and then care for the chicks until early March. Adults and fledglings fly northward to Tierra del Fuego and the Falkland Islands to escape the Antarctic winter.

Blue-eyed Shag/Imperial Cormorant (*Phalacrocorax atriceps*)
Wingspan: About 4 feet
Distinguishing marks: A swanlike neck, cobalt blue rings around its eyes, and vivid orange-yellow fleshy growths at the base of its bill.
Nests: Cliff tops on the Antarctic Peninsula and the subantarctic islands. Seaweed, feathers, moss, and bits of orange lichens are cemented into flattopped mounds with penguin guano. Cormorants, black-browed albatross, and rockhopper penguins nest in close proximity and enjoy each other's company, especially on the Falkland Islands.

Cormorants, also known as shags, do not head north as nights lengthen and days grow colder. Instead, they fly to the edge of the expanding pack ice and dive for fish in open water. At night they return to roost in cliff-side rookeries on the Antarctic Peninsula and the nearby subantarctic islands.

Shags were always a welcome sight to mariners, for it meant that land was near. One of the great moments in Antarctic history occurred in May 1916, when Frank Worsley and Ernest Shackleton spotted several cormorants and knew that their epic eight-hundred-mile journey from Elephant Island to South Georgia was over. "These birds are as sure an indication of the proximity of land as a lighthouse is," wrote Shackleton.

Each year with the approach of spring, shags return to the same mates and nest-mounds. There, the enraptured birds once again court mates and sway cheek to feathered cheek in perfect unison in an avian version of the tango. After the courtship dance ends, the pair repairs the nest with fresh-guano plaster and penguin feathers. Several eggs are laid in early November. About one month later, the hatchlings emerge from the eggs like miniature plucked chickens—naked, without even a thin coat of down to help regulate their temperature. Adults are unusually attentive for the next two to three weeks, until the chicks can maintain a constant body temperature. The young are fed small fish until they fledge in early March.

South Polar Skua (*Catharacta maccormicki*) and Antarctic Skua (*Catharacta antarctica*)
Wingspan: South polar skua: 4 feet; Antarctic skua: 5 feet
Distinguishing marks: Both species: Dark brown plumage and hawklike beaks; aggressive attitudes of superbly adapted predators; brilliant white wing patches that flash when the birds fly or posture in a threatening manner. South polar skua: Smaller in size and has lighter brown underparts.
Nests: Near piles of penguin chick bones; Antarctic skuas return to nest on the peninsula and the subantarctic islands; south polar skuas are circumpolar and nest near penguin or blue-eyed shag colonies.

Skuas may not inspire fits of rapturous prose from ornithologists, but most researchers develop a grudging admiration for these scavengers from the high southern latitudes. Away from chicks and nests, skuas are inquisitive toward people, and appraise visitors with fearless candor. If, however, a person approaches too close to a nest, encounters between human and skua are less gracious. With great wings pointing skyward from behind its head and eyes staring unflinchingly at its intended victim, the skua opens its great hooked beak and emits a high-pitched screech at rock-concert decibel levels. If the earsplitting shriek isn't enough to deter the unwary, the skua will fly toward the offender's head with breathtaking alacrity to deliver a resounding blow with its wings or feet.

The skua's scientific name, *Catharacta*, means *cleanser*; they live up to their name as scavengers and opportunistic feeders who take food on the wing, from the sea, or on the ground. Smaller petrels have learned to rush back to burrows at night to escape detection from skuas. Although krill and small fish are seized from the sea, carrion, penguin eggs, and chicks make up the bulk of the skuas' diet. For quick and easy meals, skuas pull cormorants' wing and tail feathers until the bullied birds disgorge the small fish intended for chicks.

The skuas' acute sense of smell leads them to distant carrion. During Ernest Shackleton's 1907–09 South Pole expedition, the scavengers were spotted hundreds of miles inland on the Ross Ice Shelf just hours after the first Mongolian pony was killed.

Cleaning up carrion may be the birds' first priority, but jubilant community baths at local meltwater streams and ponds are a close second. Skuas are fastidious about their feathers, and rival only penguins as to the amount of time devoted to grooming plumage.

One or two eggs are incubated in a shallow nest lined with whatever the adults can filch from other birds' nests or salvage from their rubbish heap—limpet shells, penguin and small petrel bones, indigestible feet, and bits of tattered seal hide. Any strange skua in the vicinity receives a vicious assault, for the adults guard their own chicks with admirable dedication, although they do not hesitate to snatch and devour a neighbor's hatchling.

As the chick grows, its parents feed it a variety of food gleaned from the nearby penguin metropolis, elephant seal wallows, or blue-eyed shag colonies. Other dietary staples are krill and the young rabbits that breed on more northerly subantarctic islands. Adults feed and guard the chick until it fledges in late March or early April. Young skuas will not return to nesting sites for about five years. As winter approaches, adults and juveniles fly as far north as California and Newfoundland before returning to Antarctica to breed.

Kelp Gulls (*Larus dominicanus*)
Wingspan: About 4.5 feet
Distinguishing marks: Black upper wings, white heads and chests.
Nests: Cliff ledges on the Antarctic Peninsula and subantarctic islands; nests are lined with moss, lichens, and seaweed.

Gulls plucking limpets from tide pools near penguin rookeries are a common sight along the Antarctic Peninsula and South Shetland Islands. Other choice pickings are found at research station dumpsites—more than enough to convince hundreds of gulls to remain as year-round residents, in spite of the inhospitable climate.

In spring, kelp gulls incubate a clutch of two or three eggs for about four weeks. The adults continue to feed the offspring small fish and limpets for six to eight weeks after the chicks have fledged to ensure the young birds' survival.

Arctic (*Sterna paradisaea*) and Antarctic (*Sterna vittata*) Terns
Wingspan: About 2.5 feet
Distinguishing marks: Both species: Silvery gray with black caps. Antarctic terns: Red bills and feet; forked swallowlike tail. Arctic terns: black bills and feet; molts on pack ice and rarely visits the peninsula.
Nests: Open ground scrapes lined with dried seaweed and shell fragments on the Antarctic Peninsula and subantarctic islands.

Although Arctic terns are famous for following summer from Pole to Pole, they do not breed in the Southern Hemisphere. Antarctic terns, however, must brave November blizzards and wind-driven sleet to brood eggs on rocky beaches of the Antarctic Peninsula and nearby island groups. Any breathing entity that wanders into tern territory should expect to be mobbed by furious birds. Even skuas avoid noisy tern colonies. When chicks are just four or five days old, they hide among the rocks or sparse vegetation and remain there until they fledge, about twenty-five days later. The juveniles remain with the adults for several more months.

Both tern species snatch krill and small fish during shallow dives and surface dips. They also hover near the tips of glaciers and wait for large walls of ice to calve into the sea, sending up enormous waves that are loaded with krill and other crustaceans.

Antarctic and Subantarctic Seals

Much of Antarctica's history is a tale not only of exploration but also exploitation. The desire for the fur seal's premium pelt led to the species' near extinction. When new colonies were discovered, captains recorded cryptic messages in logbooks as sealers swarmed over islands to butcher males, females, and pups. The men were sworn to secrecy. If closed mouths could not be guaranteed, the sealers were abandoned on lonely islands either to survive until the ship returned six months later or to die of starvation and exposure.

By 1830 fur seals were so rare that hunting them was unprofitable. Brief population booms once again drew sealers to the subantarctic islands in the 1870s, and the colonies were once more decimated. Twenty years later, Scottish and Norwegian ships plied Antarctic waters to hunt for right whales. When none were found, the men turned to seals for oil.

Oil rendered from seals' blubber was almost as good as that from whales, and the animals were far less dangerous to hunt. During the early 1900s, Great Britain acted swiftly and demanded expensive licensing fees for sealing and whaling on the subantarctic islands they claimed as territory. Many captains, used to the lawlessness and free-for-all atmosphere of former days, chose to ignore the new regulations. "Forty-two sea elephants were killed today," wrote Robert Cushman Murphy in

1912. "Some of them, particularly the cows, had recently hauled up from the sea and were extraordinarily fat, which means that they will yield the maximum amount of oil. The Old Man's [Captain Cleveland's] license... forbids him to kill any females or young, but he cheerfully disregards all agreements and makes the slaughter universal, even down to the winsome and playful pups." Sealing continued on South Georgia until 1964.

Public attitudes toward sealing were slow to change—until graphic films of the Canadian harp seal pup slaughter were shown on television, and gruesome photos appeared in major wildlife magazines. Man's brutality toward these animals could no longer be ignored. The public responded with a grassroots movement that changed attitudes and consumer habits. Demand for sealskin coats and other luxuries dropped dramatically.

Although the harp seals were the focus for antisealing protests, public sympathies extended toward other seal species. In 1978 the Convention for the Conservation of Antarctic Seals was adopted. Fur, elephant, and Ross seals are now completely protected, and catch limits for the other species have been determined, should sealing commence once more.

The order *Pinnipedia*, which means "fin-" or "wing-footed," consists of two large groups. The first suborder includes fur seals and sea lions, animals that rotate hind flippers under their bodies and gallop across pebbly beaches with alacrity and agility. This group also has external ear flaps.

The second group, or "true seals," are the familiar ponderous bodies that haul out on sandy beaches to bask beneath the Antarctic sun. These seals heave over land with a distinct caterpillar-like rippling motion, their sides quivering with each heavy lunge. True seals have no external ear flaps.

Differences in movement through the water also distinguish the two groups. Fur seals use paddle-shaped foreflippers like oars for deft maneuvers. True seals appear clumsy on land, but in water they twist and tumble like weightless acrobats as they pursue their prey in semidarkness.

EARED SEALS

ANTARCTIC (*Arctocephalus gazella*) AND KERGUÉLEN (*Arctocephalus tropicalis*) FUR SEALS
Distinguishing marks: Males have luxurious silvery-brown ruffs, weigh 200 to 250 pounds, and measure 6 to 7 feet from nose to tail; females are about one-third the size of males.
Rookeries: Congregate on sloping rock slabs and subantarctic island boulder beaches.

In 1931 a small number of breeding fur seals was discovered on Bird Island, near South Georgia. Once reduced to a world population of several hundred, they were protected by law and now number over one million animals.

In October males haul out and establish territories. The most highly prized sites are near the water's edge above the tidal mark, the realm of only the most dominant males. When challenged, a bull points his muzzle skyward, his long whiskers quivering with tension, and casts oblique glances toward the interloper. If disdainful looks and snorts don't persuade the other male to seek safer quarters, the beachmaster gallops toward the intruding male with the intention of slashing the upstart's hide with his sharp teeth. Usually the threat of painful wounds is enough to send the less-experienced male scampering to higher and less valued terrain. Young males with no territories are forced to bide their time in tide pools and shallows, hoping for a moment of inattention by the bull.

The pregnant females arrive several weeks later and are promptly herded into harems. The bull patrols the perimeters of his tiny kingdom, an area defined by the outermost lounging females. His expressive voice sounds like a dog's whimper for food or attention.

Two days after being rounded up, the females give birth to single pups. Even at birth, differences between male and female are evident, for male pups are larger and gain weight faster than females. When the pups are just one week old, their mothers return to the sea to forage for krill, fish, and squid. Meals for the pups average about once every three to four days until they are weaned about four months after birth.

Southern Sea Lions (*Otaria flavescens*)
Distinguishing marks: Males weigh 500–700 pounds, measure about 8 feet in length, and have luxuriant ruffs that frame their faces; females are about 6 feet long and weigh 300 pounds.
Rookeries: Congregate on rough pebble beaches and large rock slabs protected by kelp beds or reefs as far south as Tierra del Fuego and the Falkland Islands.

Like fur seals, southern sea lions can also rotate hind flippers under their bodies, which allows them great mobility on shore. In late November the bulls haul out and establish territories. Dominant males claim and defend the choicest areas nearest to the sea and force other males farther inland.

Within days of the females' arrival and roundup, the pups are born. Several weeks later the pups gather in pods while the females forage and feed on fish, octopus, and squid. The young are suckled for at least four more months before they are weaned and then abandoned.

Southern sea lions were rarely hunted; their coarse fur lacks the soft undercoat that almost doomed the fur seals to extinction. These sinuous marine animals are a pleasure to watch as they frolic like oversized otters among the long strands of kelp near shore.

Hooker's Sea Lions (*Phocarctos hookeri*)
Distinguishing marks: Males are chocolate brown; females, a rich creamy buff or pale tawny gold. Both have huge velvet-brown eyes, full of expression and curiosity. Males reach about 10 feet in length and weigh close to 1,000 pounds; Females are 6 to 7 feet with weights to 350 pounds.
Rookeries: This species is the rarest and most endangered of any seal or sea lion species. They are found only on Campbell and Enderby Islands, south of New Zealand.

Like the fur seals, Hooker's sea lions were nearly hunted to extinction. Hides were tanned into fine supple leather, and the high-quality oil rendered from their blubber was plentiful. Unlike fur seals, Hooker's sea lions have not recovered from the relentless slaughter in the early 1800s. Their estimated population is five to six thousand animals.

Males arrive at the beaches in late November. About two weeks later, pregnant females haul out and are promptly escorted toward the beachmaster's ever-expanding territory. Within seven days the pups are born; ten days later the dominant male mates with each female in his harem.

In January the beachmasters desert the harem for long leisurely feeds at sea. Although feisty younger males try to take the bulls' place as haremmasters, the females ignore them and move their pups to safer inland locations.

The seasonal influx of squid around the islands provides ample food for the sea lions—and a profitable catch for trawlers, whose nets each year drown about fifty to a hundred sea lions, the majority female. Pups face another danger—from rabbits that were released on the islands to breed and provide food for castaways. The ever-curious pups tumble into the rabbit burrows and become trapped. About 10 percent of the total number of pups are lost this way each year.

When the surviving pups are weaned, the females teach them to swim and dive. At first, the strenuous survival lessons exhaust the pups. The tolerant mothers nose the youngsters to the surface and carry them piggyback to the beach for rest and recuperation among the ferns and mosses.

EARLESS SEALS

Southern Elephant Seals (*Mirounga leonina*)

Distinguishing marks: Astounding bulk makes elephant seals easy to identify, for they are the largest seals on earth. Males weigh up to 8,000 pounds and are 18 to 20 feet in length; females average about 2,000 pounds and measure 10 to 12 feet. Males have large proboscises that they inflate to appear larger and more threatening to other males. This appendage looks like a shortened elephant's trunk.

Rookeries: Breed only on the subantarctic islands.

In late August, the first male elephant seal arrives at a subantarctic beach in peak condition and weight. His ponderous blubber quivers as he humps over the sand "like a gigantic inchworm," noted Robert Cushman Murphy in 1912. Within days, more bulls haul out and the inevitable sparring for territory begins. The most experienced beachmasters set up harems close to the shore but above the tidal mark. A few weeks later the pregnant females arrive, and are promptly persuaded to join a harem that will include fifty or more.

"In the code of bulls," noted Murphy, "the correct number of wives is just one more than you've got." Battles for females are ferocious and bloody. The combatants inflate their proboscises, open their mouths wide, and emit protracted belchlike noises that are punctuated with retches, strangles, and gurgles. Steam swirls around their bodies in the cold air.

"Next, they rear up like a pair of rocking horses," Murphy wrote. When the tension becomes unbearable, the males lunge toward each other and smack chests, their thick rolls of blubber quaking with the shock. Although they slash each other's hide with canine teeth, the bulls rarely fight to the death. The weaker male withdraws into the waves before too much damage has been inflicted on his or the beachmaster's body. However, these battles often wreak havoc throughout the harems. Casualties include pups, crushed beneath the bulls' bodies.

The young that do survive the terrible battles thrive on the world's richest milk, gaining an average of fifteen to twenty pounds a day. Just over three weeks after their birth they are abruptly weaned when the females abandon them in favor of food from the sea. They will not return to the beach until it is time for their late-season molt. Hunger drives the pups into the shallows, where they teach themselves to swim and dive. As they gain confidence, they venture farther from shore to hunt squid, fish, and octopus in deeper water.

In early December, yearlings and nonbreeding males are the first to return to the beaches to molt. Later, the females arrive and lie in great mud wallows, packed together like fat mottled sausages. On very cold days, clouds of steam linger over the wallows. During the forty-day molt, the seal's outer skin peels away in long strips. Flippers arch over chests or poke behind heads to scratch at itchy skin fragments with articulated fingerlike claws. Movement within the wallow is discouraged with prolonged, open-mouth gargles that sound like a miserable case of unrelenting indigestion. Finally, from late March to early May, the seals emerge from the mud with filthy but intact new skins.

CRABEATER SEALS (*Lobodon carcinophagus*)
Distinguishing marks: Buff to golden brown; long pointy muzzles and large brown eyes—features that remind many people of dogs with cropped ears. Males and females are 8 to 9 feet long and weigh about 500 pounds.
Rookeries: Crabeaters haul out on the pack ice to bear their pups and later to molt.

Their name is a misnomer, for crabeater seals dine exclusively on krill by trapping the small crustaceans between their overlapping teeth. Crabeaters are the world's most abundant seals. Although population estimates vary from five to thirty million, little is known about these seals since they are creatures of the pack ice. They are gregarious, but gather in small groups on floes with only a few penguins and petrels for company.

In October crabeaters haul out to bear their pups on the ice. One male defends a single female. Unlike the bull elephant or fur seal harems, male crabeaters mate with just one female. After its birth, she suckles the pup for about four weeks and then abandons it to return to the sea to feed. Leopard seals often attack the vulnerable pups when they leave the floes to feed on krill. By the time they reach adulthood, the seals' necks and hindquarters are crisscrossed with pale gash-like scars. Killer whales also prey on crabeaters, pursuing them not only in the water but also on the ice floes. The orcas smash or violently rock the floes until the resting seals tumble into the water... or the pursuers' jaws.

LEOPARD SEALS (*Hydrurga leptonyx*)
Distinguishing marks: Spotted chest and belly fur; long neck, large triangular head, massive jaws, and a fixed grimace. Its sinuous body measures 10 to 12 feet and weighs 700 to 800 pounds. Females are slightly larger.
Rookeries: Bears its young on the pack ice. Little is known about its breeding habits.

While most seals flop or hump laboriously across the ice, leopard seals glide with natural grace and slip into the sea without a splash. These solitary hunters snake among the bergs and floes to ambush crabeater seal pups and penguins.

In early spring Adélie penguins must sometimes march over thin sea ice to reach their rookery, unaware that leopard seals lurk beneath. When the seal detects the telltale slip-slap waddle walk with its accompanying shadow, it crashes through the ice and startles the plodding penguins. With a quick snap of its jaws, the seal snatches an unlucky bird and retreats to the water.

Leopard seals also patrol the waters near penguin rookeries. It is not unusual to see several dark gray heads holding steady in the surf, surveying penguin traffic to and from the colony. The seal dives, grabs a penguin by its rump or feet, and surfaces. With quick side-to-side jerks of its head, the leopard seal thwacks the bird against the water to loosen its skin. Seconds later the carcass is consumed. As predators, leopard seals are second only to killer whales for stealth.

Solitary leopard seals haul out on ice floes to recover from the rigors of the hunt or to digest a recent acquisition. Most show mild curiosity toward humans and lift their dark gray heads to monitor a small boat or two. On rare occasions a seal "treats" occupants to close-up inspections of its lobed teeth.

> Weddell Seals (*Leptonychotes weddelli*)
> Distinguishing marks: A disproportionately small head compared with the seal's 900- to 1000-pound body, second in size only to elephant seals; face looks like a smiling but earless Cheshire cat; fur is dark blue-gray, dappled with silver and light brown patches and streaks.
> Rookeries: Bear young in small colonies during September and early October on stable fast ice around the continent; most northern colony is found on South Georgia Island.

These seals were named in honor of explorer James Weddell, who in 1823 first sailed the sea that now bears his name. When he returned to Scotland, Weddell presented the Edinburgh Museum with the skull and skin of a "sea leopard" that he had killed in the South Orkney Islands. The specimen turned out to be not a leopard seal but a new species.

Weddells are the only mammals besides humans to remain as far south as 78° during the dark, inhospitable winter. Since Weddell seals haul out on the fast ice so close to shore, they are the most available seals for scientists to study. Tidal cracks in the fast ice give Weddells access to the sea to hunt the Antarctic cod they favor. Although they can remain underwater for over ninety minutes and dive to over two thousand feet, their prey is elusive and not always found within a comfortable breathing distance from the tidal cracks. To solve this problem, Weddells maintain several breathing holes by gnawing through newly formed ice with their teeth. The amount of time to grind down their molars to nubs varies from twelve to twenty years. When a Weddell can no longer chisel through the ice with its worn-down teeth, the seal drowns.

During winter Weddells rarely haul out, for the water below the sea ice is much warmer than the air above, especially when katabatic winds sweep down from the Polar Plateau. However, in early spring the females must haul out to bear their pups. Breathing holes or tidal cracks are now prime territory for the roving males, who each claim a single hole and the surrounding water. Although the male allows any females who want to use the hole to catch a breath or to haul out, he denies other males access to air and females.

In the eerie gloom of half-light, Weddells sing plaintive songs as they swim among the submerged contours of icebergs during the breeding season. High-pitched moans fade into rapid piping or clicking notes, "ending up exactly on the call-note of a bullfinch," noted Edward Wilson in 1902. "This changed to a long shrill whistle and a snort, as though [the seal] was out of breath with it all." As Wilson lay in his bunk on the *Discovery* during the long winter, he listened to the melancholy sounds emanating from beneath the ice, a strange melody that lulled him to sleep.

The pups are born in October and grow rapidly on milk rich in fat. When they are just over a week old, they enter the water after plaintive coaxing from their mothers. Six weeks later the pups are abandoned and must survive on their own.

Ross Seals (*Ommatophoca rossii*)
Distinguishing mark: Enormous dark brown eyes, large heads with short snouts, and slender bodies; dark gray fur on top and silvery gray underneath, with parallel streaks of darker fur marking throats and necks. They measure 6 to 7 feet and weigh about 400 pounds.
Rookeries: Live and breed deep within pack ice; pups are born on the ice by early November.

Explorer James Ross first studied the seals that now bear his name in 1841. Ross seals are the hardest seals to study because they live within the vast regions of semi-permanent pack ice. Fewer than one hundred sightings of the seals have been recorded since James Ross discovered them.

Like Weddells, Ross seals sing complex songs, filled with trills, whistles, moans, and clicks. One unusual characteristic is its behavior when threatened or startled. The seal rears back and inflates its throat. Looking like a bullfrog ready to croak, the seal utters a series of high-pitched sparrowlike chirps, much to the surprise of human onlookers.

Ross seals are solitary hunters and feed on fish, krill, and squid.

Whales: Back from the Brink?

Commercial whalers have hunted in Antarctic waters since the early 1900s. The number of whales killed is staggering: at least 175,000 from 1904 to 1966 just in South Georgia Island waters, the site of the first whaling station. The toll from other subantarctic islands is just as grim. Although it is difficult to estimate the number of whales that now roam the seas, most scientists believe the current protected populations are 1 to 5 percent of pre-whaling totals. Perhaps the most poignant reminders of what has been lost are found in the pages of diaries and logbooks that date from before the slaughter.

"Whales are very common in this vicinity," wrote American James Eights in 1829, near the South Shetland Islands. "Great numbers of them may be seen breaking the surface of the ocean between the numerous icebergs, sometimes sending forth volumes of spray and then elevating their huge flukes in the air."

In December 1842, near the eastern side of the Antarctic Peninsula, explorer James Ross wrote that he saw "a great number of the largest-sized black [southern right] whales, so tame that they allowed the ship sometimes almost to touch them before they would get out of the way." He concluded, "... any number of ships might procure a cargo of oil in a short time."

Fifty years later, William Bruce sailed with the Dundee whaling expedition to do just that. The ships followed Ross's route to potential riches near the eastern side of the Antarctic Peninsula. Unknown to Bruce, the southern rights were almost extinct, slaughtered at breeding sites along South American coasts and other warm water locations. The only whales he saw were the swift-swimming rorquals. "As far as the eye would reach in all directions, one could see their curved backs, and see and hear their resounding 'blasts,'" he noted.

Antarctic whaling began in 1904, when Captain Carl Anton Larsen founded Grytviken on South Georgia Island. More whaling stations appeared on other subantarctic islands. In December 1908 French explorer Jean-Baptiste Charcot visited Deception Island, one of the South Shetlands. Three whaling stations were in full operation on the island. At Pendulum Cove, Charcot anchored near a black beach where steam rose from volcanic vents and hung like misty shrouds over the landscape. "At the junction-line of snow and beach there is a regular hedge of whale skeletons," he wrote, "and the blue waters of the basin are tinged red with blood."

The slaughter continued until the mid-1960s when the "booming resonance" of the whales' blows was heard no more in the waters surrounding most of the subantarctic islands. Processing whale carcasses moved from shore-based stations to sophisticated factory ships that roamed Antarctic seas until 1986, the year an international moratorium on commercial whaling went into effect. The whaling countries embraced this goodwill gesture only after commercial extinction was declared for blue, southern right, sperm, sei, and humpback whales.

But even now, whales are still pursued and harvested for "research purposes," primarily by Japan. Whale meat is an expensive culinary novelty and served only in the most exclusive Japanese restaurants. Norway rekindled its commercial whaling interests during the 1990s and continues to hunt minke whales for human consumption. Minke whale meat is also used as commercial feed for fur-bearing animals.

On October 23, 1997, the International Whaling Commission approved a request for Russian subsistence natives of the Chukotka region to kill 120 gray whales. At the same meeting the Makah Indian tribe from northwestern United States was granted permission to resume whaling once more after a seventy-year hiatus.

BALEEN WHALES: THE MYSTICETI

Two rows of horny plates hanging from the roofs of the whales' mouths and the presence of two blowholes link the following six species. The baleen plates overlap like vertical Venetian blinds to trap krill and plankton. The whales plow through fields of these tiny creatures with mouths open and squeeze enormous quantities of water between the baleen plates to trap the bits of food.

Baleen is made from keratin, the same protein material that forms fingernails and hooves. During the 1800s it was the most prized part of the whale. With its plasticlike flexibility, baleen was used for buggy whips, hoop skirts, corsets, and collar stays. Other nineteenth-century products included chimney sweeps' brooms, magicians' wands, fishing rods, chair seats, and shoehorns. Twentieth-century automobiles, sensible clothing, the invention of plastics, and steel for springs in machinery eliminated the market for baleen.

Whale oil, however, remained indispensable. Earlier in the twentieth century, manufacturers of explosives used glycerin from whale oil in their products, and tanners rubbed whale oil into hides to turn them into high-grade leathers. Paints, varnishes, soaps, inks, and face lotions were also manufactured with whale oil. Hydrogenation, the process of hardening or saturating oil, turned glycerin into margarine, crayons, candles, shaving creams, and cosmetics. With the ability to withstand high heat and pressure, sperm oil was used until the 1970s to lubricate turbojet engines. Its use subsided when oil from the jojoba plant, native to North American deserts, was found to be an excellent substitute.

In addition to humpbacks and southern rights, blue, sei, fin, and minke whales form a subcategory called the rorquals. This Norwegian word means a furrow or groove and refers to the accordion-like folds under the jaw that expand as the whale feeds. Until the late 1800s whalers ignored the rorquals because they could not get close enough to kill them, but harpoon guns and swifter whale-catcher boats soon devastated the rorqual population.

> Blue Whales (*Balaenoptera musculus*)
> Distinguishing marks: Spectacular, frothy plume-shaped blow projected twenty-five to thirty-five feet upward; named for its mottled blue-gray back.

Blue whales are the largest mammals ever to live on Earth, surpassing even brontosauruses in length and tonnage. In their prime, blue whales average 80–100 feet in length and weigh 120 tons.

To replenish their stores of blubber after the rigors of mating and bearing calves, blue whales migrate to Antarctic waters and feed on swarms of krill. Each whale strains four or more tons of tiny crustaceans from the ocean each day. The largest blue whales may cruise as far south as the edge of the pack ice in the early Antarctic summer.

Nineteenth-century whalers could do little but admire these magnificent animals. "Old sulfur-bottom," or so they called this whale for the yellow film of diatoms on its belly, was too fast for hand-thrown harpoons; but the whales' survival tactics failed when confronted with harpoon guns and speedy whale-catcher boats. The number of blue whales dwindled to less than one percent of an estimated two hundred thousand. In 1965 the species was granted international protection.

Blue whales have a distinct diving sequence. First, the whale shows its head followed by an impressive vertical plume of mist. The great head sinks and the long back arches slightly and rolls above the slate-colored sea. Its stubby dorsal fin appears and moments later the flukes are held at a shallow angle before disappearing into the sea.

Fin whales (*Balaenoptera physalis*)
Distinguishing marks: Dark grayish-brown on its sides and back; the right side of its head is white. The blow is a substantial plume, twenty feet high. Fin whales rarely show flukes when they dive.

Fins are the second-largest whales, after the blues, and reach eighty to ninety feet in length. The asymmetrical coloration of the whale's head may be connected with how the fin tilts its head as it scoops up krill or plankton. They are swift swimmers, and have been observed to sprint for short bursts at about twenty miles per hour. Fins like to travel and feed in small groups of three to seven whales. They often showed marked curiosity toward whaling ships.

After World War II demand for whale oil to lubricate airplane parts skyrocketed; the pursuit of fin whales was relentless. Although they were fully protected in 1976, it is uncertain whether these whales will ever recover their former numbers.

Sei Whales (*Balaenoptera borealis*)
Distinguishing marks: Dark gray-blue backs and ash-colored throat grooves; smaller than fins, ranging in size from about fifty to seventy feet in length. Conical-shaped blow rises about ten feet.

Sei (pronounced "sigh") whales swim no further south than the Antarctic Convergence, although occasionally one is spotted near the peninsula. They are skittish and can race through the water at speeds up to thirty-five miles per hour when disturbed.

Minke Whales (*Balaenoptera acutorostrata*)
Distinguishing marks: Long, pointy triangular head; a single sharp ridge connects the two blowholes to the tip of the snout; black upper body parts with white underneath. The minke is the smallest of the rorqual whales.

Minkes are the most commonly sighted whales in Antarctic seas. They are, however, skittish and tend to remain at a safe distance from ships. They travel and feed in small pods of three to eight whales. In areas of dense plankton or krill concentrations, several hundred or more may gather to feed. Each whale's fifty to seventy skin folds turn deep rosy pink when expanded.

Until recently, whalers have ignored the thirty-foot, eight-ton minkes in favor of larger and more profitable rorquals. But with the population collapse of each great whale species, minkes looked more promising over the years. Japan kills over four hundred minkes per year for research purposes and restaurant resources. Although this small number does not endanger the overall minke population, purposeful killing of whales is still a controversial issue.

HUMPBACK WHALES (*Megaptera novaeangliae*)
Distinguishing marks: A pronounced hump in front of its stubby dorsal fin; long side flippers look like enormous paddles. Humpbacks measure 48–55 feet and weigh 25–30 tons.

Humpbacks are not as sleek and speedy as the rorquals. In profile these whales have a decidedly chunky appearance. The broad knobby flukes look like teeth on a saw. Black and white patterns on the underside of the flukes vary with each individual, and are like whale fingerprints that help experts to track specific animals. The haunting songs that male humpbacks sing at breeding sites also differ from whale to whale, but by the time the humpbacks reach Antarctic waters, breeding is over and the males are silent.

Humpbacks feed almost exclusively on krill and have developed several unique methods to concentrate the crustaceans. Several whales corral the krill by swimming in circles around the krill swarm. Then the humpbacks dive and come up in the middle of the horde with mouths wide open.

Another technique is to create a bubble curtain around the krill by releasing a stream of air below and around the teeming mass. As the bubbles rise to the surface, they form a barrier. The humpbacks slowly rise through the center, capturing hundreds of pounds of krill in their baleen.

Humpbacks seem to have a playful streak. They will suddenly rise from the depths and leap out of the water like a monstrous salmon. The humpbacks twist in the air with long flippers stretched outward from their bodies, like great black and white wings. This behavior is called *breaching* and is common to all whale species—humpbacks just seem to do it more often. Scientists have suggested that breaching could be a form of courtship display, a show of agility and strength, or a signal to the group. The enormous splash that follows the leap may be a way to stun small prey or loosen the parasites on their skin.

Humpbacks also *lobtail*, or smack, the water forcefully with their flukes. Experts believe that lobtailing is aggressive behavior and may express extreme irritation. The resounding slaps could also be attention-getting maneuvers directed toward other whales or onlookers. Humpbacks and other whales also *spy-hop* by maneuvering their bulky bodies into a vertical position, heads above the water. They turn slowly like mammoth periscopes and survey the ocean's surface or nearby land.

> SOUTHERN RIGHT WHALES (*Eubalaena australis* or *Balaena glacialis*)
> Distinguishing marks: A distinctive V-shaped double blow; many thick parasitic calluses on its body that look like flowering heads of cauliflower. The placements of the callosities help scientists identify and track individual whales. A right whale measures 55–65 feet and weighs 70–80 tons.

Southern right whales were the "right" ones to hunt, for they were slow swimmers and easily harpooned. The high percentage of oil in the right whale's blubber acted as a very effective buoy and kept the animal afloat long after death. The baleen hanging from the roof of its mouth was worth a fortune during the 1800s. As a consequence, right whales were hunted to commercial extinction. The few remaining whales were finally protected in 1935. Experts believe that these whales are slowly increasing their numbers from 5 to 7 percent per year.

This whale lacks a dorsal fin. Its flukes are broad and graceful with pointed tips and a deep middle notch. Like humpbacks, southern right whales have a reputation for playfulness. They seem to wave their flukes or breach many consecutive times for no discernible reason other than for sheer joy of the activity.

Although southern rights rarely venture beyond 55° south in summer, a few have been spotted in South Georgia's bays and inlets. They have also been reported at Îles Kerguélen and other subantarctic islands, but are usually alone.

TOOTHED WHALES: THE ODONTOCETI

These whales are generally smaller than the rorquals, ranging in size from sixty-foot sperm whales to six-foot dolphins and porpoises. Toothed whales are very social and travel to feeding and breeding grounds in large pods. Since the species in this division have at least a few teeth or rudimentary stumps, they are able to grasp slippery fish and squid. Beside these culinary staples, one species, orca, has developed a taste for seals and other warm-blooded animals.

> SPERM WHALES (*Physeter macrocephalus*)
> Distinguishing marks: Huge, square "Moby Dick" head that is one-third the length of its 36-to-60-foot body; a bushy forward projecting blow that is very loud, like a sudden release of hot steam from a large vat. Sperm whale skins are wrinkled like prunes. Dorsal fins are absent; instead, a trail of knuckles extends from the hump down the back to the small fluke.

Only mature males visit Antarctic waters, to feed on the giant squid that dwell in very deep water at the edges of steep continental shelves. The sperm whale's enormous head is boxy and dwarfs the beaklike lower jaw with its row of conical teeth. The distorted appearance of the head is the result of a huge chamber filled with spermaceti, a clear liquid wax that may help the animal intensify and concentrate sound waves. If directed at its prey, the sonic "blast" stuns and immobilizes the next meal. If the squid is not debilitated, it may put up a fight of epic proportions. Circular scars sometimes seen on a sperm whale's head are from giant squid suckers.

"There was a violent commotion in the sea right where the moon's rays were concentrated," wrote Frank T. Bullen, who worked on a whaling ship in the late 1800s. At first he thought a massive volcano had suddenly formed and was lifting its top out of the water. However, within minutes he realized he was watching something most unusual. "A very large sperm whale was locked in deadly conflict with a cuttle-fish, or squid, almost as large as himself...The head of the whale especially seems a perfect network of writhing arms." The whale suddenly gripped the squid's hind section in its mouth and chewed through the writhing body. "Then the conflict ceased, the sea resumed its placid calm, and nothing remained to tell of the fight but a strong odor of fish."

> KILLER WHALES (*Orcinus orca*)
> Distinguishing marks: Striking black and white markings, including white elliptical patches above the eyes and behind the dorsal fin on the sides. The five-to-six-foot-high dorsal fin makes the orca easy to recognize. Females have smaller and more curved fins.

Orcas travel in family pods and usually remain with the same group for life. The dozen or more members create their own unique calls, clicks, and chortles that are common only to the pod, like a regional slang. Males are about twenty-two to thirty feet long; females, about four feet less.

Orcas were called "sea wolves" for their predatory skills. Early explorers noted that they work cooperatively when they hunt, like their four-legged namesakes, and have been known to attack and kill prey that is many times larger than themselves, including blue whales. Several may work together to dislodge a resting seal by rocking the ice floe. The sea's bounty provides orcas with seals, penguins, dolphins, squid, fish, and sea turtles. Although there have been several vivid close encounters with orcas, there is no record of a human ever being killed by the whales.

In April 1916 Ernest Shackleton and the crew of the ill-fated *Endurance* rowed for their lives in three tiny whaleboats toward Elephant Island, in the South Shetlands. Darkness fell and the men camped on the sea ice, hoping for rest among the broken floes. "There was no sleep for any of us," Shackleton wrote. "The killer whales were blowing in the sea lanes all around as we waited for daylight and watched for signs of cracks in the ice. The hours passed with laggard feet as we stood huddled together." Surgeon Macklin found the whales' presence "companionable and rather comforting." Their hissing noises, like the persistent release of random safety valves, were at least as tolerable as the notorious snoring of his friend, Thomas Orde-Lees.

Photographer Frank Hurley worried throughout the long cold night. "We had seen the killers charge and upset heavy masses of ice on which luckless seals basked, and we had little doubt that these voracious monsters would appreciate a variation in the diet." The men survived that night and all the nights until August 30, the day that Shackleton returned to Elephant Island to rescue and take his men home.

Part III of this book began with the physical history of Antarctica, from its pivotal role in Gondwana to the white wilderness it is today. Layered rock sediments and fossils have revealed their secrets, written with mineral compositions and imprints of life that long ago disappeared into the shadowy eons of geologic time. More recently, another type of "rock" is contributing to scientists' understanding of Earth's past climates. That rock is ice.

A three-mile-thick sheet of ice covers all but 2 percent of the continent. Scientists have drilled and retrieved deep ice cores, reaching down 10,000 feet and 450,000 years into the past. These compressed seasonal ice layers are Antarctica's weather archives. Like geologists who study earth's history in layers of rocks, glaciologists "read" ice core strata that are encoded with traces of nuclear fallout, factory and auto emissions, chemical sulfates from volcanic eruptions, and carbon dioxide from the burning of fossil fuels.

Deeper ice layers originated as snow falling on the Polar Plateau when Captain Cook circumnavigated *Terra Incognita*; farther down, Herodotus wandered through Asia recording his histories. Deeper still, an air bubble—smaller than a pinprick—holds clues to the climate at a time when an unknown prehistoric artist stroked ocher on a cave wall. Near the bottom strata, humans discovered fire and burned wood to push back the darkness. By analyzing the ancient bubbles of air deep within the ice, scientists may discover the warning signs for accelerating global warming or the onset of another Ice Age.

The Antarctic Peninsula is 4.5° F warmer now than fifty years ago, when temperatures were first systematically recorded. Ice shelves, such as the Larsen and the Wordie, have deteriorated and calved mammoth tabular icebergs. Several smaller shelves have completely disappeared. Two species of flowering plants, pearlwort and hairgrass, continue to spread rapidly as more boulder-strewn lands are exposed by the retreating ice. Whether this warming trend is accelerating or represents just a minor blip in the long continuum of Earth's geologic history is the question that scientists strive to answer. Perhaps Antarctica's ice will be as sensitive a harbinger as the coal miner's canary.

Early in spring, the pack ice is laced with leads that provide routes and resting places for the return of Antarctica's wildlife. Penguins follow the channels to ancient breeding sites on top of fossilized guano. Crabeater seals rest on jagged floes. Overhead, cape petrels soar and bank as one toward reddish-brown cliffs to roost. In the distance the misty blow of a lone whale disappears against the white face of a freshly calved iceberg. The black body moves closer to the decaying edge of the pack ice to feed on swarms of krill. As the days continue to lengthen, prodigious concentrations of wildlife species converge on Antarctica, just as they have done for past millennia and will do for millennia to come.

TRAVEL RESOURCES

For more information about travel to Antarctica, contact any of the following tour operators:

Abercrombie & Kent Inc.
1520 Kensington Road, Oak Brook, IL 60523-2141
Telephone: (800) 323-7308
Fax: (630) 954-3324
Web: www.abercrombiekent.com

Adventure Associates
P.O. Box 612, Bondi Junction, NSW 1355, Australia
Telephone: 61-2-9389-7466
Fax: 61-2-9369-1853
Web: www.adventureassociates.com

Adventure Network International
15a The Broadway, Penn Road, Buckinghamshire HP9 2PD, UK
Telephone: 44-1494671-808
Fax: 44-1494671-725
Web: www.adventure-network.com

Aurora Expeditions
182A Cumberland St., The Rocks, NSW 2000, Australia
Telephone: 02-9252-1033
Fax: 02-9252-1373
Web: www.auroraexpeditions.com.au

Clipper Cruise Line
7711 Bonhomme Avenue, St. Louis, MO 63105-1956
Telephone: (314) 727-2929, (800) 325-0100
Web: www.clippercruise.com

Expeditions, Inc.
550 Industrial Way, Suite 27, Bend, OR 97702
Telephone: (541) 330-2454
Fax: (541) 330-2456
Web: www.expeditioncruises.com

Hapag-Lloyd Cruises GmbH
Ballindamm 25, D-20095 Hamburg, Germany
Telephone: 49-40-3001-4600
Fax: 49-40-3001-4601
Web: www.hapag-lloyd.com (English), www.hlfk.de (German)

Heritage Expeditions
P.O. Box 6282, Christchurch, New Zealand
Telephone: 64-3-338-9944
Web: www.heritage-expeditions.com

Lindblad Expeditions
720 Fifth Avenue, New York, NY 10019
Telephone: (212) 765-7740, (800) 397-3348
Web: www.expeditions.com

Marine Expeditions
890 Yonge Street, Third Floor, Toronto, Ontario M4W 3P4, Canada
Telephone: (416) 964-5751, (800) 263-9147
Outside North America: (416) 964-9069
Web: www.marineex.com

Mountain Travel-Sobek
6420 Fairmount Avenue, El Cerrito, CA 94530-3606
Telephone: (510) 527-8100, (888) 687-6235 (MTSOBEK)
Web: www.mtsobek.com

Natural Habitat Adventures
2945 Center Green Court, Suite H, Boulder, CO 80301
Telephone: (800) 543-8917
Web: www.nathab.com

Orient Lines
1510 SE 17th Street, Suite 400, Fort Lauderdale, FL 33316
Telephone: (954) 527-6660, (800) 333-7300
Web: www.orientlines.com

Quark Expeditions
980 Post Road, Darien, CT 06820-4509
Telephone: (203) 656-0499
Web: www.quark-expeditions.com

Society Expeditions
2001 Western Avenue, Suite 300, Seattle, WA 98121-2114
Telephone: (206) 728-9400
Web: www.societyexpeditions.com

Zegrahm Expeditions
192 Nickerson Street, Suite 200, Seattle, WA 98109-1632
Telephone: (206) 285-4000, (800) 628-8747
Web: www.zeco.com

FLIGHTS OVER ANTARCTICA

Croydon Travel
34 Main Street, Croydon Victoria 3136, Australia
Telephone: 03-9725-8555
Web: www.antarcticaflights.com.au

GLOSSARY

ANTARCTIC CONVERGENCE: Fluctuating boundary at approximately 50° to 63° south latitude where cold, north-flowing Antarctic surface water meets warm tropical water. Also called the *Polar Front*.

AURORA AUSTRALIS: Shimmering curtains of light that occur about sixty miles above the earth when sun-emitted electric particles are captured by earth's magnetic field, exciting molecules of ionized gas which produce the luminosity; Maoris of New Zealand described the southern lights as *Tahu-Nui-A-Rangi*, or the great burning in the sky.

BEACHMASTER: Dominant male seal that controls the choicest breeding territory, usually land nearest the sea, and maintains the most females.

BLIZZARD: A fierce windstorm that fills the air with dense clouds of fine dry snowflakes; visibility reduced to near zero.

BLUE ICE: Oldest, purest, and most compressed ice from the bottommost layers of glaciers, ice tongues, and ice sheets.

BRASH ICE: Fragments of ice representing the final stages of iceberg deterioration.

CAIRN: A tower of stones or ice blocks built as a marker for directions, food, messages, or graves.

CALVING: The sudden splitting of a mass of ice from the front of a glacier, ice shelf, or large iceberg.

CRÈCHE: A group of penguin chicks protected by a few adults, or "aunties," while parents procure food at sea.

CREVASSE: A crack produced by the variation in flow rates between different parts of the glacier.

DIAMOND DUST: Frozen water vapor that refracts light like millions of miniature prisms.

DRIFT ICE: Broken fragments of pack ice in areas where there is more water than solid ice.

DRIFT SNOW: Fine, windblown, rounded granules of snow.

FAST ICE: Sea ice that is attached to the shore; also sea ice between two or more grounded icebergs.

FLENSE: To cut and remove the skin and/or blubber of a whale or seal.

FLOE: An irregular section of broken pack ice.

FLUKE: One of the lobes of a whale's tail.

FRAZIL ICE: Thick slushy clumps of flattened, elongated ice crystals that form below the surface in agitated but freezing waters.

GLACIER: River of ice flowing down a valley.

GLACIER TONGUE: A narrow, floating, seaward extension of a glacier; may extend twenty-five or more miles.

GROUNDED ICEBERG: An iceberg that has drifted into shallow water and run aground.

GROWLER: Small piece of dense blue or green ice that float level with the surface of the sea, making it difficult to spot; dangerous to ships.

HAREM: Group of female seals controlled by a dominant male seal or beachmaster.

HOOSH: A thick, hearty slurry composed of dried meat, lard, hard biscuits, and water; consumed on sledging expeditions by the early explorers.

HUMMOCKY ICE: Massive blocks or ridges of buckled pack ice caused either by the natural expansion of the pack ice or by the action of wind and currents.

IAATO: International Association of Antarctica Tour Operators.

ICEBERG: A piece of ice that breaks from an ice shelf or glacier.

ICE BLINK: A bright section above the horizon illuminating the underside of stratus clouds; caused by intense reflections from pack ice.

ICE SHEET: Slow-moving mass of compressed snow and ice with depths to three miles; the East Antarctica Ice Sheet and West Antarctica Ice Sheet cover most of Antarctica.

ICE SHELF: A floating extension of an ice sheet attached to the coast; height varies from 6 to 180 feet above sea level.

KATABATIC WINDS: Powerful cold winds that roll down from the Polar Plateau toward the sea; also called *gravity winds*.

KRILL: A tiny, shrimplike crustacean that feeds on plankton and forms the foundation of the Antarctic food chain; a Norwegian word for "whale food."

LEAD: An expanding fracture in pack ice; navigable by animals and/or vessels.

MELT POOL: A pond of frozen fresh water on the surface of a glacier or iceberg.

NUNATAK: An isolated mountain peak, or outcrop of rock that pokes through the ice sheet.

PACK ICE: Seasonal ice formed on the surface of the sea, broken into sections and re-frozen in response to storms and temperature changes.

POLYNYA: Any irregularly shaped body of water, pond-sized or larger, within the pack ice.

PEMMICAN: Dried meat mixed with lard; expedition food consumed by the early explorers.

POLAR PLATEAU: The broad, flat region of the East Antarctic Ice Sheet that includes the South Pole.

PRESSURE ICE: Pack ice and floes that have been thrust upward, or *rafted*, by pressure and movement from wind and currents.

PRESSURE RIDGES: Walls or columns of ice that are formed by movement of pack ice against a shore.

SASTRUGI: A corrugated surface formed when prevailing winds carve ridges and furrows on a snow plain.

SEA ICE: Frozen seawater stretching out from land into open water.

SNOWDRIFTS: Accumulations of windblown snow deposited in the lee, or protected, areas of obstructions.

SPA: Specially Protected Area; a protected region which cannot be entered without a permit or permission.

SSSI: Site of Special Scientific Interest; a protected area which cannot be entered without a permit or permission.

STAVES: Narrow strips of wood placed edge to edge to form the sides or lining of a boat.

TABULAR ICEBERG: A massive flat-topped and straight-sided iceberg calved from an ice shelf or glacier tongue.

WHITEOUT: A condition that occurs when low clouds merge with the snow or ice surface, resulting in diffused light and the absence of shadows and perspective.

SELECTED BIBLIOGRAPHY

Allen, Reginald E., editor. *Greek Philosophy: Thales to Aristotle*. New York: Macmillan Publishing Company, Inc., 1966.

Amundsen, Roald. *The South Pole: An Account of the Norwegian Expedition in the Ship Fram, 1910–1912.* Translated from the Norwegian by A. G. Chater. New York: Lee Keedick, 1929.

———. *My Life as an Explorer.* New York: Doubleday, 1928.

Back, June Debenham, editor. *The Quiet Land: The Antarctic Diaries of Frank Debenham*. Norfork, VA: Bluntisham Books, Erskine Press, 1992.

Barron, William. *An Apprentice's Reminiscences of Whaling in Davis Strait.* Boston, MA: Hull: N. Waller, 1890.

Baughman, T. H. *Before the Heroes Came.* Lincoln, NE: University of Nebraska Press, 1994.

Beaglehole, J. C., editor. *The Journals of Captain James Cook on His Voyages of Discovery*. Cambridge, MA: Hakluyt Society, 1955.

———. *The Life of Captain James Cook*. Stanford, CA: Stanford University Press, 1974.

Bechervaise, John. *Antarctica: The Last Horizon*. Sidney: Cassell Australia Limited, 1979.

Bernacchi, L. C. *To the South Polar Regions*. London: Hurst and Blackett, 1901.

———. *Saga of the Discovery.* London: Hurst and Blackett, 1938.

Bertrand, Kenneth J. *Americans in Antarctica 1775–1948*. New York: American Geographical Society (Special Publication #39), 1971.

Bonner, W. Nigel. *The Natural History of Seals*. New York: Facts on File Inc., 1990.

Borchgrevink, Carsten E. *First on the Antarctic Continent.* Montreal: McGill-Queen's University Press, 1980.

Brewster, Barney. *Antarctica: Wilderness at Risk.* Melbourne: A. H. & A. W. Reed Ltd., 1982.

Bruce, William S. *The Log of the Scotia.* Edited by Peter Speak. Edinburgh: Edinburgh University Press, 1992.

Bull, Henryk J. *The Cruise of the Antarctic to the South Polar Regions.* London: Edward Arnold, 1896.

Bullen, Frank T. *The Cruise of the Cachalot.* Reprint. Christchurch, New Zealand: Capper Press, 1976.

Burney, James. *With Captain James Cook in the Antarctic and Pacific: The Private Journal of James Burney, Second Lieutenant of the Adventure on Cook's Second Voyage, 1772–1773.* Canberra: National Library of Australia, 1975.

Burton, Robert. *The Life and Death of Whales.* New York: Universe Books, 1980.

Byrd, Richard E. *Alone.* New York: G. P. Putnam's Sons, 1938.

Campbell, David G. *The Crystal Desert.* Boston: Houghton Mifflin Company, 1992.

Carwardine, Mark. *Whales, Dolphins, and Porpoises.* New York: Dorling Kindersley, 1995.

Chapman, Walter. *The Loneliest Continent.* Greenwich, CT: New York Graphic Society Publishers, 1964.

———, editor. *Antarctic Conquest: The Great Explorers in Their Own Words.* Indianapolis, IN: Bobbs-Merrill Company, Inc., 1965.

Charcot, Jean-Baptiste. *The Voyage of the* Pourquoi-Pas? Reprint. Toronto: Hurst and Company, 1978.

Cherry-Garrard, Apsley. *The Worst Journey in the World.* Second edition. New York: Carrol & Graf Publishers, Inc., 1997.

Christie, E. W. Hunter. *The Antarctic Problem.* London: George Allen & Unwin Ltd., 1951.

Cook, Frederick A. *Through the First Antarctic Night 1898–1899.* New York: Doubleday and McClure, 1900.

Debenham, Frank, editor. Translated by E. Bullough. *The Voyage of Captain Bellingshausen to the Antarctic Seas, 1819–1821.* London: Hakluyt Society, 1945.

———. *Antarctica: The Story of a Continent*. London: Herbert Jenkins Limited, 1959.

Evans, E. R. G. R. *South with Scott*. London: Collins, 1924.

Fanning, Edmund. *Voyages to the South Seas, Indian and Pacific Oceans, China Sea, Northwest Coast, Feejee Islands, South Shetlands, &, &*. [sic] Fairfield: Ye Galleon Press, 1970.

Filchner, Wilhelm. *To the Sixth Continent: The German South Polar Expedition, 1911–1913*. Translated by William Barr. Norfork, VA: Bluntisham Books, The Erskine Press, 1994.

Fisher, Margery and James Fisher. *Shackleton and the Antarctic*. London: James Barrie Books Ltd., 1957.

Fox, Robert. *Antarctica and the South Atlantic*. London: British Broadcasting Corporation, 1985.

Fraser, Conon. *Beyond the Roaring Forties: New Zealand's Subantarctic Islands*. Auckland: Government Printing Office, New Zealand, 1986.

Giaever, John. *The White Desert: The Official Account of the Norwegian–British–Swedish Antarctic Expedition*. Translated by E. M. Huggard. London: Chatto & Windus, 1954.

Gorman, James. *The Total Penguin*. Eaglewood Cliffs, NJ: Prentice Hall Press, 1990.

Harrowfield, David. *Icy Heritage: Historic Sites of the Ross Sea Region*. Christchurch: Antarctic Heritage Trust, 1995.

Headland, Robert. *The Island of South Georgia*. Cambridge, MA: Cambridge University Press, 1992.

———, compiler. *Chronological List of Antarctic Expeditions and Related Historical Events*. Cambridge, MA: Cambridge University Press, 1990.

Hillary, Edmund, and Vivian Fuchs. *The Crossing of Antarctica*. New York: Macmillan Publishing, 1958.

Hoare, Michael M., editor. *The Resolution Journal of Johann Reinhold Forster*. London: Hakluyt Society, 1982.

Hohman, Elmo P. *The American Whaleman*. Reprint. Philadelphia: Augustus M. Kelley, Publishers, 1972.

Huntford, Roland. *Scott and Amundsen*. London: Hodder & Stoughton, 1979.

———. *Shackleton*. New York: Atheneum, 1986.

Hurley, Frank. *Shackleton's Argonauts*. Sidney: Angus and Robertson, 1948.

Huxley, Leonard, editor. *Scott's Last Expedition*. London: John Murray, 1913.

King, H. G. R. *The Antarctic*. New York: Arco Publishing Company, Inc. 1969.

———, editor. *Diary of the Terra Nova Expedition to the Antarctic, 1910–1913 by Edward Wilson*. London: Blandford Press Ltd., 1972.

Kirk, G. S., and J. E. Raven. *The Presocratic Philosophers*. Cambridge, England: Cambridge University Press, 1957.

Kirwan, L. P. *A History of Polar Exploration*. New York: W. W. Norton & Company, Inc., 1960.

Knox, George A., compiler. *The Biology of the Southern Ocean*. Cambridge, England: Cambridge University Press, 1994.

Land, Barbara. *The New Explorers: Women in Antarctica*. New York: Dodd & Mead, 1981.

Lansing, Alfred. *Endurance: Shackleton's Incredible Voyage*. New York: McGraw-Hill, 1959.

Limb, Sue, and Patrick Cordingley. *Captain Oates, Soldier and Explorer*. London: B. T. Batsford Ltd., 1984.

Lovering, J. F., and J. R. V. Prescott. *Last of Lands: Antarctica*. Melbourne: Melbourne University Press, 1979.

Low, Charles R., editor. *Three Voyages Round the World Made by Captain James Cook, R.N.* London: A. L. Burt, Publisher, 1922.

Mawson, Douglas. *The Home of the Blizzard: Being the Story of the Australian Antarctic Expedition 1911–1914*. London: Heinemann, 1915.

Maxtone-Graham, John Charles. *Safe Return Doubtful: The Heroic Age of Polar Exploration*. New York: Charles Scribner's Sons, Macmillan Publishing Company, 1988.

May, John. *Greenpeace Book of Antarctica*. New York: Doubleday, 1988.

McNally, Robert. *So Remorseless a Havoc: Of Dolphins, Whales, & Men*. New York: Little, Brown & Company, 1981.

Mickleburgh, Edwin. *Beyond the Frozen Sea: Visions of Antarctica*. New York: St. Martin's Press, 1987.

Mill, Robert Hugh. *The Siege of the South Pole*. New York: Frederick A. Stokes, 1905.

Mitterling, Philip I. *America in the Antarctic to 1840*. Urbana, IL: University of Illinois Press, 1959.

Moss, Sanford, and Lucia de Leiris. *Natural History of the Antarctic Peninsula*. New York: Columbia University Press, 1988.

Mossman, R. C., J. H. Harvey Pirie, and R. N. Rudmose Brown. *The Voyage of the Scotia: Being the Record of a Voyage of Exploration in Antarctic Seas*. Edinburgh: William Blackwood and Sons, 1906.

Murdoch, W. Burn. *From Edinburgh to the Antarctic: An Artist's Notes and Sketches During the Dundee Antarctic Expedition of 1892–93*. Reprinted. Edinburgh: The Paradigm Press in 1984.

Murphy, Robert Cushman. *Logbook for Grace*. New York: The Macmillan Company, 1947.

Murray, George, editor. *The Antarctic Manual for the Use of the Expedition of 1901*. London: Royal Geographical Society, 1901.

Naveen, Ron, Colin Monteath, Tui de Roy, and Mark Jones. *Wild Ice*. Washington, DC: Smithsonian Institution Press, 1991.

Neider, Charles, editor. *Antarctica*. New York: Random House, Inc., 1972.

———. *Edge of the World, Ross Island, Antarctica*. New York: Doubleday, Garden City, 1974.

Noble, John. *The Mapmakers: The Story of the Great Pioneers in Cartography from Antiquity to the Space Age*. New York: Vintage Books, 1982.

Nordenskjöld, Otto, and John Gunnar. *Antarctica, or Two Years Amongst the Ice of the South Pole*. London: Hurst & Blackett, Ltd., 1905.

Paige, Paula Spurlin, translator. *The Voyage of Magellan and The Journal of Antonio Pigafetta*. Eaglewood Cliffs, NJ: Prentice-Hall, Inc., 1969.

Peterson, Roger Tory. *Penguins*. Boston, MA: Houghton Mifflin Company, 1979.

Poesch, Jessie, editor. *Titian Ramsey Peale 1799–1885 and His Journals of the Wilkes Expedition*. Philadelphia, PA: American Philosophical Society, 1961.

Ponting, Herbert G. *Great White South*. London: Duckworth, 1921. Reprinted, 1932.

Priestley, Raymond E. *Antarctic Adventure: Scott's Northern Party*. Reprinted. London: Hurst & Co., 1974.

Pyne, Stephen J. *The Ice: A Journey to Antarctica*. Iowa City, IA: University of Iowa Press, 1987.

Quigg, Philip W. *A Pole Apart*. New York: McGraw-Hill Book Company, 1983.

Reader's Digest. *Antarctica: Great Stories from the Frozen Continent*. Sidney: Reader's Digest Services Pty. Limited, 1990.

Rehbock, Philip F., editor. *At Sea with the Scientifics: The Challenger Letters of Joseph Matkin*. Honolulu, HI: University of Hawaii Press, 1992.

Rosenman, Helen, editor and translator. *Two Voyages to the South Seas by Jules-Sebastien-Cesar Dumont d'Urville*. Honolulu, HI: University of Hawaii Press, 1988.

Ross, Captain Sir James Clark. *A Voyage of Discovery and Research in the Southern and Antarctic Regions, During the Years 1839–43*. Reprinted. London: Latimer, Trend & Company Limited, London: Whitstable for David & Charles (Holdings) Limited, 1969.

Ross, M. J. *Ross in the Antarctic*. London: Caedmon of Whitby, 1982.

Rymill, John. *Southern Lights: The Official Account of the British Graham Land Expedition 1934–1937*. London: Chatto & Windus, 1938.

Sage, Bryan. *Antarctic Wildlife*. New York: Facts on File, Inc., 1982.

Savours, Ann, editor. *Edward Wilson: Diary of the Discovery Expedition to the Antarctic 1901–1904*. London: Blandford Press Ltd., 1966.

———, editor. *Scott's Last Voyage Through the Camera of Herbert Ponting*. London: Praeger, 1975.

Scott, Robert F. *The Voyage of the Discovery*. Two volumes. New York: Charles Scribner's Sons. London: Smith Elder, 1907.

Seaver, George. *Edward Wilson of the Antarctic*. London: John Murray, 1935.

———. *Birdie Bowers of the Antarctic*. London: John Murray, 1938. Reprinted in 1951.

Shackleton, Ernest H. *Heart of the Antarctic*. Philadelphia: J. B. Lippincott Company, 1909.

———. *South*. London: Heinemann, 1919. Reprinted, 1936.

Shelvocke, George. *A Voyage Round the World*. London: F. Senex: 1726. Reprinted. New York: Jonathan Cape & Harrison Smith, 1928.

Simpson, George Gaylord. *Penguins Past and Present, Here and There*. New Haven: Yale University Press, 1976.

Smith, Geoffry Hattersley, editor. *The Norwegian with Scott: Tryggve Gran's Antarctic Diary 1910–1913*. Translated by Ellen Johanne McGhie. London: Her Majesty's Stationery Office, 1984.

Small, George L. *The Blue Whale*. New York: Columbia University Press, 1971.

Smith, Thomas W. *Narrative of the Life, Travels, and Sufferings of Thomas W. Smith . . . Written by Himself*. Boston, MA: William C. Hill, 1844.

Steward, John. *Antarctica: An Encyclopedia*. Jefferson, NC: McFarland & Company, 1990.

Spufford, Francis. *I May Be Some Time: Ice and the English Imagination*. London: Faber and Faber, 1996.

Stackpole, Edouard A. *The Sea-Hunters: The New England Whalemen During Two Centuries, 1635–1835*. Philadelphia: J. B. Lippincott Company, 1953.

Strange, Ian J. *A Field Guide to the Wildlife of the Falkland Islands and South Georgia*. New York: Harper Collins Publishers, 1992.

Stonehouse, Bernard. *Animals of the Antarctic: The Ecology of the Far South*. London: Peter Lowe, 1972.

Sullivan, Walter. *Quest for a Continent*. New York: McGraw-Hill, 1957.

Thomson, George Malcolm. *Sir Francis Drake*. New York: William Morrow & Company, Inc., 1972.

Tingey, Robert J., editor. *The Geology of Antarctica*. New York: Oxford University Press, 1991.

Tod, Frank. *Whaling in Southern Waters*. Dunedin: New Zealand Tablet Co., 1982.

Tooley, R. V. *Maps and Map-Makers*. London: B. T. Batsford Ltd., 1988.

Tyler, David B. *The Wilkes Expedition*. Philadelphia, PA.: The American Philosophical Society, 1968.

Victor, Paul-Emile. *Man and the Conquest of the Poles*. Translated by Scott Sullivan. New York: Simon and Schuster, 1963.

Villiers, Alan J. *Whaling in the Frozen South*. Indianapolis, IN: Bobbs-Merrill, 1925.

Viola, Herman J., and Carolyn Margolis, editors. *Magnificent Voyagers: The U.S. Exploring Expedition, 1838–1842*. Washington, DC: Smithsonian Institution Press, 1985.

Weddell, James. *A Voyage Towards the South Pole Performed in the Years 1822–24*. Reprinted by David & Charles Reprints, London, 1970.

Wilkes, Charles. *Narrative of the United States Exploring Expedition*. Philadelphia, PA: Lea and Blanchard, 1845.

Wilson, Derek. *The World Encompassed: Drake's Great Voyage 1577–1580*. London: Hamish Hamilton Ltd., 1977.

Wood, William. *Elizabethan Sea-Dogs*. New Haven, CT: Yale University Press, 1918.

Worsley, Frank. *Shackleton's Boat Journey*. New York: W. W. Norton & Company, Inc., 1977.

Young, Neal W., editor. *Antarctica: Weather and Climate*. Melbourne: Melbourne University Press, 1981.

INDEX

PICTURE CREDITS

ENDSHEETS Antarctica. *Permission granted by the Surveyor-General, Land Information New Zealand*

PAGE X Antarctic Expedition map. *Permission granted by the Surveyor-General, Land Information New Zealand*

PAGE 2 Ice floes near the Antarctic Peninsula. *Marilyn J. Landis*

PAGE 26 Lone Adélie penguin near the Antarctic Peninsula. *Marilyn J. Landis*

PAGE 48 Weddell seal near Cape Evans, Ross Sea sector. *Marilyn J. Landis*

PAGE 62 Dumont d'Urville's *Astrolabe* trapped in the Weddell Sea pack ice. *The George Peabody Library of The Johns Hopkins University*

PAGE 108 Abandoned Grytviken Whaling Station, South Georgia Island. *Marilyn J. Landis*

PAGE 124 At the South Pole. Standing: Wilson, Scott, Oates. Sitting: Bowers and Evans. *Scott Polar Research Institute*

PAGE 166 Richard E. Byrd cooking his meal at Bolling Advance Weather Station. *National Archives and Records Service*

PAGE 190 A little penguin music: Piper Kerr with an emperor penguin. *University of Glasgow Archives & Business Records Centre*

PAGE 210 The doomed *Endurance*, Shackleton's ship on his 1914–1916 expedition. *Scott Polar Research Institute*

PAGE 244 Antarctica: Evening glow on mountains and ice. *Marilyn J. Landis*

PAGE 264 East Antarctica: Cliffs of ice. *Marilyn J. Landis*

PAGE 278 Beak to beak: King penguins on South Georgia Island. *Marilyn J. Landis*

PAGE 304 Reflections of ice and sky. *Marilyn J. Landis*

PAGE 316 Adélie penguins claim Paulet Island hut from the Swedish Antarctic Expedition. *Marilyn J. Landis*

Additional photographs in the time line and Chapter 14 by Marilyn J. Landis.

EAST
ANTARCTICA
South Geomagnetic Pole
VOSTOK (RUSSIA)
POLAR PLATEAU
AMUNDSEN-SCOTT (USA)
QUEEN MARY LAND
WILHELM II LAND
PRINCESS ELIZABETH LAND
Law Dome 1395
SABRINA COAST
SCOTT (NZ)
MCMURDO (USA)
DRY VALLEYS
ROYAL SOCIETY RA
COOK MTS
DARWIN MTS
BRITTANIA RA
CHURCHILL MTS
HILLARY COAST
SHACKLETON COAST
QUEEN ALEXANDRA RANGE
Mt Kirkpatrick 4528
HUGHES RA
ROSS
Pole of Relative Inaccessibility
GROVE MTS
INGRID CHRISTENSEN COAST
ZHONGSHAN (CHINA)
MAWSON ESCARPMENT
NANSENISEN
WEGENERISEN
SØR RONDANE
ASUKA (JAPAN)
PRINSESSE RAGNHILD KYST
FIMBULHEIMEN
WOHLTHAT MASSIVET
ORVINFJELLA
MÜLIG-HOFMANFJELLA
NOVOLAZAREVSKAYA (RUSSIA)
PRINSESSE ASTRID KYST
MAITRI (INDIA)
80°
85°
75°
70°
LOCATION MAP
1:88 000 000
Azimuthal Equal - Area Projection
NEW ZEALAND
Auckland
Sydney
Wellington
Christchurch
Chatham Is
Melbourne
Stewart I
Bounty Is
Auckland Is
Antipodes
TASMANIA
Campbell Is
Macquarie I
3835 kilometres
Balleny Is
McMurdo
1353 km
South Pole
South Pacific Ocean
Southern Ocean
South Atlantic Ocean
SOUTH AMERICA
Falkland Is
South Georgia
Glen Crozet
Marion Is
Prince Edward Is
Bouvetoya Is
AFRICA
120° W
90° W
60° W
80° S
70° S
60° S
50° S
40° S